A Concise Introduction to MATLAB

A Concise Introduction to MATLAB

William J. Palm III

University of Rhode Island

**McGraw-Hill
Higher Education**

Boston   Burr Ridge, IL   Dubuque, IA   New York   San Francisco   St. Louis
Bangkok   Bogotá   Caracas   Kuala Lumpur   Lisbon   London   Madrid   Mexico City
Milan   Montreal   New Delhi   Santiago   Seoul   Singapore   Sydney   Taipei   Toronto

McGraw-Hill
Higher Education

A CONCISE INTRODUCTION TO MATLAB

Published by McGraw-Hill, a business unit of The McGraw-Hill Companies, Inc., 1221 Avenue of the Americas, New York, NY 10020.

Some ancillaries, including electronic and print components, may not be available to customers outside the United States.

This book is printed on acid-free paper.

5 6 7 8 9 DOC 12 11

ISBN 978–0–07–338583–9
MHID 0–07–338583–2

Global Publisher: *Raghothaman Srinivasan*
Executive Editor: *Michael Hackett*
Senior Sponsoring Editor: *Bill Stenquist*
Director of Development: *Kristine Tibbetts*
Developmental Editor: *Lora Kalb*
Executive Marketing Manager: *Michael Weits*
Project Manager: *Joyce Watters*
Senior Production Supervisor: *Laura Fuller*
Associate Media Producer: *Christina Nelson*
Associate Design Coordinator: *Brenda A. Rolwes*
Cover Designer: *Studio Montage, St. Louis, Missouri*
Compositor: *Carlisle Publishing Services*
Typeface: 10/12 *Times Roman*
Printer: *R. R. Donnelley Crawfordsville, IN*
(USE) Cover Image: *Air plane lands at runway:* © *Ilene MacDonald/ Alamy RF; Cruise ship off the Antarctic coast:* © *McGraw-Hill Companies/Ian Coles RF; Astoria Bridge and Train:* © *Royalty-Free/CORBIS*

MATLAB® and Simulink® are trademarks of The MathWorks, Inc. and are used with permission. The MathWorks does not warrant the accuracy of the text or exercises in this book. This book's use or discussion of MATLAB® and Simulink® software or related products does not constitute endorsement or sponsorship byThe MathWorks of a particular pedagogical approach or particular use of the MATLAB® and Simulink® software.

Library of Congress Cataloging-in-Publication Data

Palm, William J.
  A concise introduction to MATLAB / William J. Palm, III. -- 1st ed.
    p. cm.
  Includes bibliographical references and index.
  ISBN 978-0-07-338583-9 — ISBN 0-07-338583-2 (hard copy : alk. paper) 1. MATLAB. 2. Numerical analysis--Data processing.
3. Signal processing--Data processing. I. Title.
QA297.P319 2008
620.001'51--dc22
                          2007036050

www.mhhe.com

To my sisters, Linda and Chris, and to my parents, Lillian and William

ABOUT THE AUTHOR

William J. Palm III is Professor of Mechanical Engineering and Applied Mechanics at the University of Rhode Island. In 1966 he received a B.S. from Loyola College in Baltimore, and in 1971 a Ph.D. in Mechanical Engineering and Astronautical Sciences from Northwestern University in Evanston, Illinois.

During his 36 years as a faculty member, he has taught 19 courses. One of these is a freshman MATLAB course, which he helped develop. He has authored nine textbooks dealing with modeling and simulation, system dynamics, control systems, vibration, and MATLAB. These include *System Dynamics* (McGraw-Hill, 2005) and *Mechanical Vibration* (Wiley, 2007). He wrote a chapter on control systems in the *Mechanical Engineers' Handbook* (M. Kutz, ed., Wiley, 1999), and was a special contributor to the fifth editions of *Statics* and *Dynamics,* both by J. L. Meriam and L. G. Kraige (Wiley, 2002).

Professor Palm's research and industrial experience are in control systems, robotics, vibrations, and system modeling. He was the Director of the Robotics Research Center at the University of Rhode Island from 1985 to 1993, and is the coholder of a patent for a robot hand. He served as Acting Department Chair from 2002 to 2003. His industrial experience is in automated manufacturing; modeling and simulation of naval systems, including underwater vehicles and tracking systems; and design of control systems for underwater-vehicle engine-test facilities.

CONTENTS

PREFACE

Formerly used mainly by specialists in signal processing and numerical analysis, MATLAB* in recent years has achieved widespread and enthusiastic acceptance throughout the engineering, mathematics, and scientific communities. Many schools now require a course based entirely or in part on MATLAB early in the curriculum. MATLAB is programmable and has the same logical, relational, conditional, and loop structures as other programming languages, such as Fortran, C, BASIC, and Pascal. Thus it can be used to teach programming principles. In most schools a MATLAB course has replaced the traditional Fortran course, and MATLAB is the principal computational tool used throughout the curriculum.

The popularity of MATLAB is partly due to its long history, and thus it is well developed and well tested. People trust its answers. Its popularity is also due to its user interface, which provides an easy-to-use interactive environment that includes extensive numerical computation and visualization capabilities. Its compactness is a big advantage. For example, you can solve a set of many linear algebraic equations with just three lines of code, a feat that is impossible with traditional programming languages. MATLAB is also extensible; currently more than 20 "toolboxes" in various application areas can be used with MATLAB to add new commands and capabilities.

MATLAB is available for MS Windows and Macintosh personal computers and for other operating systems. It is compatible across all these platforms, which enables users to share their programs, insights, and ideas. This text is based on MATLAB Version 7.4 (R 2007a). Some of the material in Chapter 7 is based on the Control System Toolbox, Version 8.0. Chapter 8 is based on Version 3.2 of the Symbolic Math Toolbox.

TEXT OBJECTIVES AND PREREQUISITES

This text is intended as a stand-alone introduction to MATLAB. It can be used in an introductory course, as a self-study text, or as a supplementary text. The text's material is based on the author's experience in teaching a required two-credit semester course devoted to MATLAB for engineering freshmen. In addition, the text can serve as a reference for later use. The text's many tables, and its referencing system in an appendix and a three-part index each, have been designed with this purpose in mind.

The reader is assumed to have some knowledge of algebra and trigonometry; knowledge of calculus is not required for the first six chapters. Some knowledge

---

*MATLAB is a registered trademark of The MathWorks, Inc.

of high school chemistry and physics, primarily simple electrical circuits and basic statics and dynamics, is required to understand some of the examples.

This text is a condensed version of *Introduction to MATLAB 7 for Engineers* (McGraw-Hill, 2005), using the same pedagogy and instructional style. For this text, we have removed some of the lengthier examples and the background material in mathematics that may not be needed for some readers or for some courses. *Introduction to MATLAB 7 for Engineers* also contains a chapter on Simulink,[†] which is a graphical interface for dynamic systems simulation.

TEXT ORGANIZATION

The text consists of eight chapters. The first chapter gives an overview of MATLAB features, including its windows and menu structures. Chapter 2 introduces the concept of an array, which is the fundamental data element in MATLAB, and describes how to use numeric arrays, cell arrays, and structure arrays for basic mathematical operations. Chapter 2 also covers the solution of linear algebraic equations, which arise in many applications.

Chapter 3 discusses the use of functions and files. MATLAB has an extensive number of built-in math functions, and users can define their own functions and save them as a file for reuse.

Chapter 4 shows how to create decision-making programs with MATLAB, and it covers relational and logical operators, conditional statements, for and while loops, and the switch structure.

Chapter 5 treats two-dimensional plots in greater detail, as well as three-dimensional plots. Function discovery, which uses data plots to discover a mathematical description of the data, is a common application of plotting, and a separate section is devoted to this topic. The chapter also treats polynomial and multiple linear regression as part of its modeling coverage.

Chapter 6 reviews basic statistics and probability and shows how to use MATLAB to generate histograms, perform calculations with the normal distribution, and create random number simulations. The chapter concludes with linear and cubic-spline interpolation.

Chapter 7 covers numerical methods for calculus and differential equations. Numerical integration and differentiation methods are treated. Ordinary differential equation solvers in the core MATLAB program are covered, as well as the linear-system solvers in the Control System toolbox.

Chapter 8 covers symbolic methods for manipulating algebraic expressions and for solving algebraic and transcendental equations, calculus, differential equations, and matrix algebra problems. The calculus applications include integration and differentiation, optimization, Taylor series, series evaluation, and limits. Laplace transform methods for solving differential equations are also introduced. This chapter requires the use of the Symbolic Math toolbox or the Student Edition of MATLAB.

[†]Simulink is a registered trademark of The MathWorks, Inc.

Appendix A contains a guide to the commands and functions introduced in the text. Appendix B is a list of references. Answers to selected problems and a three-part index appear at the end of the text.

All figures, tables, equations, and exercises have been numbered according to their chapter and section. For example, Figure 3.4–2 is the second figure in Chapter 3, Section 4. This system is designed to help the reader locate these items. The end-of-chapter problems are the exception to this numbering system. They are numbered 1, 2, 3, and so on to avoid confusion with the in-chapter exercises. The problems are grouped according to the relevant chapter section.

The first four chapters constitute a course in the essentials of MATLAB. The remaining four chapters are independent of each other, and may be covered in any order, or may be omitted if necessary. These chapters provide additional coverage and examples of plotting and model building, probability and statistics, calculus and differential equations, and symbolic processing, respectively.

SPECIAL REFERENCE FEATURES

The text has the following special features, which have been designed to enhance its usefulness as a reference.

- Throughout each of the chapters, numerous tables summarize the commands and functions as they are introduced.
- Appendix A is a complete summary of all the commands and functions described in the text, grouped by category, along with the number of the page on which they are introduced.
- At the end of each chapter is a list of the key terms introduced in the chapter, with the page number referenced.
- Key terms have been placed in the margin or in section headings where they are introduced.
- The index has three sections: a listing of symbols, an alphabetical list of MATLAB commands and functions, and an alphabetical list of topics.

PEDAGOGICAL AIDS

The following pedagogical aids have been included:

- Each chapter begins with an overview.
- **Test Your Understanding** exercises appear throughout the chapters near the relevant text. These relatively straightforward exercises allow readers to assess their grasp of the material as soon as it is covered. In most cases the answer to the exercise is given with the exercise.
- Each chapter ends with numerous problems, grouped according to the relevant section.
- Each chapter contains numerous practical examples. The major examples are numbered. A guide to these examples appears on the inside front cover.
- Each chapter has a summary section that reviews the chapter's objectives.

■ Answers to many end-of-chapter problems appear at the end of the text. These problems are denoted by an asterisk next to their number (for example, **15***).

An Instructor's Manual is available online for instructors who have adopted this text for a course. This manual contains the complete solutions to all the **Test Your Understanding** exercises and to all the chapter problems. The text website (at http:// www.mhhe.com/palm) also has downloadable files containing PowerPoint slides keyed to the text.

ACKNOWLEDGMENTS

Many individuals are due credit for this text. Working with faculty at the University of Rhode Island in developing and teaching a freshman course based on MATLAB has greatly influenced this text. Email from many users contained useful suggestions. The following people, as well as several anonymous reviewers, suggested many helpful corrections and additions.

Spyros Andreou, *Georgia Southern University*

David Arnold, *College of the Redwoods*

Kirk Breitenbach, *NASA-JPL*

Steven Ciccarelli, *Rochester Institute of Technology*

Dwight Davy, *Case Western Reserve University*

Mike Ecker, *Medtronic Inc.*

Michael Gustafson, *Duke University*

Yueh-Jaw Lin, *The University of Akron*

Armando Rodriquez, *Arizona State University*

Don Smith, *Texas A&M University*

Thomas Sullivan, *Carnegie Mellon University*

Daniel Valentine, *Clarkson University*

Susan Vandiver, *Southern Methodist University*

Elizabeth Wyler, *Thomas Nelson Community College*

Richard Zaccone, *Bucknell University*

The MathWorks, Inc. has always been very supportive of educational publishing. I especially want to thank Naomi Fernandes of The MathWorks, Inc. for her help. Bill Stenquist, Lora Kalb, and Joyce Watters of McGraw-Hill efficiently guided the text through production.

My sisters, Linda and Chris, and my mother Lillian, have always been there, cheering my efforts. My father was always there for support before he passed away. Finally, I want to thank my wife, Mary Louise, and my children, Aileene, Bill, and Andy, for their understanding and support of this project.

William J. Palm III

Kingston, Rhode Island

May 2007

An Overview of MATLAB®*

OUTLINE

This is the most important chapter in the book. By the time you have finished this chapter, you will be able to use MATLAB to solve many kinds of problems. Section 1.1 provides an introduction to MATLAB as an interactive calculator. Section 1.2 covers the main menus and toolbar. Section 1.3 introduces arrays, files, and plots. Section 1.4 discusses how to create, edit, and save MATLAB programs. Section 1.5 introduces the extensive MATLAB Help System.

How to Use This Book

The book's chapter organization is flexible enough to accommodate a variety of users. However, it is important to cover at least the first four chapters, in that order. Chapter 2 covers *arrays,* which are the basic building blocks in MATLAB. Chapter 3 covers file usage, functions built into MATLAB, and user-defined functions. Chapter 4 covers programming using relational and logical operators, conditional statements, and loops.

Chapters 5 through 8 are independent chapters that can be covered in any order. They contain in-depth discussions of how to use MATLAB to solve several

*MATLAB is a registered trademark of The MathWorks, Inc.

common types of problems. Chapter 5 covers two- and three-dimensional plots in more detail, and shows how to use plots to build mathematical models from data. Chapter 6 covers probability, statistics, and interpolation applications. Chapter 7 introduces numerical methods for calculus and ordinary differential equations. Chapter 8 covers symbolic processing in MATLAB, with applications to algebra, calculus, differential equations, linear algebra, and transforms.

Reference and Learning Aids

The book has been designed as a reference as well as a learning tool. The special features useful for these purposes are as follows.

- Throughout each chapter margin notes identify where new terms are introduced.
- Throughout each chapter short Test Your Understanding exercises appear. Where appropriate, answers immediately follow the exercise so you can measure your mastery of the material.
- Homework exercises conclude each chapter. These usually require more effort than the Test Your Understanding exercises.
- Each chapter contains tables summarizing the MATLAB commands introduced in that chapter.
- At the end of each chapter is:
  - A summary of what you should be able to do after completing that chapter and
  - A list of key terms you should know.
- Appendix A contains tables of MATLAB commands, grouped by category, with the appropriate page references.
- Two indexes are included. The first is an index of MATLAB commands and symbols; the second is an index of topics.

1.1 MATLAB Interactive Sessions

We now show how to start MATLAB, how to make some basic calculations, and how to exit MATLAB.

Conventions

In this text we use `typewriter font` to represent MATLAB commands, any text that you type in the computer, and any MATLAB responses that appear on the screen, for example, `y = 6*x`. Variables in normal mathematics text appear in italics; for example, $y = 6x$. We use boldface type for three purposes: to represent vectors and matrices in normal mathematics text (for example, **Ax = b**), to represent a key on the keyboard (for example, **Enter**), and to represent the name of a screen menu or an item that appears in such a menu (for example, **File**). It is assumed that you press the **Enter** key after you type a command. We do not show this action with a separate symbol.

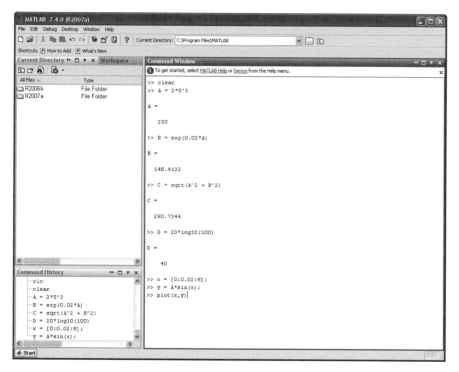

Figure 1.1–1 The default MATLAB Desktop.

Starting MATLAB

To start MATLAB on a MS Windows system, double-click on the MATLAB icon. You will then see the MATLAB *Desktop*. The Desktop manages the Command window and a Help Browser, as well as other tools. The default appearance of the Desktop is shown in Figure 1.1–1. Three windows appear. These are the Command window, the Command History window, and the Current Directory window. Across the top of the Desktop are a row of menu names, and a row of icons called the *toolbar*. To the right of the toolbar is a box showing the directory where MATLAB looks for and saves files. We will describe the menus, toolbar, and directories later in this chapter.

DESKTOP

You use the Command window to communicate with the MATLAB program, by typing instructions of various types called *commands, functions,* and *statements*. Later we will discuss the differences between these types, but for now, to simplify the discussion, we will call the instructions by the generic name *commands*. MATLAB displays the prompt (≫) to indicate that it is ready to receive instructions. Before giving MATLAB instructions, make sure the cursor is located just after the prompt. If it is not, use the mouse to move the cursor.

COMMAND WINDOW

The prompt in the Student Edition looks like EDU ≫. We will use the normal prompt symbol ≫ to illustrate commands in this text.

Three other windows appear in the default Desktop. The Current Directory window is much like a file manager window; you can use it to access files. Double-clicking on a file name with the extension .m will open that file in the MATLAB Editor. The Editor is discussed in Section 1.4.

Underneath the Current Directory window is the Workspace window. To activate it, click on its tab to the right of the Current Directory window. The Workspace window displays the variables created in the Command window. Double-click on a variable name to open the Array Editor, which is discussed in Chapter 2.

The fourth window in the default Desktop is the Command History window. This window shows all the previous keystrokes you entered in the Command window. It is useful for keeping track of what you typed. You can click on a keystroke and drag it to the Command window or the Editor. Double-clicking on a keystroke executes it in the Command window.

You can alter the appearance of the Desktop if you wish. For example, to eliminate a window, just click on its Close-window button (✕) in its upper right-hand corner. To undock, or separate the window from the Desktop, click on the button containing a curved arrow. You can manipulate other windows in the same way. To restore the default configuration, click on the **Desktop** menu, then click on **Desktop Layout,** and select **Default.**

Entering Commands and Expressions

To see how simple it is to use MATLAB, try entering a few commands on your computer. If you make a typing mistake, just press the **Enter** key until you get the prompt, and then retype the line. Or, because MATLAB retains your previous keystrokes in a command file, you can use the up-arrow key (↑) to scroll back through the commands. Press the key once to see the previous entry, twice to see the entry before that, and so on. Use the down-arrow key (↓) to scroll forward through the commands. When you find the line you want, you can edit it using the left- and right-arrow keys (← and →), and the **Backspace** key, and the **Delete** key. Press the **Enter** key to execute the command. This technique enables you to correct typing mistakes quickly.

Note that you can see your previous keystrokes displayed in the Command History window. You can copy a line from this window to the Command window by highlighting the line with the mouse, holding down the left mouse button, and dragging the line to the Command window.

Make sure the cursor is at the prompt in the Command window. To divide 8 by 10, type 8 / 10 and press **Enter** (the symbol / is the MATLAB symbol for division). Your entry and the MATLAB response looks like the following on the screen (we call this interaction between you and MATLAB an *interactive session,* or simply *a session*). Remember, the symbol ≫ automatically appears on the screen; you do not type it.

SESSION

```
>> 8/10
ans =
    0.8000
```

MATLAB indents the numerical result. MATLAB uses high precision for its computations, but by default it usually displays its results using four decimal places except when the result is an integer.

MATLAB assigns the most recent answer to a variable called `ans`, which is an abbreviation for *answer*. A *variable* in MATLAB is a symbol used to contain a value. You can use the variable `ans` for further calculations; for example, using the MATLAB symbol for multiplication (*), we obtain

VARIABLE

```
>> 5*ans
ans =
    4
```

Note that the variable `ans` now has the value 4.

You can use variables to write mathematical expressions. Instead of using the default variable `ans`, you can assign the result to a variable of your own choosing, say `r`, as follows:

```
>> r=8/10
r =
    0.8000
```

Spaces in the line improve its readability; for example, you can put a space before and after the = sign if you want. MATLAB ignores these spaces when making its calculations. It also ignores spaces surrounding + and − signs.

If you now type `r` at the prompt and press **Enter,** you will see

```
>> r
r =
    0.8000
```

thus verifying that the variable `r` has the value 0.8. You can use this variable in further calculations. For example,

```
>> s=20*r
s =
    16
```

A common mistake is to forget the multiplication symbol * and type the expression as you would in algebra, as $s = 20r$. If you do this in MATLAB, you will get an error message.

MATLAB has hundreds of functions available. One of these is the *square root* function, `sqrt`. A pair of parentheses is used after the function's name to enclose the value—called the function's *argument*—that is operated on by the function. For example, to compute the square root of 9, and assign its value to

ARGUMENT

Table 1.1–1 Scalar arithmetic operations

Symbol	Operation	MATLAB form
^	exponentiation: a^b	a^b
*	multiplication: ab	a*b
/	right division: $a/b = \frac{a}{b}$	a/b
\	left division: $a\backslash b = \frac{b}{a}$	a\b
+	addition: $a + b$	a+b
−	subtraction: $a - b$	a−b

the variable r, you type `r = sqrt(9)`. Note that the previous value of r has been replaced by 3.

Order of Precedence

A *scalar* is a single number. A *scalar variable* is a variable that contains a single number. MATLAB uses the symbols + − * / ^ for addition, subtraction, multiplication, division, and exponentiation (power) of scalars. These are listed in Table 1.1–1. For example, typing `x = 8 + 3*5` returns the answer `x = 23`. Typing `2^3-10` returns the answer `ans = -2`. The *forward slash* (/) represents *right division*, which is the normal division operator familiar to you. Typing `15/3` returns the result `ans = 5`.

MATLAB has another division operator, called *left division,* which is denoted by the *backslash* (\). The left division operator is useful for solving sets of linear algebraic equations, as we will see. A good way to remember the difference between the right and left division operators is to note that the slash slants toward the denominator. For example, $7/2 = 2\backslash 7 = 3.5$.

The mathematical operations represented by the symbols + − * / \, and ^ follow a set of rules called *precedence*. Mathematical expressions are evaluated starting from the left, with the exponentiation operation having the highest order of precedence, followed by multiplication and division with equal precedence, followed by addition and subtraction with equal precedence. Parentheses can be used to alter this order. Evaluation begins with the innermost pair of parentheses, and proceeds outward. Table 1.1–2 summarizes these rules. For example, note the effect of precedence on the following session.

Table 1.1–2 Order of precedence

Precedence	Operation
First	Parentheses, evaluated starting with the innermost pair.
Second	Exponentiation, evaluated from left to right.
Third	Multiplication and division with equal precedence, evaluated from left to right.
Fourth	Addition and subtraction with equal precedence, evaluated from left to right.

```
>>8 + 3*5
ans =
      23
>>(8 + 3)*5
ans =
      55
>>4^2-12- 8/4*2
ans =
      0
>>4^2-12- 8/(4*2)
ans =
      3
>>3*4^2 + 5
ans =
      53
>>(3*4)^2 + 5
ans =
      149
>>27^(1/3) + 32^(0.2)
ans =
      5
>>27^(1/3) + 32^0.2
ans =
      5
>>27^1/3 + 32^0.2
ans =
      11
```

To avoid mistakes, feel free to insert parentheses wherever you are unsure of the effect precedence will have on the calculation. Use of parentheses also improves the readability of your MATLAB expressions. For example, parentheses are not needed in the expression $8 + (3*5)$, but they make clear our intention to multiply 3 by 5 before adding 8 to the result.

Test Your Understanding

T1.1–1 Use MATLAB to compute the following expressions.

    a. $6\left(\dfrac{10}{13}\right) + \dfrac{18}{5(7)} + 5(9^2)$

    b. $6(35^{1/4}) + 14^{0.35}$

    (Answers: *a.* 410.1297 *b.* 17.1123.)

The Assignment Operator

The = sign in MATLAB is called the *assignment* or *replacement* operator. It works differently than the equals sign you know from mathematics. When you type x = 3, you tell MATLAB to assign the value 3 to the variable x. This usage is no different than in mathematics. However, in MATLAB we can also type something like this: x = x + 2. This tells MATLAB to add 2 to the current value of x, and to replace the current value of x with this new value. If x originally had the value 3, its new value would be 5. This usage of the = operator is different from its use in mathematics. For example, the mathematics equation $x = x + 2$ is invalid because it implies that $0 = 2$.

In MATLAB the variable on the *left-hand* side of the = operator is replaced by the value generated by the *right-hand* side. Therefore, one variable, and only one variable, must be on the left-hand side of the = operator. Thus in MATLAB you cannot type 6 = x. Another consequence of this restriction is that you cannot write in MATLAB expressions like the following:

```
>>x+2=20
```

The corresponding equation $x + 2 = 20$ is acceptable in algebra, and has the solution $x = 18$, but MATLAB cannot solve such an equation without additional commands (these commands are available in the Symbolic Math toolbox, which is described in Chapter 8).

Another restriction is that the right-hand side of the = operator must have a computable value. For example, if the variable y has not been assigned a value, then the following will generate an error message in MATLAB.

```
>>x = 5 + y
```

In addition to assigning known values to variables, the assignment operator is very useful for assigning values that are not known ahead of time, or for changing the value of a variable by using a prescribed procedure. The following example shows how this is done.

EXAMPLE 1.1–1   Volume of a Circular Cylinder

The volume of a circular cylinder of height *h* and radius *r* is given by $V = \pi r^2 h$. A particular cylindrical tank is 15 m tall and has a radius of 8 m. We want to construct another cylindrical tank with a volume 20 percent greater but having the same height. How large must its radius be?

■ Solution
First solve the cylinder equation for the radius *r*. This gives

$$r = \sqrt{\frac{V}{\pi h}}$$

The session is shown below. First we assign values to the variables r and h representing the radius and height. Then we compute the volume of the original cylinder, and increase the volume by 20 percent. Finally we solve for the required radius. For this problem we can use the MATLAB built-in constant `pi`.

```
>>r = 8;
>>h = 15;
>>V = pi*r^2*h;
>>V = V + 0.2*V;
>>r = sqrt(V/(pi*h))
r =
    8.7636
```

Thus the new cylinder must have a radius of 8.7636 m. Note that the original values of the variables `r` and `V` are replaced with the new values. This is acceptable as long as we do not wish to use the original values again. Note how precedence applies to the line `V = pi*r^2*h;`. It is equivalent to `V = pi*(r^2)*h;`.

Variable Names

The term *workspace* refers to the names and values of any variables in use in the current work session. Variable names must begin with a letter; the rest of the name can contain letters, digits, and underscore characters. MATLAB is case-sensitive. Thus the following names represent five different variables: `speed`, `Speed`, `SPEED`, `Speed_1`, and `Speed_2`. In MATLAB 7, variable names can be no longer than 63 characters.

WORKSPACE

Managing the Work Session

Table 1.1–3 summarizes some commands and special symbols for managing the work session. A semicolon at the end of a line suppresses printing the results to the screen. If a semicolon is not put at the end of a line, MATLAB displays the results of the line on the screen. Even if you suppress the display with the semicolon, MATLAB still retains the variable's value.

Table 1.1–3 Commands for managing the work session

Command	Description
`clc`	Clears the Command window.
`clear`	Removes all variables from memory.
`clear var1 var2`	Removes the variables `var1` and `var2` from memory.
`exist('name')`	Determines if a file or variable exists having the name 'name'.
`quit`	Stops MATLAB.
`who`	Lists the variables currently in memory.
`whos`	Lists the current variables and sizes, and indicates if they have imaginary parts.
`:`	Colon; generates an array having regularly spaced elements.
`,`	Comma; separates elements of an array.
`;`	Semicolon; suppresses screen printing; also denotes a new row in an array.
`...`	Ellipsis; continues a line.

You can put several commands on the same line if you separate them with a comma—if you want to see the results of the previous command—or semicolon if you want to suppress the display. For example,

```
>>x=2;y=6+x,x=y+7
y =
    8
x =
   15
```

Note that the first value of x was not displayed. Note also that the value of x changed from 2 to 15.

If you need to type a long line, you can use an *ellipsis,* by typing three periods, to delay execution. For example,

```
>>NumberOfApples = 10; NumberOfOranges = 25;
>>NumberOfPears = 12;
>>FruitPurchased = NumberOfApples + NumberOfOranges ...
+NumberOfPears
FruitPurchased =
       47
```

Use the arrow, tab, and control keys to recall, edit, and reuse functions and variables you typed earlier. For example, suppose you mistakenly enter the line

```
>>volume = 1 + sqr(5)
```

MATLAB responds with an error message because you misspelled sqrt. Instead of retyping the entire line, press the up-arrow key (↑) once to display the previously typed line. Press the left-arrow key (←) several times to move the cursor and add the missing t, then press **Enter.** Repeated use of the up-arrow key recalls lines typed earlier.

Tab and Arrow Keys

You can use the *smart recall* feature to recall a previously typed function or variable whose first few characters you specify. For example, after you have entered the line starting with volume, typing vol and pressing the up-arrow key (↑) once recalls the last-typed line that starts with the function or variable whose name begins with vol. This feature is case-sensitive.

You can use the *tab completion* feature to reduce the amount of typing. MATLAB automatically completes the name of a function, variable, or file if you type the first few letters of the name and press the **Tab** key. If the name is unique, it is automatically completed. For example, in the session listed earlier, if you type Fruit and press **Tab**, MATLAB completes the name and displays FruitPurchased. Press **Enter** to display the value of the variable, or continue editing to create a new executable line that uses the variable FruitPurchased.

If there is more than one name that starts with the letters you typed, MATLAB displays these names when you press the **Tab** key. Use the mouse to select the desired name from the pop-up list by double-clicking on its name.

The left-arrow (←) and right-arrow (→) keys move left and right through a line one *character* at a time. To move through one *word* at a time, press **Ctrl** and → simultaneously to move to the *right;* press **Ctrl** and ← simultaneously to move to the *left.* Press **Home** to move to the beginning of a line; press **End** to move to the end of a line.

Deleting and Clearing

Press **Del** to delete the character at the cursor; press **Backspace** to delete the character before the cursor. Press **Esc** to clear the entire line; press **Ctrl** and **k** simultaneously to delete (*kill*) to the end of the line.

MATLAB retains the last value of a variable until you quit MATLAB or clear its value. Overlooking this fact commonly causes errors in MATLAB. For example, you might prefer to use the variable x in a number of different calculations. If you forget to enter the correct value for x, MATLAB uses the last value, and you get an incorrect result. You can use the clear function to remove the values of *all* variables from memory, or you can use the form clear var1 var2 to clear the variables named var1 and var2. The effect of the clc command is different; it clears the Command window of everything in the window display, but the values of the variables remain.

You can type the name of a variable and press **Enter** to see its current value. If the variable does not have a value (i.e., if it does not exist), you see an error message. You can also use the exist function. Type exist ('x') to see if the variable x is in use. If a 1 is returned, the variable exists; a 0 indicates that it does not exist. The who function lists the names of all the variables in memory, but does not give their values. The form who var1 var2 restricts the display to the variables specified. The wildcard character * can be used to display variables that match a pattern. For instance, who A* finds all variables in the current workspace that start with A. The whos function lists the variable names and their sizes, and indicates whether or not they have nonzero imaginary parts.

The difference between a function and a command or a statement is that functions have their arguments enclosed in parentheses. Commands, such as clear, need not have arguments, but if they do, they are not enclosed in parentheses; for example, clear x. Statements cannot have arguments; for example, clc and quit are statements.

Press Ctrl-C to cancel a long computation without terminating the session. You can quit MATLAB by typing quit. You can also click on the **File** menu, and then click on **Exit MATLAB.**

Predefined Constants

MATLAB has several predefined special constants, such as the built-in constant pi we used in Example 1.1–1. Table 1.1–4 lists them. The symbol Inf stands

Table 1.1–4 Special variables and constants

Command	Description
ans	Temporary variable containing the most recent answer.
eps	Specifies the accuracy of floating point precision.
i,j	The imaginary unit $\sqrt{-1}$.
Inf	Infinity.
NaN	Indicates an undefined numerical result.
pi	The number π.

for ∞, which in practice means a number so large that MATLAB cannot represent it. For example, typing 5/0 generates the answer Inf. The symbol NaN stands for "not a number." It indicates an undefined numerical result such as that obtained by typing 0/0. The symbol eps is the smallest number which, when added to 1 by the computer, creates a number greater than 1. We use it as an indicator of the accuracy of computations.

The symbols i and j denote the imaginary unit, where $i = j = \sqrt{-1}$. We use them to create and represent complex numbers, such as x = 5 + 8i.

Try not to use the names of special constants as variable names. Although MATLAB allows you to assign a different value to these constants, it is not good practice to do so.

Complex Number Operations

MATLAB handles complex number algebra automatically. For example, the number $c_1 = 1 - 2i$ is entered as follows: c1 = 1-2i. You can also type c1 = Complex(1, -2).

> **Caution:** Note that an asterisk is not needed between i or j and a number, although it is required with a variable, such as c2 = 5 - i*c1. This convention can cause errors if you are not careful. For example, the expressions y = 7/2*i and x = 7/2i give two different results: $y = (7/2)i = 3.5i$ and $x = 7/(2i) = -3.5i$.

Addition, subtraction, multiplication, and division of complex numbers are easily done. For example,

```
>>s = 3+7i;w = 5-9i;
>>w+s
ans =
    8.0000 - 2.0000i
>>w*s
ans =
   78.0000 + 8.0000i
>>w/s
ans =
   -0.8276 - 1.0690i
```

Table 1.1–5 Numeric display formats

Command	Description and example
`format short`	Four decimal digits (the default); 13.6745.
`format long`	16 digits; 17.27484029463547.
`format short e`	Five digits (four decimals) plus exponent; 6.3792e+03.
`format long e`	16 digits (15 decimals) plus exponent; 6.379243784781294e−04.
`format bank`	Two decimal digits; 126.73.
`format +`	Positive, negative, or zero; +.
`format rat`	Rational approximation; 43/7.
`format compact`	Suppresses some blank lines.
`format loose`	Resets to less compact display mode.

Test Your Understanding

T1.1–2 Given $x = -5 + 9i$ and $y = 6 - 2i$, use MATLAB to show that $x + y = 1 + 7i$, $xy = -12 + 64i$, and $x/y = -1.2 + 1.1i$.

Formatting Commands

The `format` command controls how numbers appear on the screen. Table 1.1–5 gives the variants of this command. MATLAB uses many significant figures in its calculations, but we rarely need to see all of them. The default MATLAB display format is the `short` format, which uses four decimal digits. You can display more by typing `format long`, which gives 16 digits. To return to the default mode, type `format short`.

   You can force the output to be in scientific notation by typing `format short e`, or `format long e`, where e stands for the number 10. Thus the output `6.3792e+03` stands for the number 6.3792×10^3. The output `6.3792e-03` stands for the number 6.3792×10^{-3}. Note that in this context e does *not* represent the number e, which is the base of the natural logarithm. Here e stands for "exponent." It is a poor choice of notation, but MATLAB follows conventional computer programming standards that were established many years ago.

   Use `format bank` only for monetary calculations; it does not recognize imaginary parts.

1.2 Menus and the Toolbar

The Desktop manages the Command window and other MATLAB tools. The default appearance of the Desktop is shown in Figure 1.1–1 on page 3. Across the top of the Desktop are a row of menu names, and a row of icons called the *toolbar*. To the right of the toolbar is a box showing the *current directory,* where MATLAB looks for files.

CURRENT DIRECTORY

   Other windows appear in a MATLAB session, depending on what you do. For example, a graphics window containing a plot appears when you use the plotting functions; an editor window, called the Editor/Debugger, appears for use in creating program files. Each window type has its own menu bar, with one or more menus, at the top. *Thus the menu bar will change as you change windows.* To activate, or select, a menu, click on it. Each menu has several items. Click on an item to select it. *Keep in mind that menus are context-sensitive. Thus their contents change, depending on which features you are currently using.*

The Desktop Menus

Most of your interaction will be in the Command window. When the Command window is active, the default MATLAB 7 Desktop (shown in Figure 1.1–1) has six menus: **File, Edit, Debug, Desktop, Window,** and **Help.** Note that these menus change depending on what window is active. Every item on a menu can be selected with the menu open either by clicking on the item or by typing its underlined letter. Some items can be selected without the menu being open by using the shortcut key listed to the right of the item. Those items followed by three dots (**. . .**) open a submenu or another window containing a dialog box.

The three most useful menus are the **File, Edit,** and **Help** menus. The **Help** menu is described in Section 1.5. The **File** menu in MATLAB 7 contains the following items, which perform the indicated actions when you select them.

The File Menu in MATLAB 7

New Opens a dialog box that allows you to create a new program file, called an M-file, using a text editor called the Editor/Debugger, or a new Figure or Model file (a file type used by Simulink).

Open. . . Opens a dialog box that allows you to select a file for editing.

Close Command Window Closes the Command window.

Import Data. . . Starts the Import Wizard which enables you to import data easily.

Save Workspace As. . . Opens a dialog box that enables you to save a file.

Set Path. . . Opens a dialog box that enables you to set the MATLAB search path.

Preferences. . . Opens a dialog box that enables you to set preferences for such items as fonts, colors, tab spacing, and so forth.

Page Setup Opens a dialog box that enables you to format printed output.

Print. . . Opens a dialog box that enables you to print all of the Command window.

Print Selection. . . Opens a dialog box that enables you to print selected portions of the Command window.

File List Contains a list of previously used files, in order of most recently used.

Exit MATLAB Closes MATLAB.

The **Edit** menu contains the following items.

The Edit Menu in MATLAB 7

Undo Reverses the previous editing operation.

Redo Reverses the previous Undo operation.

Cut Removes the selected text and stores it for pasting later.

Copy Copies the selected text for pasting later, without removing it.

Paste Inserts any text on the clipboard at the current location of the cursor.

Paste to Workspace. . . Inserts the contents of the clipboard into the workspace as one or more variables.

Select All Highlights all text in the Command window.

Delete Clears the variable highlighted in the Workspace Browser.

Find. . . Finds and replaces phrases.

Find Files. . . Finds files.

Clear Command Window Removes all text from the Command window.

Clear Command History Removes all text from the Command History window.

Clear Workspace Removes the values of all variables from the workspace.

You can use the **Copy** and **Paste** selections to copy and paste commands appearing on the Command window. However, an easier way is to use the up-arrow key to scroll through the previous commands, and press **Enter** when you see the command you want to retrieve.

Use the **Debug** menu to access the Debugger, which is discussed in Chapter 4. Use the **Desktop** menu to control the configuration of the Desktop and to display toolbars. The **Window** menu has one or more items, depending on what you have done thus far in your session. Click on the name of a window that appears on the menu to open it. For example, if you have created a plot and not closed its window, the plot window will appear on this menu as **Figure 1.** However, there are other ways to move between windows (such as pressing the **Alt** and **Tab** keys simultaneously if the windows are not docked).

The toolbar, which is below the menu bar, provides buttons as shortcuts to some of the features on the menus. Clicking on the button is equivalent to clicking on the menu, then clicking on the menu item; thus the button eliminates one click of the mouse. The first seven buttons from the left correspond to the **New M-File, Open File, Cut, Copy, Paste, Undo,** and **Redo.** The eighth button activates Simulink, which is a program built on top of MATLAB. The ninth button activates the Profiler, which can be used to optimize program performance. The tenth button activates the GUIDE Quick Start, which is used to create and edit graphical user interfaces (GUIs). The eleventh button (the one with the question mark) accesses the Help System.

Below the toolbar is a button that accesses help for adding shortcuts to the toolbar and a button that accesses a list of the features added since the previous release.

1.3 Arrays, Files, and Plots

This section introduces arrays, which are the basic building blocks in MATLAB, and shows how to handle files and generate plots.

Arrays

MATLAB has hundreds of functions, which we will discuss throughout the text. For example, to compute sin x, where x has a value in radians, you type `sin(x)`. To compute cos x, type `cos(x)`. The exponential function e^x is computed from `exp(x)`. The natural logarithm, ln x, is computed by typing `log(x)`. (Note the spelling difference between mathematics text, ln, and MATLAB syntax, `log`.) You compute the base 10 logarithm by typing `log10(x)`. The inverse sine, or arcsine, is obtained by typing `asin(x)`. It returns an answer in radians, not degrees.

ARRAY

    One of the strengths of MATLAB is its ability to handle collections of numbers, called *arrays,* as if they were a single variable. A numerical array is an ordered collection of numbers (a set of numbers arranged in a specific order). An example of an array variable is one that contains the numbers 0, 4, 3, and 6, in that order. We can use square brackets to define the variable x to contain this collection by typing x = [0, 4, 3, 6]. The elements of the array may also be separated by spaces, but commas are preferred to improve readability and avoid mistakes. Note that the variable y defined as y = [6, 3, 4, 0] is not the same as x because the order is different.

    We can add the two arrays x and y to produce another array z by typing the single line z = x + y. To compute z, MATLAB adds all the corresponding numbers in x and y to produce z. The resulting array z contains the numbers 6, 7, 7, 6.

    You need not type all the numbers in the array if they are regularly spaced. Instead, you type the first number and the last number, with the spacing in the middle, separated by colons. For example, the numbers 0, 0.1, 0.2, . . . , 10 can be assigned to the variable u by typing u = [0:0.1:10].

    To compute $w = 5 \sin u$ for $u = 0, 0.1, 0.2, \ldots, 10$, the session is:

```
>>u = [0:0.1:10];
>>w = 5*sin(u);
```

The single line w = 5*sin(u) computed the formula $w = 5 \sin u$ 101 times, once for each value in the array u, to produce an array z that has 101 values.

ARRAY INDEX

    You can see all the u values by typing u after the prompt or, for example, you can see the seventh value by typing u(7). The number 7 is called an *array index,* because it points to a particular element in the array.

```
>>u(7)
ans =
     0.6000
>>w(7)
ans =
     2.8232
```

You can use the `length` function to determine how many values are in an array. For example, continue the previous session as follows:

```
>>m = length(w)
m =
   101
```

Arrays that display on the screen as a single row of numbers with more than one column are called *row arrays*. You can create *column* arrays, which have more than one row, by using a semicolon to separate the rows.

Polynomial Roots

We can describe a polynomial in MATLAB with an array whose elements are the polynomial's coefficients, *starting with the coefficient of the highest power of x*. For example, the polynomial $4x^3 - 8x^2 + 7x - 5$ would be represented by the array `[4,-8,7,-5]`. The *roots* of the polynomial $f(x)$ are the values of x such that $f(x) = 0$. Polynomial roots can be found with the `roots(a)` function, where `a` is the polynomial's coefficient array. The result is a *column* array that contains the polynomial's roots. For example, to find the roots of $x^3 - 7x^2 + 40x - 34 = 0$, the session is

```
>>a = [1,-7,40,-34];
>>roots(a)
ans =
     3.0000 + 5.000i
     3.0000 - 5.000i
     1.0000
```

The roots are $x = 1$ and $x = 3 \pm 5i$. The two commands could have been combined into the single command `roots([1,-7,40,-34])`.

Test Your Understanding

T1.3–1 Use MATLAB to determine how many elements are in the array `[cos(0):0.02:log10(100)]`. Use MATLAB to determine the 25th element. (Answer: 51 elements and 1.48.)

T1.3–2 Use MATLAB to find the roots of the polynomial $290 - 11x + 6x^2 + x^3$. (Answer: $x = -10, 2 \pm 5i$.)

Built-in Functions

We have seen several of the functions built into MATLAB, such as the `sqrt` and the `sin` functions. Table 1.3–1 lists some of the commonly used built-in functions. Chapter 3 gives extensive coverage of the built-in functions. MATLAB users can create their own functions for their special needs. Creation of user-defined functions is covered in Chapter 3.

Table 1.3–1 Some commonly used mathematical functions

Function	MATLAB syntax*
e^x	exp(x)
$\sqrt{x}$	sqrt(x)
$\ln x$	log(x)
$\log_{10} x$	log10(x)
$\cos x$	cos(x)
$\sin x$	sin(x)
$\tan x$	tan(x)
$\cos^{-1} x$	acos(x)
$\sin^{-1} x$	asin(x)
$\tan^{-1} x$	atan(x)

*The MATLAB trigonometric functions listed here use radian measure. Trigonometric functions ending in d, such as sind(x) and cosd(x), take the argument x in degrees.

Working with Files

MATLAB uses several types of files that enable you to save programs, data, and session results. As we will see in Section 1.4, MATLAB function files and program files are saved with the extension .m, and thus are called M-files. *MAT-files* have the extension .mat and are used to save the names and values of variables created during a MATLAB session.

Because they are *ASCII files,* M-files can be created using just about any word processor. MAT-files are *binary* files that are generally readable only by the software that created them. MAT-files contain a machine signature that allows them to be transferred between machine types such as MS windows and Macintosh machines.

The third type of file we will be using is a data file, specifically an ASCII *data file,* that is, one created according to the ASCII format. You may need to use MATLAB to analyze data stored in such a file created by a spreadsheet program, a word processor, or a laboratory data acquisition system or in a file you share with someone else.

MAT-FILES

ASCII FILES

DATA FILE

Saving and Retrieving Your Workspace Variables

If you want to continue a MATLAB session at a later time, you must use the save and load commands. Typing save causes MATLAB to save the workspace variables, that is, the variable names, their sizes, and their values, in a binary file called matlab.mat, which MATLAB can read. To retrieve your workspace variables, type load. You can then continue your session as before. To save the workspace variables in another file named filename.mat, type save filename. To load the workspace variables, type load filename. If the saved MAT-file filename contains the variables A, B, and C, then loading the file filename places these variables back into the workspace and overwrites any existing variables having the same name.

To save just some of your variables, say, `var1` and `var2`, in the file `filename.mat`, type `save filename var1 var2`. You need not type the variable names to retrieve them; just type `load filename`.

Directories and Path It is important to know the location of the files you use with MATLAB. File location frequently causes problems for beginners. Suppose you use MATLAB on your home computer and save a file to a removable disk, as discussed later in this section. If you bring that disk to use with MATLAB on another computer, say, in a school's computer lab, you must make sure that MATLAB knows how to find your files. Files are stored in *directories,* called *folders* on some computer systems. Directories can have subdirectories below them. For example, suppose MATLAB was installed on drive c: in the directory `c:\matlab`. Then the `toolbox` directory is a subdirectory under the directory `c:\matlab`, and `symbolic` is a subdirectory under the `toolbox` directory. The *path* tells us and MATLAB how to find a particular file. For example, the file `solve.m` is a function in the Symbolic Math toolbox. The path to this file is `c:\matlab\toolbox\symbolic`. The full name of a file consists of its path and its name, for example, `c:\matlab\toolbox\symbolic\solve.m`.

PATH

Working with Removable Disks In Section 1.4 you will learn how to create and save M-files. Suppose you have saved the file `problem1.m` in the directory `\homework` on a disk, which you insert in drive a:. The path for this file is `a:\homework`. As MATLAB is normally installed, when you type `problem1`,

1. MATLAB first checks to see if `problem1` is a variable and if so, displays its value.
2. If not, MATLAB then checks to see if `problem1` is one of its own commands, and executes it if it is.
3. If not, MATLAB then looks in the current directory for a file named `problem1.m` and executes `problem1` if it finds it.
4. If not, MATLAB then searches the directories in its *search path,* in order, for `problem1.m` and then executes it if found.

SEARCH PATH

You can display the MATLAB search path by typing `path`. If `problem1` is on the disk only and if directory a: is not in the search path, MATLAB will not find the file and will generate an error message, unless you tell it where to look. You can do this by typing `cd a:\homework`, which stands for "change directory to a:\homework." This will change the current directory to `a:\homework` and force MATLAB to look in that directory to find your file. The general syntax of this command is `cd dirname`, where `dirname` is the full path to the directory.

An alternative to this procedure is to copy your file to a directory on the hard drive that is in the search path. However, there are several pitfalls with this approach: (1) if you change the file during your session, you might forget to copy the revised file back to your disk; (2) the hard drive becomes cluttered (this is a problem in public computer labs, and you might not be permitted to save your file on the hard drive);

Table 1.3–2 System, directory, and file commands

Command	Description
`addpath dirname`	Adds the directory `dirname` to the search path.
`cd dirname`	Changes the current directory to `dirname`.
`dir`	Lists all files in the current directory.
`dir dirname`	Lists all the files in the directory `dirname`.
`path`	Displays the MATLAB search path.
`pathtool`	Starts the Set Path tool.
`pwd`	Displays the current directory.
`rmpath dirname`	Removes the directory `dirname` from the search path.
`what`	Lists the MATLAB-specific files found in the current working directory. Most data files and other non-MATLAB files are not listed. Use `dir` to get a list of all files.
`what dirname`	Lists the MATLAB-specific files in directory `dirname`.
`which item`	Displays the pathname of `item` if `item` is a function or file. Identifies `item` as a variable if so.

(3) the file might be deleted or overwritten if MATLAB is reinstalled; and (4) someone else can access your work!

You can determine the current directory (the one where MATLAB looks for your file) by typing `pwd`. To see a list of all the files in the current directory, type `dir`. To see the files in the directory `dirname`, type `dir dirname`.

The `what` command displays a list of the MATLAB-specific files in the current directory. The `what dirname` command does the same for the directory `dirname`. Type `which item` to display the full pathname of the function `item` or the file `item` (include the file extension). If `item` is a variable, then MATLAB identifies it as such.

You can add a directory to the search path by using the `addpath` command. To remove a directory from the search path, use the `rmpath` command. The Set Path tool is a graphical interface for working with files and directories. Type `pathtool` to start the browser. To save the path settings, click on **Save** in the tool. To restore the default search path, click on **Default** in the browser.

These commands are summarized in Table 1.3–2.

Plotting with MATLAB

MATLAB contains many powerful functions for easily creating plots of several different types, such as rectilinear, logarithmic, surface, and contour plots. As a simple example, let us plot the function $y = 5 \sin x$ for $0 \leq x \leq 6$. We choose to use an increment of 0.02 to generate a large number of x values in order to produce a smooth curve. The function `plot(x,y)` generates a plot with the x values on the horizontal axis (the abscissa) and the y values on the vertical axis (the ordinate). The session is:

```
>>x = [0:0.02:6];
>>y = 5*sin(x);
>>plot(x,y),xlabel('x'),ylabel('y')
```

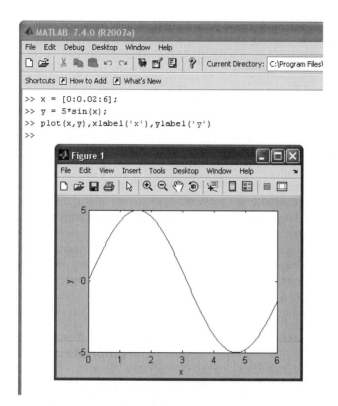

Figure 1.3–1 A graphics window showing a plot.

The plot appears on the screen in a *graphics window,* named **Figure No. 1,** as shown in Figure 1.3–1. The `xlabel` function places the text in single quotes as a label on the horizontal axis. The `ylabel` function performs a similar function for the vertical axis. When the plot command is successfully executed, a graphics window automatically appears. If a hard copy of the plot is desired, the plot can be printed by selecting **Print** from the **File** menu on the graphics window. The window can be closed by selecting **Close** on the **File** menu in the graphics window. You will then be returned to the prompt in the Command window.

GRAPHICS WINDOW

  Other useful plotting functions are `title` and `gtext`. These functions place text on the plot. Both accept text within parentheses and single quotes, as with the `xlabel` function. The `title` function places the text at the top of the plot; the `gtext` function places the text at the point on the plot where the cursor is located when you click the left mouse button.

  You can create multiple plots—called *overlay plots*—by including another set or sets of values in the `plot` function. For example, to plot the functions $y = 2\sqrt{x}$ and $z = 4 \sin 3x$ for $0 \le x \le 5$ on the same plot, the session is

OVERPLAY PLOT

```
>>x = [0:0.01:5];
>>y = 2*sqrt(x);
>>z = 4*sin(3*x);
>>plot(x,y,x,z),xlabel('x'),gtext('y'),gtext('z')
```

After the plot appears on the screen, the program waits for you to position the cursor and click the mouse button, once for each `gtext` function used. Use the `gtext` function to place the labels *y* and *z* next to the appropriate curves.

You can also distinguish curves from one another by using different line types for each curve. For example, to plot the *z* curve using a dashed line, replace the `plot(x,y,x,z)` function in the above session with `plot(x,y,x,z, '− −')`. Other line types can be used. These are discussed in Chapter 5.

Sometimes it is useful or necessary to obtain the coordinates of a point on a plotted curve. The function `ginput` can be used for this purpose. Place it at the end of all the plot and plot formatting statements, so that the plot will be in its final form. The command `[x,y] = ginput(n)` gets *n* points and returns the *x* and *y* coordinates in the vectors `x` and `y`, which have a length *n*. Position the cursor using a mouse, and press the mouse button. The returned coordinates have the same scale as the coordinates on the plot.

DATA MARKER

In cases where you are plotting *data,* as opposed to functions, you should use *data markers* to plot each data point (unless there are very many data points). To mark each point with a plus sign +, the required syntax for the `plot` function is `plot(x,y,'+')`. You can connect the data points with lines if you wish. In that case, you must plot the data twice, once with a data marker, and once without a marker.

For example, suppose the data for the independent variable is `x  = [15:2:23]`, and the dependent variable values are `y = [20, 50, 60, 90, 70]`. To plot the data with plus signs use the following session:

```
>>x = [15:2:23];
>>y = [20, 50, 60, 90, 70];
>>plot(x,y,'+',x,y),xlabel('x'),ylabel('y'), grid
```

The `grid` command puts grid lines on the plot. Other data markers are available. These are discussed in Chapter 5.

Table 1.3–3 summarizes these plotting commands. We will discuss other plotting functions, and the Plot Editor, in Chapter 5.

Table 1.3–3 Some MATLAB plotting commands

Command	Description
`[x,y] = ginput(n)`	Enables the mouse to get *n* points from a plot, and returns the *x* and *y* coordinates in the vectors `x` and `y`, which have a length *n*.
`grid`	Puts grid lines on the plot.
`gtext('text')`	Enables placement of text with the mouse.
`plot(x,y)`	Generates a plot of the array `y` versus the array `x` on rectilinear axes.
`title('text')`	Puts text in a title at the top of the plot.
`xlabel('text')`	Adds a text label to the horizontal axis (the abscissa).
`ylabel('text')`	Adds a text label to the vertical axis (the ordinate).

Test Your Understanding

T1.3–3 Use MATLAB to plot the function $s = 2\sin(3t + 2) + \sqrt{5t + 1}$ over the interval $0 \le t \le 5$. Put a title on the plot, and properly label the axes. The variable s represents speed in feet per second; the variable t represents time in seconds.

T1.3–4 Use MATLAB to plot the functions $y = 4\sqrt{6x + 1}$ and $z = 5e^{0.3x} - 2x$ over the interval $0 \le x \le 1.5$. Properly label the plot and each curve. The variables y and z represent force in newtons; the variable x represents distance in meters.

1.4 Script Files and the Editor/Debugger

**EDITOR/
DEBUGGER**

You can perform operations in MATLAB in two ways:

1. In the interactive mode, in which all commands are entered directly in the Command window, or
2. By running a MATLAB program stored in *script* file. This type of file contains MATLAB commands, so running it is equivalent to typing all the commands—one at a time—at the Command window prompt. You can run the file by typing its name at the Command window prompt.

When the problem to be solved requires many commands, a repeated set of commands, or has arrays with many elements, the interactive mode is inconvenient. Fortunately, MATLAB allows you to write your own programs to avoid this difficulty. You write and save MATLAB programs in M-files, which have the extension `.m`; for example, `program1.m`.

MATLAB uses two types of M-files: *script files* and *function* files. You can use the Editor/Debugger built into MATLAB to create M-files. Because they contain commands, script files are sometimes called *command* files. Function files are discussed in Chapter 3.

SCRIPT FILE

Creating and Using a Script File

The symbol `%` designates a *comment,* which is not executed by MATLAB. Comments are used mainly in script files for the purpose of documenting the file. The comment symbol may be put anywhere in the line. MATLAB ignores everything to the right of the `%` symbol. For example, consider the following session.

COMMENT

```
>>% This is a comment.
>>x = 2+3 % So is this.
x =
   5
```

Note that the portion of the line before the `%` sign is executed to compute `x`.

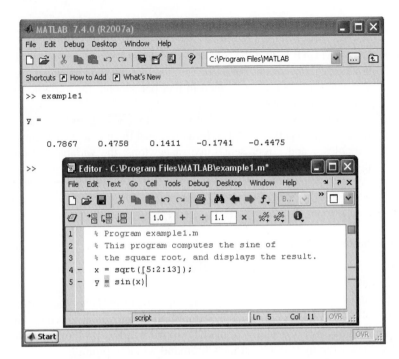

Figure 1.4–1 The MATLAB Command window with the Editor/Debugger open.

Here is a simple example that illustrates how to create, save, and run a script file, using the Editor/Debugger built into MATLAB. However, you may use another text editor to create the file. The sample file is shown below. It computes the sine of the square root of several numbers and displays the results on the screen.

```
% Program example1.m
% This program computes the sine of
% the square root and displays the result.
x = sqrt([5:2:13]);
y = sin(x)
```

To create this new M-file in the Command window select **New** from the **File** menu, then select **M-file.** You will then see a new edit window. This is the Editor/Debugger window as shown in Figure 1.4–1. Type in the file as shown above. You can use the keyboard and the **Edit** menu in the Editor/Debugger as you would in most word processors to create and edit the file. When finished, select **Save** from the **File** menu in the Editor/Debugger. In the dialog box that appears, replace the default name provided (usually named Untitled) with the name example1, and click on **Save.** The Editor/Debugger will automatically provide the extension .m and save the file in the MATLAB current directory, which for now we will assume to be on the hard drive.

Once the file has been saved, in the MATLAB Command window type the script file's name example1 to execute the program. You should see the result displayed in the Command window. Figure 1.4–1 shows a screen containing the resulting Command window display and the Editor/Debugger opened to display the script file.

Effective Use of Script Files

Create script files to avoid the need to retype lengthy and commonly used procedures. Here are some other things to keep in mind when using script files:

1. The name of a script file must follow the MATLAB convention for naming variables.

2. Recall that typing a variable's name at the Command window prompt causes MATLAB to display the value of that variable. Thus, do not give a script file the same name as a variable it computes because MATLAB will not be able to execute that script file more than once, unless you clear the variable.

3. Do not give a script file the same name as a MATLAB command or function. You can check to see if a command, function or file name already exists by using the `exist` command. For example, to see if a variable `example1` already exists, type `exist('example1');` this will return a 0 if the variable does not exist, and a 1 if it does. To see if an M-file `example1.m` already exists, type `exist('example1.m','file')` *before* creating the file; this will return a 0 if the file does not exist, and a 2 if it does. Finally, to see if a built-in function `example1` already exists, type `exist 'example1', 'builtin')` *before* creating the file; this will return a 0 if the built-in function does not exist, and a 5 if it does.

Note that not all functions supplied with MATLAB are built-in functions. For example, the function `mean.m` is supplied but is not a built-in function. The command `exist('mean.m', 'file')` will return a 2, but the command `exist('mean', 'builtin')` will return a 0. You may think of built-in functions as primitives that form the basis for other MATLAB functions. You cannot view the entire file of a built-in function in a text editor, only the comments.

Debugging Script Files

Debugging a program is the process of finding and removing the "bugs," or errors, in a program. Such errors usually fall into one of the following categories.

DEBUGGING

1. Syntax errors such as omitting a parenthesis or comma, or spelling a command name incorrectly. MATLAB usually detects the more obvious errors and displays a message describing the error and its location.

2. Errors due to an incorrect mathematical procedure, called *runtime errors*. They do not necessarily occur every time the program is executed; their occurrence often depends on the particular input data. A common example is division by zero.

To locate an error, try the following:

1. Always test your program with a simple version of the problem, whose answers can be checked by hand calculations.

2. Display any intermediate calculations by removing semicolons at the end of statements.

3. Use the debugging features of the Editor/Debugger, which are introduced in Chapter 4. However, one advantage of MATLAB is that it requires relatively simple programs to accomplish many types of tasks. Thus you probably will not need to use the Debugger for the problems encountered in this text.

Programming Style

Comments may be put anywhere in the script file. However, because the first comment line before any executable statement is the line searched by the `lookfor` command, discussed later in this chapter, consider putting key words that describe the script file in this first line (called the H1 line). A suggested structure for a script file is the following.

1. *Comments section* In this section put comment statements to give:
   a. The name of the program and any key words in the first line.
   b. The date created, and the creators' names in the second line.
   c. The definitions of the variable names for every input and output variable. Divide this section into at least two subsections, one for input data, and one for output data. A third, optional section may include definitions of variables used in the calculations. *Be sure to include the units of measurement for all input and all output variables!*
   d. The name of every user-defined function called by the program.

2. *Input section* In this section put the input data and/or the input functions that enable data to be entered. Include comments where appropriate for documentation.

3. *Calculation section* Put the calculations in this section. Include comments where appropriate for documentation.

4. *Output section* In this section put the functions necessary to deliver the output in whatever form required. For example, this section might contain functions for displaying the output on the screen. Include comments where appropriate for documentation.

The programs in this text often omit some of these elements to save space. Here the text discussion associated with the program provides the required documentation.

Controlling Input and Output

MATLAB provides several useful commands for obtaining input from the user and for formatting the output (the results obtained by executing the MATLAB commands). Table 1.4–1 summarizes these commands.

The `disp` function (short for "display") can be used to display the value of a variable but not its name. Its syntax is `disp(A)`, where A represents a MATLAB variable name. The `disp` function can also display text such as a message to the user. You enclose the text within single quotes. For example, the command `disp('The predicted speed is:')` causes the message to appear on

Table 1.4–1 Input/output commands

Command	Description
disp(A)	Displays the contents, but not the name, of the array A.
disp('text')	Displays the text string enclosed within single quotes.
format	Controls the screen's output display format (see Table 1.1–5).
x = input('text')	Displays the text in quotes, waits for user input from the keyboard, and stores the value in x.
x = input('text','s')	Displays the text in quotes, waits for user input from the keyboard, and stores the input as a string in x.
k=menu('title','option1', 'option2',...	Displays a menu whose title is in the string variable 'title' and whose choices are 'option1','option2', and so on.

the screen. This command can be used with the first form of the disp function in a script file as follows (assuming the value of Speed is 63):

```
disp('The predicted speed is:')
disp(Speed)
```

When the file is run, these lines produce the following on the screen:

```
The predicted speed is:
    63
```

The input function displays text on the screen, waits for the user to enter something from the keyboard, and then stores the input in the specified variable. For example, the command x = input('Please enter the value of x:') causes the message to appear on the screen. If you type 5 and press **Enter,** the variable x will have the value 5.

A *string variable* is composed of text (alphanumeric characters). If you want to store a text input as a string variable, use the other form of the input command. For example, the command Calendar = input('Enter the day of the week:','s') prompts you to enter the day of the week. If you type Wednesday, this text will be stored in the string variable Calendar.

STRING VARIABLE

Use the menu function to generate a menu of choices for user input. Its syntax is

```
k = menu('title','option1','option2',...)
```

The function displays the menu whose title is in the string variable 'title', and whose choices are string variables 'option1', 'option2', and so on. The returned value of k is 1, 2, ... depending on whether you click on the button for option1, option2, and so forth. For example, the following script uses a menu to select the data marker for a graph, assuming that the arrays x and y already exist.

```
k = menu('Choose a data marker','o','*','x');
type = ['o','*','x'];
plot(x,y,x,y,type(k))
```

Table 1.4–2 Steps for developing a computer solution

1. State the problem concisely.
2. Specify the data to be used by the program. This is the "input."
3. Specify the information to be generated by the program. This is the "output."
4. Work through the solution steps by hand or with a calculator; use a simpler set of data if necessary.
5. Write and run the program.
6. Check the output of the program with your hand solution.
7. Run the program with your input data and perform a "reality check" on the output. Does it make sense? Estimate the range of the expected result and compare it with your answer.
8. If you will use the program as a general tool in the future, test it by running it for a range of reasonable data values; perform a reality check on the results.

Test Your Understanding

T1.4–1 The surface area A of a sphere depends on its radius r as follows: $A = 4\pi r^2$. Write a script file that prompts the user to enter a radius, computes the surface area, and displays the result.

Steps for Obtaining a Computer Solution

If you use a program such as MATLAB to solve a problem, follow the steps shown in Table 1.4–2.

1.5 The MATLAB Help System

To explore the more advanced features of MATLAB not covered in this book, you will need to know how to use effectively the MATLAB Help System. MATLAB has these options to get help for using MathWorks products.

1. *Help Browser* This graphical user interface helps you find information and view online documentation for your MathWorks products.
2. *Help Functions* The functions help, lookfor, and doc can be used to display syntax information for a specified function.
3. *Other Resources* For additional help, you can run demos, contact technical support, search documentation for other MathWorks products, view a list of other books, and participate in a newsgroup.

The Help Browser

To open the Help Browser, select MATLAB **Help** from the **Help** menu, or click the question mark button in the toolbar. The Help Browser contains two window "panes": the Help Navigator pane on the left and the Display pane on the right (see Figure 1.5–1). The Help Navigator contains four tabs:

■ *Contents:* a contents listing tab,
■ *Index:* a global index tab,

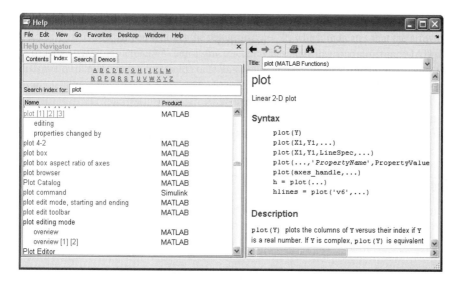

Figure 1.5–1 The MATLAB Help Browser.

- *Search:* a search tab having a find function and full text search features, and
- *Demos:* a bookmarking tab to start built-in demonstrations.

Use the tabs in the Help Navigator to find documentation. You view documentation in the Display pane. To open the Help Navigator pane from the display pane, click on **Help Navigator** in the **View** menu.

Viewing Documentation

After finding documentation with the Help Navigator, view the documentation in the Display pane. While viewing a page of documentation, you can:

- Scroll to see contents not currently visible in the window.
- View the previous or next page in the document by clicking the left or right arrow at the top of the page.
- View the previous or next item in the index by clicking the left or right arrow at the bottom of the page.
- Find a phrase in the currently displayed page by clicking on the binoculars icon and typing it in the **Find what:** box in the Help Browser toolbar and pressing the **Enter** key.

Using the Contents Tab

Click the **Contents** tab in the Help Navigator to list the titles and table of contents for all product documentation. To expand the listing for an item, click the

+ to the left of the item. To collapse the listings for an item, click the − to the left of the item, or double-click the item. Click on an item to select it. The first page of that document appears in the Display pane. Double-clicking an item in the contents listing expands the listing for that item and shows the first page of that document in the Display pane.

The Contents pane is synchronized with the Display pane. By default, the item selected in the Contents pane always matches the documentation appearing in the Display pane. Thus, the contents tree is synchronized with the displayed document.

Using the Index Tab

Click the **Index** tab in the Help Navigator pane to find specific index entries (keywords) from all of your MathWorks documentation. Type a word or words in the "Search index for" box. As you type, the index highlights the matching entries. Scroll down in the Help Navigator pane to see more matching entries. Click on an entry to display the corresponding page. If you do not find a matching index entry or if the corresponding page does not contain the information you seek, try a less specific topic by using only part of the wording, or use the **Search** tab.

Using the Search Tab

Click the **Search** tab in the Help Navigator pane to find all MATLAB documents containing a specified phrase. Type the phrase in the "Search for" box. Then click the **Go** button. The list of documents and the heading under which the phrase is found in that document then appear in the Help Navigator pane. Select an entry from the list of results to view that document in the Display pane.

Help Functions

Three MATLAB functions can be used for accessing online information about MATLAB functions.

The `help` Function   The `help` function is the most basic way to determine the syntax and behavior of a particular function. For example, typing `help log10` in the Command window produces the following display:

```
LOG10 Common (base 10) logarithm.
  LOG10(X) is the base 10 logarithm of the elements of X.
  Complex results are produced if X is not positive.

  See also LOG, LOG2, EXP, LOGM.
```

Note that the display describes what the function does, warns about any unexpected results if nonstandard argument values are used, and directs the user to other related functions.

All the MATLAB functions are organized into logical groups, upon which the MATLAB directory structure is based. For instance, all elementary mathematical functions such as `log10` reside in the `elfun` directory, and the polynomial functions reside in the `polyfun` directory. To list the names of all the functions in that directory, with a brief description of each, type `help polyfun`. If you are unsure of what directory to search, type `help` to obtain a list of all the directories, with a description of the function category each represents.

Typing `helpwin topic` displays the help text for the specified `topic` inside the Desktop Help Browser window. Links are created to functions referenced in the "See Also" line of the help text. You can also access the Help window by selecting the **Help** option under the **Help** menu, or by clicking the question mark button on the toolbar.

The `lookfor` Function   The `lookfor` function allows you to search for functions on the basis of a key word. It searches through the first line of help text, known as the H1 line, for each MATLAB function, and returns all the H1 lines containing a specified key word. For example, MATLAB does not have a function named `sine`. So the response from `help sine` is

```
sine.m not found
```

However, typing `lookfor` sine produces over a dozen matches, depending on which toolboxes you have installed. For example, you will, see among others,

```
ACOS        Inverse cosine, result in radians
ACOSD       Inverse cosine, result in degrees
ACOSH       Inverse hyperbolic cosine
ASIN        Inverse sine, result in radians
...
SIN         Sine of argument in radians
...
```

From this list you can find the correct name for the sine function. Note that all words containing sine are returned, such as cosine. Adding `-all` to the `lookfor` function searches the entire help entry, not just the H1 line.

The `doc` Function   Typing `doc function` displays the documentation for the MATLAB function `function`. Typing `doc toolbox/function` displays the documentation for the specified toolbox function. Typing `doc toolbox` displays the documentation road map page for the specified toolbox.

Table 1.5–1 summarizes the MATLAB Help functions.

1.6 Summary

You should now be familiar with basic operations in MATLAB. These include

- Starting and exiting MATLAB,
- Computing simple mathematical expressions, and
- Managing variables.

Table 1.5–1 MATLAB Help functions

Function	Use
`doc`	Displays the start page of the documentation in the Help Browser.
`doc function`	Displays the documentation for the MATLAB function `function`.
`doc toolbox/ function`	Displays the documentation for the specified toolbox function.
`doc toolbox`	Displays the documentation road map page for the specified toolbox.
`help`	Displays a list of all the function directories, with a description of the function category each represents.
`help function`	Displays in the Command window a description of the specified function `function`.
`helpwin topic`	Displays the help text for the specified `topic` inside the desktop Help Browser window.
`lookfor topic`	Displays in the Command window a brief description for all functions whose description includes the specified key word `topic`.
`type filename`	Displays the M-file `filename` without opening it with a text editor.

You should also be familiar with the MATLAB menu and toolbar system.

The chapter gives an overview of the various types of problems MATLAB can solve. These include

■ Using arrays and polynomials,
■ Creating plots, and
■ Creating script files

The following chapters give more details on these topics.

Key Terms with Page References

Problems

Answers to problems marked with an asterisk are given at the end of the text.

Section 1.1

1. Make sure you know how to start and quit a MATLAB session. Use MATLAB to make the following calculations, using the values: $x = 10$, $y = 3$. Check the results using a calculator.

   a. $u = x + y$ *b.* $v = xy$ *c.* $w = x/y$

   d. $z = \sin x$ *e.* $r = 8 \sin y$ *f.* $s = 5 \sin (2y)$

2.* Suppose that $x = 2$ and $y = 5$. Use MATLAB to compute the following.

   a. $\dfrac{yx^3}{x - y}$ *b.* $\dfrac{3x}{2y}$ *c.* $\dfrac{3}{2}xy$ *d.* $\dfrac{x^5}{x^5 - 1}$

3. Suppose that $x = 3$ and $y = 4$. Use MATLAB to compute the following, and check the results with a calculator.

   a. $\left(1 - \dfrac{1}{x^5}\right)^{-1}$ *b.* $3\pi x^2$ *c.* $\dfrac{3y}{4x - 8}$ *d.* $\dfrac{4(y - 5)}{3x - 6}$

4. Evaluate the following expressions in MATLAB for the given value of x. Check your answers by hand.

   a. $y = 6x^3 + \dfrac{4}{x}$, $x = 2$ *b.* $y = \dfrac{x}{4}3$, $x = 8$

   c. $y = \dfrac{(4x)^2}{25}$, $x = 10$ *d.* $y = 2\dfrac{\sin x}{5}$, $x = 2$

   e. $y = 7(x^{1/3}) + 4x^{0.58}$, $x = 20$

5. Assuming that the variables a, b, c, d, and f are scalars, write MATLAB statements to compute and display the following expressions. Test your statements for the values $a = 1.12$, $b = 2.34$, $c = 0.72$, $d = 0.81$, $f = 19.83$.

   $$x = 1 + \frac{a}{b} + \frac{c}{f^2} \qquad\qquad s = \frac{b - a}{d - c}$$

   $$r = \frac{1}{\frac{1}{a} + \frac{1}{b} + \frac{1}{c} + \frac{1}{d}} \qquad\qquad y = ab\,\frac{1}{c}\frac{f^2}{2}$$

6. Use MATLAB to calculate

   a. $\dfrac{3}{4}(6)(7^2) + \dfrac{4^5}{7^3 - 145}$ *b.* $\dfrac{48.2(55) - 9^3}{53 + 14^2}$

   c. $\dfrac{27^2}{4} + \dfrac{319^{4/5}}{5} + 60(14)^{-3}$

   Check your answers with a calculator.

7. The volume of a sphere is given by $V = 4\pi r^3/3$, where r is the radius. Use MATLAB to compute the radius of a sphere having a volume 30 percent greater than that of a sphere of radius 5 ft.

8.* Suppose that $x = -7 - 5i$ and $y = 4 + 3i$. Use MATLAB to compute

    a. $x + y$ *b.* xy *c.* x/y

9. Use MATLAB to compute the following. Check your answers by hand.

    a. $(3 + 6i)(-7 - 9i)$ *b.* $\dfrac{5 + 4i}{5 - 4i}$

    c. $\dfrac{3}{2}i$ *d.* $\dfrac{3}{2i}$

10. Evaluate the following expressions in MATLAB, for the values $x = 5 + 8i$, $y = -6 + 7i$. Check your answers by hand.

    a. $u = x + y$ *b.* $v = xy$ *c.* $w = x/y$

    d. $z = e^x$ *e.* $r = \sqrt{y}$ *f.* $s = xy^2$

11. The *ideal gas law* provides one way to estimate the pressure exerted by a gas in a container. The law is

$$P = \frac{nRT}{V}$$

More accurate estimates can be made with the *van der Waals* equation:

$$P = \frac{nRT}{V - nb} - \frac{an^2}{V^2}$$

where the term nb is a correction for the volume of the molecules, and the term an^2/V^2 is a correction for molecular attractions. The values of a and b depend on the type of gas. The gas constant is R, the *absolute* temperature is T, the gas volume is V, and the number of gas molecules is indicated by n. If $n = 1$ mol of an ideal gas were confined to a volume of $V = 22.41$ L at 0°C (273.2 K), it would exert a pressure of 1 atmosphere. In these units, $R = 0.08206$.

    For chlorine (Cl_2), $a = 6.49$ and $b = 0.0562$. Compare the pressure estimates given by the ideal gas law and the van der Waals equation for 1 mol of Cl_2 in 22.41 L at 273.2 K. What is the main cause of the difference in the two pressure estimates: the molecular volume or the molecular attractions?

12. The *ideal gas law* relates the pressure P, volume V, absolute temperature T, and amount of gas n. The law is

$$P = \frac{nRT}{V}$$

where R is the gas constant.

An engineer must design a large natural gas storage tank to be expandable to maintain the pressure constant at 2.2 atmospheres. In December when the temperature is 4°F (-15°C), the volume of gas in the tank is 28,500 ft^3. What will the volume of the same quantity of gas be in July when the temperature is 88°F (31°C)? (Hint: Use the fact that n, R, and P are constant in this problem. Note also that K = °C + 273.2.)

Section 1.3

13. Suppose x takes on the values $x = 1, 1.2, 1.4, \ldots, 5$. Use MATLAB to compute the array `y` that results from the function $y = 7 \sin(4x)$. Use MATLAB to determine how many elements are in the array `y`, and the value of the third element in the array `y`.

14. Use MATLAB to determine how many elements are in the array `[sin(-pi/2):0.05:cos(0)]`. Use MATLAB to determine the 10th element.

15. Use MATLAB to calculate

    a. $e^{(-2.1)^3} + 3.47 \log(14) + \sqrt[4]{287}$ *b.* $(3.4)^7 \log(14) + \sqrt[4]{287}$

    c. $\cos^2\left(\dfrac{4.12\pi}{6}\right)$ *d.* $\cos\left(\dfrac{4.12\pi}{6}\right)^2$

    Check your answers with a calculator.

16. Use MATLAB to calculate

    a. $6\pi \tan^{-1}(12.5) + 4$ *b.* $5 \tan[3 \sin^{-1}(13/5)]$

    c. $5 \ln(7)$ *d.* $5 \log(7)$

    Check your answers with a calculator.

17. The Richter scale is a measure of the intensity of an earthquake. The energy E (in joules) released by the quake is related to the magnitude M on the Richter scale as follows.

$$E = 10^{4.4} 10^{1.5M}$$

How much more energy is released by a magnitude 7.3 quake than a 5.5 quake?

18.* Use MATLAB to find the roots of $13x^3 + 182x^2 - 184x + 2503 = 0$.

19. Use MATLAB to find the roots of the polynomial $36x^3 + 12x^2 - 5x + 10$.

20. Determine which search path MATLAB uses on your computer. If you use a lab computer as well as a home computer, compare the two search paths. Where will MATLAB look for a user-created M-file on each computer?

21. Use MATLAB to plot the function $T = 6 \ln t - 7e^{0.2t}$ over the interval $1 \leq t \leq 3$. Put a title on the plot and properly label the axes. The variable T represents temperature in degrees Celsius; the variable t represents time in minutes.

22. Use MATLAB to plot the functions $u = 2 \log_{10}(60x + 1)$ and $v = 3 \cos(6x)$ over the interval $0 \le x \le 2$. Properly label the plot and each curve. The variables u and v represent speed in miles per hour; the variable x represents distance in miles.

23. The Fourier series is a series representation of a periodic function in terms of sines and cosines. The Fourier series representation of the function

$$f(x) = \begin{cases} 1 & 0 < x < \pi \\ -1 & -\pi < x < 0 \end{cases}$$

is

$$\frac{4}{\pi}\left(\frac{\sin x}{1} + \frac{\sin 3x}{3} + \frac{\sin 5x}{5} + \frac{\sin 7x}{7} + \cdots \right)$$

Plot on the same graph the function $f(x)$ and its series representation using the four terms shown.

24. A *cycloid* is the curve described by a point P on the circumference of a circular wheel of radius r rolling along the x axis. The curve is described in parametric form by the equations

$$x = r(\phi - \sin \phi)$$
$$y = r(1 - \cos \phi)$$

Use these equations to plot the cycloid for $r = 10$ inches and $0 \le \phi \le 4\pi$.

Section 1.4

25. A fence around a field is shaped as shown in Figure P25. It consists of a rectangle of length L and width W, and a right triangle that is symmetrical about the central horizontal axis of the rectangle. Suppose the width W is known (in meters), and the enclosed area A is known (in square meters). Write a MATLAB script file in terms of the given variables W and A to determine the length L required so that the enclosed area is A. Also determine the total length of fence required. Test your script for the values $W = 6$ m and $A = 80$ m².

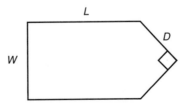

Figure P25

26. The four-sided figure shown in Figure P26 consists of two triangles having a common side a. The law of cosines for the top triangle states that

$$a^2 = b_1^2 + c_1^2 - 2b_1 c_1 \cos A_1$$

and a similar equation can be written for the bottom triangle. Develop a procedure for computing the length of side c_2 if you are given the lengths of sides b_1, b_2, and c_1, and the angles A_1 and A_2 in degrees. Write a script file to implement this procedure. Test your script using the following values: $b_1 = 180$ m, $b_2 = 165$ m, $c_1 = 115$ m, $A_1 = 120°$ and $A_2 = 100°$.

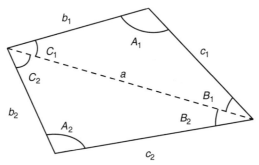

Figure P26

Section 1.5

27. Use the MATLAB help facilities to find information about the following topics and symbols: plot, label, cos, cosine, :, and *.

28. Use the MATLAB help facilities to determine what happens if you use the sqrt function with a negative argument.

29. Use the MATLAB help facilities to determine what happens if you use the exp function with an imaginary argument.

2

C H A P T E R

Numeric, Cell, and Structure Arrays

OUTLINE

One of the strengths of MATLAB is the capability to handle collections of items, called *arrays,* as if they were a single entity. The array-handling feature means that MATLAB programs can be very short.

The array is the basic building block in MATLAB. The following classes of arrays are available in MATLAB 7:

Array

numeric	character	logical	cell	structure	function handle	Java

So far we have used only numeric arrays, which are arrays containing only numeric values. Within the numeric class are the subclasses *single* (single precision), *double* (double precision), *int8*, *int16*, and *int32* (signed 8-bit, 16-bit, and 32-bit integers), and *uint8*, *uint16*, and *uint32* (unsigned 8-bit, 16-bit, and 32-bit integers). A character array is an array containing strings. The elements of logical arrays are "true" or "false," which, although represented by the symbols 1 and 0, are not numeric quantities. We will study the logical arrays in Chapter 4. Cell arrays and structure arrays are covered in Sections 2.7 and 2.8 of this chapter. Function handles are treated in Chapter 3. The Java class is not covered in this text.

The first four sections of this chapter treat concepts that are essential to understanding MATLAB and therefore must be covered. Sections 2.5 and 2.6 treat specific applications that may not be of interest to all readers. Sections 2.7 and 2.8 introduce two types of arrays that are useful for some specialized applications.

2.1 One- and Two-Dimensional Numeric Arrays

We can represent the location of a point in three-dimensional space by three Cartesian coordinates x, y, and z. These three coordinates specify a *vector* **p.** (In mathematical text we often use boldface type to indicate vectors.) The set of *unit vectors* **i, j, k,** whose lengths are 1 and whose directions coincide with the x, y, and z axes, respectively, can be used to express the vector mathematically as follows: $\mathbf{p} = x\mathbf{i} + y\mathbf{j} + z\mathbf{k}.$ The unit vectors enable us to associate the vector components x, y, z with the proper coordinate axes; therefore, when we write $\mathbf{p} = 5\mathbf{i} + 7\mathbf{j} + 2\mathbf{k},$ we know that the x, y, and z coordinates of the vector are 5, 7, and 2, respectively. We can also write the components in a specific order, separate them with a space, and identify the group with brackets, as follows: [5 7 2]. As long as we agree that the vector components will be written in the order x, y, z, we can use this notation instead of the unit-vector notation. In fact, MATLAB uses this style for vector notation. MATLAB allows us to separate the components with commas for improved readability if we desire so that the equivalent way of writing the preceding vector is [5, 7, 2]. This expression is a *row vector*, which is a horizontal arrangement of the elements.

ROW VECTOR

We can also express the vector as a *column vector*, which has a vertical arrangement. A vector can have only one column, or only one row. Thus, a vector is a one dimensional array. In general, arrays can have more than one column and more than one row.

COLUMN VECTOR

Creating Vectors in MATLAB

The concept of a vector can be generalized to any number of components. In MATLAB a vector is simply a list of scalars, whose order of appearance in the list might be significant, as it is when specifying xyz coordinates. As another example, suppose we measure the temperature of an object once every hour. We can represent the measurements as a vector, and the 10th element in the list is the temperature measured at the 10th hour.

To create a row vector in MATLAB, you simply type the elements inside a pair of square brackets, separating the elements with a space or a comma. Brackets are required for arrays in some cases, but not all. To improve readability, we will always use them. The choice between a space or comma is a matter of personal preference, although the chance of an error is less if you use a comma. (You can also use a comma followed by a space for maximum readability.)

To create a column vector, you can separate the elements by semicolons; alternatively, you can create a row vector and then use the *transpose* notation (′), which converts a row vector into a column vector, or vice versa. For example:

TRANSPOSE

```
>>g = [3;7;9]
g =
     3
     7
     9
>>g = [3,7,9]'
g =
     3
     7
     9
```

The third way to create a column vector is to type a left bracket ([ ) and the first element, press **Enter,** type the second element, press **Enter,** and so on until you type the last element followed by a right bracket ( ]) and **Enter.** On the screen this sequence looks like

```
>>g = [3
7
9]
g =
     3
     7
     9
```

Note that MATLAB displays row vectors horizontally and column vectors vertically.

You can create vectors by "appending" one vector to another. For example, to create the row vector u whose first three columns contain the values of r = [2,4,20] and whose fourth, fifth, and sixth columns contain the values of w = [9,-6,3], you type u = [r,w]. The result is the vector u = [2,4,20,9,-6,3].

The colon operator (:) easily generates a large vector of regularly spaced elements. Typing

```
>>x = [m:q:n]
```

creates a vector x of values with a spacing q. The first value is m. The last value is n if m − n is an integer multiple of q. If not, the last value is less than n. For example, typing x = [0:2:8] creates the vector x = [0,2,4,6,8], whereas typing x = [0:2:7] creates the vector x = [0,2,4,6]. To create a row vector z consisting of the values from 5 to 8 in steps of 0.1, you type z = [5:0.1:8]. If the increment q is omitted, it is presumed to be 1. Thus y = [-3:2] produces the vector y = [-3,-2,-1,0,1,2].

The increment q can be negative. In this case m should be greater than n. For example, u = [10:-2:4] produces the vector [10,8,6,4].

The linspace command also creates a linearly spaced row vector, but instead you specify the number of values rather than the increment. The syntax is linspace(x1,x2,n), where x1 and x2 are the lower and upper limits and n is the number of points. For example, linspace(5,8,31) is equivalent to [5:0.1:8]. If n is omitted, the spacing is 1.

The logspace command creates an array of logarithmically spaced elements. Its syntax is logspace(a,b,n), where n is the number of points between 10^a and 10^b. For example, x = logspace(-1,1,4) produces the vector x = [0.1000, 0.4642, 2.1544, 10.000]. If n is omitted, the number of points defaults to 50.

Two-Dimensional Arrays

An array having rows and columns is a two-dimensional array that is sometimes called a *matrix*. In mathematical text if possible, vectors are usually denoted by boldface lowercase letters and matrices by boldface uppercase letters. An example of a matrix having three rows and two columns is

MATRIX

$$M = \begin{bmatrix} 2 & 5 \\ -3 & 4 \\ -7 & 1 \end{bmatrix}$$

We refer to the *size* of an array by the number of rows and the number of columns. For example, an array with 3 rows and 2 columns is said to be a 3×2 array. *The number of rows is always stated first!* We sometimes represent a matrix **A** as $[a_{ij}]$ to indicate its elements a_{ij}. The subscripts i and j—called *indices*—indicate the row and column location of the element a_{ij}. *The row number must always come first!* For example, the element a_{32} is in row 3, column 2. Two matrices **A** and **B** are equal if they have the same size and if all their corresponding elements are equal; that is, $a_{ij} = b_{ij}$ for every value of i and j.

ARRAY SIZE

Creating Matrices

The most direct way to create a matrix is to type the matrix row by row, separating the elements in a given row with spaces or commas and separating the rows with semicolons. For example, typing

```
>>A = [2,4,10;16,3,7];
```

creates the following matrix:

$$A = \begin{bmatrix} 2 & 4 & 10 \\ 16 & 3 & 7 \end{bmatrix}$$

If the matrix has many elements, you can press **Enter** and continue typing on the next line. MATLAB knows you are finished entering the matrix when you type the closing bracket ( ]).

You can append a row vector to another row vector to create either a third row vector or a matrix (if both vectors have the same number of columns). Note the difference between the results given by [a,b] and [a;b] in the following session:

```
>>a = [1,3,5];
>>b = [7,9,11];
>>c = [a,b]
c =
     1 3 5 7 9 11
>> D = [a;b]
D =
     1 3 5
     7 9 11
```

Matrices and the Transpose Operation

The transpose operation interchanges the rows and columns. In mathematics text we denote this operation by the superscript T. For an $m \times n$ matrix **A** with m rows and n columns, $\mathbf{A}^T$ (read "A transpose") is an $n \times m$ matrix.

$$\mathbf{A} = \begin{bmatrix} -2 & 6 \\ -3 & 5 \end{bmatrix} \qquad \mathbf{A}^T = \begin{bmatrix} -2 & -3 \\ 6 & 5 \end{bmatrix}$$

If $\mathbf{A}^T = \mathbf{A}$, the matrix **A** is *symmetric.* Note that the transpose operation converts a row vector into a column vector, and vice versa.

If the array contains complex elements, the transpose operator (') produces the *complex conjugate transpose;* that is, the resulting elements are the complex conjugates of the original array's transposed elements. Alternatively, you can use the *dot transpose* operator (. ') to transpose the array without producing complex conjugate elements, for example, A. '. If all the elements are real, the operators ' and . ' give the same result.

ARRAY ADDRESSING

Array Addressing

Array indices are the row and column numbers of an element in an array and are used to keep track of the array's elements. For example, the notation v(5) refers to the fifth element in the vector v, and A(2,3) refers to the element in row 2, column 3 in the matrix A. *The row number is always listed first!* This notation enables you to correct entries in an array without retyping the entire array. For example, to change the element in row 1, column 3 of a matrix **D** to 6, you can type D(1,3) = 6.

The colon operator selects individual elements, rows, columns, or "subarrays" of arrays. Here are some examples:

■ v(:) represents all the row or column elements of the vector v.

■ v(2:5) represents the second through fifth elements; that is v(2), v(3), v(4), v(5).

- $A(:,3)$ denotes all the elements in the third *column* of the matrix A.
- $A(3,:)$ denotes all the elements in the third *row* of A.
- $A(:,2:5)$ denotes all the elements in the second through fifth columns of A.
- $A(2:3,1:3)$ denotes all the elements in the second and third rows that are also in the first through third columns.
- $v = A(:)$ creates a vector v consisting of all the columns of A stacked from first to last.
- $A(end,:)$ denotes the last row in A. $A(:,end)$ denotes the last column.

You can use array indices to extract a smaller array from another array. For example, if you create the array **B**

$$\mathbf{B} = \begin{bmatrix} 2 & 4 & 10 & 13 \\ 16 & 3 & 7 & 18 \\ 8 & 4 & 9 & 25 \\ 3 & 12 & 15 & 17 \end{bmatrix} \tag{2.1-1}$$

by typing

```
>>B = [2,4,10,13;16,3,7,18;8,4,9,25;3,12,15,17];
```

and then type

```
>>C = B(2:3,1:3);
```

you can produce the following array:

$$\mathbf{C} = \begin{bmatrix} 16 & 3 & 7 \\ 8 & 4 & 9 \end{bmatrix}$$

The *empty array* contains no elements and is expressed as []. Rows and columns can be deleted by setting the selected row or column equal to the empty array. This step causes the original matrix to collapse to a smaller one. For example, $A(3,:) = []$ deletes the third row in A, while $A(:,2:4) = []$ deletes the second through fourth columns in A. Finally, $A([1\ 4],:) = []$ deletes the first and fourth rows of A.

Suppose we type $A = [6,9,4;1,5,7]$ to define the following matrix:

$$\mathbf{A} = \begin{bmatrix} 6 & 9 & 4 \\ 1 & 5 & 7 \end{bmatrix}$$

Typing $A(1,5) = 3$ changes the matrix to

$$\mathbf{A} = \begin{bmatrix} 6 & 9 & 4 & 0 & 3 \\ 1 & 5 & 7 & 0 & 0 \end{bmatrix}$$

Because **A** did not have five columns, its size is automatically expanded to accept the new element in column 5. MATLAB adds zeros to fill out the remaining elements.

EMPTY ARRAY

MATLAB does not accept negative or zero indices, but you can use negative increments with the colon operator. For example, typing B = A(:,5:-1:1) reverses the order of the columns in **A** and produces

$$\mathbf{B} = \begin{bmatrix} 3 & 0 & 4 & 9 & 6 \\ 0 & 0 & 7 & 5 & 1 \end{bmatrix}$$

Suppose that C = [-4,12,3,5,8]. Then typing B(2,:) = C replaces row 2 of B with C. Thus **B** becomes

$$\mathbf{B} = \begin{bmatrix} 3 & 0 & 4 & 9 & 6 \\ -4 & 12 & 3 & 5 & 8 \end{bmatrix}$$

Suppose that D = [3,8,5;4,-6,9]. Then typing E = D([2,2,2],:) repeats row 2 of D three times to obtain

$$\mathbf{E} = \begin{bmatrix} 4 & -6 & 9 \\ 4 & -6 & 9 \\ 4 & -6 & 9 \end{bmatrix}$$

Using `clear` to Avoid Errors

You can use the `clear` command to protect yourself from accidentally reusing an array that has the wrong dimension. Even if you set new values for an array, some previous values might still remain. For example, suppose you had previously created the 2 × 2 array A = [2, 5; 6, 9], and you then create the 5 × 1 arrays x = [1:5]' and y = [2:6]'. Suppose you now redefine A so that its columns will be x and y. If you then type A(:,1) = x to create the first column, MATLAB displays an error message telling you that the number of rows in A and x must be the same. MATLAB thinks A should be a 2 × 2 matrix because A was previously defined to have only two rows and its values remain in memory. The `clear` command wipes A and all other variables from memory and avoids this error. To clear A only, type `clear` A before typing A(:,1) = x.

Some Useful Array Functions

MATLAB has many functions for working with arrays (see Table 2.1–1). Here is a summary of some of the more commonly used functions.

The max(A) function returns the algebraically greatest element in **A** if **A** is a vector having all real elements. It returns a row vector containing the greatest elements in each *column* if **A** is a matrix containing all real elements. If *any* of the elements are complex, max(A) returns the element that has the largest magnitude. The syntax [x,k] = max(A) is similar to max(A), but it stores the maximum values in the row vector **x** and their indices in the row vector **k**.

If A and B have the same size, C = max(A,B) creates an array the same size, having the maximum value from each corresponding location in A and B. For example, the following **A** and **B** matrices give the **C** matrix shown.

Table 2.1–1 Basic syntax of array functions*

Command	Description
`find(x)`	Computes an array containing the indices of the nonzero elements of the array **x.**
`[u,v,w] = find(A)`	Computes the arrays **u** and **v**, containing the row and column indices of the nonzero elements of the matrix A, and the array **w**, containing the values of the nonzero elements. The array **w** may be omitted.
`length(A)`	Computes either the number of elements of **A** if **A** is a vector or the largest value of m or n if **A** is an $m \times n$ matrix.
`linspace(a,b,n)`	Creates a row vector of n regularly spaced values between a and b.
`logspace(a,b,n)`	Creates a row vector of n logarithmically spaced values between a and b.
`max(A)`	Returns the algebraically largest element in **A** if **A** is a vector. Returns a row vector containing the largest elements in each column if **A** is a matrix. If any of the elements are complex, `max(A)` returns the elements that have the largest magnitudes.
`[x,k] = max(A)`	Similar to `max(A)` but stores the maximum values in the row vector **x** and their indices in the row vector **k.**
`min(A)`	Same as `max(A)` but returns minimum values.
`[x,k] = min(A)`	Same as `[x,k] = max(A)` but returns minimum values.
`norm(x)`	Computes a vector's geometric length $\sqrt{x_1^2 + x_2^2 + \cdots + x_n^2}$.
`size(A)`	Returns a row vector `[m n]` containing the sizes of the $m \times n$ array **A.**
`sort(A)`	Sorts each column of the array **A** in ascending order and returns an array the same size as **A.**
`sum(A)`	Sums the elements in each column of the array **A** and returns a row vector containing the sums.

*Many of these functions have extended syntax. See the text and MATLAB help for more discussion.

$$A = \begin{bmatrix} 1 & 6 & 4 \\ 3 & 7 & 2 \end{bmatrix} \quad B = \begin{bmatrix} 3 & 4 & 7 \\ 1 & 5 & 8 \end{bmatrix} \quad C = \begin{bmatrix} 3 & 6 & 7 \\ 3 & 7 & 8 \end{bmatrix}$$

The functions `min(A)` and `[x,k] = min(A)` are the same as `max(A)` and `[x,k] = max(A)` except that they return minimum values.

The function `size(A)` returns a row vector `[m n]` containing the sizes of the $m \times n$ array **A.** The `length(A)` function computes either the number of elements of **A** if A is a vector or the largest value of m or n if **A** is an $m \times n$ matrix.

For example, if

$$A = \begin{bmatrix} 6 & 2 \\ -10 & -5 \\ 3 & 0 \end{bmatrix}$$

then `max(A)` returns the vector `[6,2]`; `min(A)` returns the vector `[-10, -5]`; `size(A)` returns `[3,2]`; and `length(A)` returns 3.

The `sum(A)` function sums the elements in each *column* of the array **A** and returns a row vector containing the sums. The `sort(A)` function sorts each *column* of the array **A** in ascending order and returns an array the same size as **A.**

If A has one or more complex elements, the `max`, `min`, and `sort` functions act on the absolute values of the elements and return the element that has the largest magnitude.

For example, if

$$A = \begin{bmatrix} 6 & 2 \\ -10 & -5 \\ 3 + 4i & 0 \end{bmatrix}$$

then `max(A)` returns the vector `[-10,-5]` and `min(A)` returns the vector `[3+4i,0]`. (The magnitude of $3 + 4i$ is 5.)

The sort will be done in descending order if the form `sort(A, 'descend')` is used. The `min`, `max`, and `sort` functions can be made to act on the rows instead of the columns by transposing the array.

The complete syntax of the `sort` function is `sort(A, dim, mode)`, where `dim` selects a dimension along which to sort, and `mode` selects the direction of the sort, `'ascend'` for ascending order and `'descend'` for descending order. So, for example, `sort(A,2, 'descend')` would `sort` the elements in each row of **A** in descending order.

The `find(x)` command computes an array containing the indices of the *nonzero* elements of the vector **x.** The syntax `[u,v,w] = find(A)` computes the arrays **u** and **v,** containing the row and column indices of the nonzero elements of the matrix **A,** and the array **w,** containing the values of the nonzero elements. The array **w** may be omitted.

For example, if

$$A = \begin{bmatrix} 6 & 0 & 3 \\ 0 & 4 & 0 \\ 2 & 7 & 0 \end{bmatrix}$$

then the session

```
>>A = [6, 0, 3; 0, 4, 0; 2, 7, 0];
>>[u, v, w] = find(A)
```

returns the vectors

$$\mathbf{u} = \begin{bmatrix} 1 \\ 3 \\ 2 \\ 3 \\ 1 \end{bmatrix} \qquad \mathbf{v} = \begin{bmatrix} 1 \\ 1 \\ 2 \\ 2 \\ 3 \end{bmatrix} \qquad \mathbf{w} = \begin{bmatrix} 6 \\ 2 \\ 4 \\ 7 \\ 3 \end{bmatrix}$$

MAGNITUDE

LENGTH

ABSOLUTE VALUE

The vectors **u** and **v** give the (row, column) indices of the nonzero values, which are listed in **w.** For example, the second entries in **u** and **v** give the indices (3, 1), which specifies the element in row 3, column 1 of **A,** whose value is 2.

These functions are summarized in Table 2.1–1.

Magnitude, Length, and Absolute Value of a Vector

The terms *magnitude, length,* and *absolute value* are often loosely used in everyday language, but you must keep their precise meaning in mind when using MATLAB.

The MATLAB `length` command gives the number of elements in the vector. The *magnitude* of a vector **x** having real elements $x_1, x_2, \ldots, x_n$ is a scalar, given by $\sqrt{x_1^2 + x_2^2 + \cdots + x_n^2}$, and is the same as the vector's geometric length. The *absolute value* of a vector **x** is a vector whose elements are the absolute values of the elements of **x**. For example, if x = [2,-4,5], its length is 3; its magnitude is $\sqrt{2^2 + (-4)^2 + 5^2} = 6.7082$; and its absolute value is [2,4,5]. The length, magnitude, and absolute value of x are computed by `length(x)`, `norm(x)`, and `abs(x)`, respectively.

Test Your Understanding

T2.1–1 For the matrix **B**, find the array that results from the operation [B;B']. Use MATLAB to determine what number is in row 5, column 3 of the result.

$$\mathbf{B} = \begin{bmatrix} 2 & 4 & 10 & 13 \\ 16 & 3 & 7 & 18 \\ 8 & 4 & 9 & 25 \\ 3 & 12 & 15 & 17 \end{bmatrix}$$

T2.1–2 For the same matrix **B**, use MATLAB to (a) find the largest and smallest element in **B** and their indices and (b) sort each column in **B** to create a new matrix **C**.

The Array Editor

The MATLAB Workspace Browser provides a graphical interface for managing the workspace. You can use it to view, save, and clear workspace variables. It includes the *Array Editor,* a graphical interface for working with arrays. To open the Workspace Browser, type `workspace` at the Command window prompt. The browser appears as shown in Figure 2.1–1.

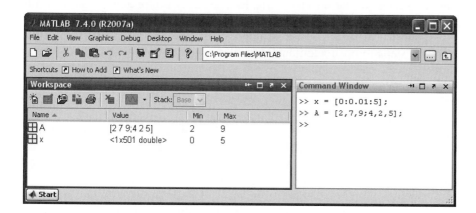

Figure 2.1–1 The Workspace Browser.

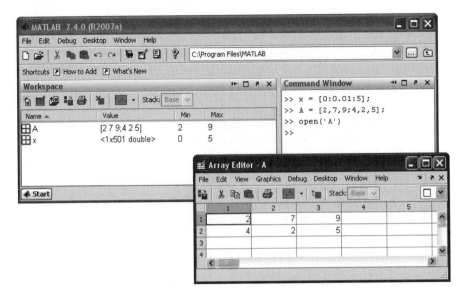

Figure 2.1–2 The Array Editor.

Keep in mind that the Desktop menus are context-sensitive. Thus their contents will change depending on which features of the Browser and Array Editor you are currently using. The Workspace Browser shows the name of each variable, its value, array size, and class. The icon for each variable illustrates its class.

From the Workspace Browser you can open the Array Editor to view and edit a visual representation of two-dimensional numeric arrays, with the rows and columns numbered. To open the Array Editor from the Workspace Browser, double-click on the variable you want to open. The Array Editor opens, displaying the values for the selected variable. The Array Editor appears as shown in Figure 2.1–2.

To open a variable, you can also right-click it and use the **Context** menu. Repeat the steps to open additional variables into the Array Editor. In the Array Editor, access each variable via its tab at the bottom of the window, or use the **Window** menu. You can also open the Array Editor directly from the Command window by typing open('var'), where var is the name of the variable to be edited. Once an array is displayed in the Array Editor, you can change a value in the array by clicking on its location, typing in the new value, and pressing **Enter.**

Right-clicking on a variable brings up the **Context** menu, which can be used to edit, save, or clear the selected variable, or to plot the rows of the variable versus its columns (this type of plot is discussed in Chapter 5).

You can also clear a variable from the Workspace Browser by first highlighting it in the Browser, then clicking on **Delete** in the **Edit** menu.

2.2 Multidimensional Numeric Arrays

MATLAB supports multidimensional arrays. For more information, type `help datatypes`.

A three-dimensional array has the dimension $m \times n \times q$. A four-dimensional array has the dimension $m \times n \times q \times r$, and so forth. The first two dimensions are the row and column, as with a matrix. The higher dimensions are called *pages*. You can think of a three-dimensional array as layers of matrices. The first layer is page 1; the second layer is page 2, and so on. If A is a $3 \times 3 \times 2$ array, you can access the element in row 3, column 2 of page 2 by typing A(3,2,2). To access all of page 1, type A(:,:,1). To access all of page 2, type A(:,:,2). The `ndims` command returns the number of dimensions. For example, for the array A just described, `ndims(A)` returns the value 3.

You can create a multidimensional array by first creating a two-dimensional array and then extending it. For example, suppose you want to create a three-dimensional array whose first two pages are

$$\begin{bmatrix} 4 & 6 & 1 \\ 5 & 8 & 0 \\ 3 & 9 & 2 \end{bmatrix} \quad \begin{bmatrix} 6 & 2 & 9 \\ 0 & 3 & 1 \\ 4 & 7 & 5 \end{bmatrix}$$

To do so, first create page 1 as a 3×3 matrix and then add page 2, as follows:

```
>>A = [4,6,1;5,8,0;3,9,2];
>>A(:,:,2) = [6,2,9;0,3,1;4,7,5];
```

Another way to produce such an array is with the `cat` command. Typing `cat(n,A,B,C,...)` creates a new array by concatenating the arrays A, B, C, and so on along the dimension n. Note that `cat(1,A,B)` is the same as [A;B] and that `cat(2,A,B)` is the same as [A,B]. For example, suppose we have the 2×2 arrays **A** and **B**:

$$\mathbf{A} = \begin{bmatrix} 8 & 2 \\ 9 & 5 \end{bmatrix} \quad \mathbf{B} = \begin{bmatrix} 4 & 6 \\ 7 & 3 \end{bmatrix}$$

Then C = cat(3,A,B) produces a three-dimensional array composed of two layers; the first layer is the matrix A, and the second layer is the matrix B. The element C(m,n,p) is located in row m, column n, and layer p. Thus the element C(2,1,1) is 9, and the element C(2,2,2) is 3.

Multidimensional arrays are useful for problems that involve several parameters. For example, if we have data on the temperature distribution in a rectangular object, we could represent the temperatures as an array T with three dimensions.

2.3 Element-by-Element Operations

To increase the magnitude of a vector, multiply it by a scalar. For example, to double the magnitude of the vector r = [3,5,2], multiply each component by two to obtain [6,10,4]. In MATLAB you type v = 2*r.

Multiplying a matrix **A** by a scalar w produces a matrix whose elements are the elements of **A** multiplied by w. For example:

$$3\begin{bmatrix} 2 & 9 \\ 5 & -7 \end{bmatrix} = \begin{bmatrix} 6 & 27 \\ 15 & -21 \end{bmatrix}$$

This multiplication is performed in MATLAB as follows:

```
>>A = [2,9;5,-7];
>>3*A
```

**ARRAY
OPERATIONS**

**ELEMENT-
BY-ELEMENT
OPERATIONS**

Thus multiplication of an array by a scalar is easily defined and easily carried out. However, multiplication of two *arrays* is not so straightforward. In fact, MATLAB uses two definitions of multiplication: (1) array multiplication and (2) matrix multiplication. Division and exponentiation must also be carefully defined when you are dealing with operations between two arrays. MATLAB has two forms of arithmetic operations on arrays. In this section we introduce one form, called *array operations,* which are also called *element-by-element* operations. In the next section we introduce *matrix* operations. Each form has its own applications, which we illustrate by examples.

Array Addition and Subtraction

Array addition can be done by adding the corresponding components. To add the arrays r = [3,5,2] and v = [2,-3,1] to create w in MATLAB, you type w = r + v. The result is w = [5,2,3].

When two arrays have identical size, their sum or difference has the same size and is obtained by adding or subtracting their corresponding elements. Thus **C = A + B** implies that $c_{ij} = a_{ij} + b_{ij}$ if the arrays are matrices. The array **C** has the same size as **A** and **B**. For example:

$$\begin{bmatrix} 6 & -2 \\ 10 & 3 \end{bmatrix} + \begin{bmatrix} 9 & 8 \\ -12 & 14 \end{bmatrix} = \begin{bmatrix} 15 & 6 \\ -2 & 17 \end{bmatrix} \tag{2.3–1}$$

Array subtraction is performed in a similar way.

The addition shown in equation 2.3–1 is performed in MATLAB as follows:

```
>>A = [6,-2;10,3];
>>B = [9,8;-12,14]
>>A+B
ans =
      15    6
      -2   17
```

Array addition and subtraction are associative and commutative. For addition these properties mean that

$$(\mathbf{A} + \mathbf{B}) + \mathbf{C} = \mathbf{A} + (\mathbf{B} + \mathbf{C}) \tag{2.3–2}$$

$$\mathbf{A} + \mathbf{B} + \mathbf{C} = \mathbf{B} + \mathbf{C} + \mathbf{A} = \mathbf{A} + \mathbf{C} + \mathbf{B} \tag{2.3–3}$$

Table 2.3–1 Element-by-element operations

Symbol	Operation	Form	Example
+	Scalar-array addition	A + b	[6,3]+2=[8,5]
−	Scalar-array subtraction	A − b	[8,3]−5=[3,−2]
+	Array addition	A + B	[6,5]+[4,8]=[10,13]
−	Array subtraction	A − B	[6,5]−[4,8]=[2,−3]
.*	Array multiplication	A.*B	[3,5].*[4,8]=[12,40]
./	Array right division	A./B	[2,5]./[4,8]=[2/4,5/8]
.\	Array left division	A.\B	[2,5].\[4,8]=[2\4,5\8]
.^	Array exponentiation	A.^B	[3,5].^2=[3^2,5^2]
			2.^[3,5]=[2^3,2^5]
			[3,5].^[2,4]=[3^2,5^4]

Array addition and subtraction require that both arrays have the same size. The only exception to this rule in MATLAB occurs when we add or subtract a *scalar* to or from an array. In this case the scalar is added or subtracted from each element in the array. Table 2.3–1 gives examples.

Element-by-Element Multiplication

MATLAB defines element-by-element multiplication only for arrays that have the same size. The definition of the product x.*y, where x and y each have n elements, is

$$x.*y. = [x(1)y(1), \quad x(2)y(2) \quad . . . \quad , \quad x(n)y(n)]$$

if x and y are row vectors. For example, if

$$\mathbf{x} = [2, 4, -5] \qquad \mathbf{y} = [-7, 3, -8] \qquad (2.3–4)$$

then z = x.*y gives

$$\mathbf{z} = [2(-7), 4(3), -5(-8)] = [-14, 12, 40]$$

This type of multiplication is sometimes called *array* multiplication.

If u and v are column vectors, the result of u.*v is a column vector.

Note that x′ is a column vector with size 3 × 1 and thus does not have the same size as y, whose size is 1 × 3. Thus for the vectors x and y the operations x′.*y and y.*x′ are not defined in MATLAB and will generate an error message. With element-by-element multiplication, it is important to remember that the dot (.) and the asterisk (*) form *one* symbol (.*). It might have been better to have defined a single symbol for this operation, but the developers of MATLAB were limited by the selection of symbols on the keyboard.

The generalization of array multiplication to arrays with more than one row or column is straightforward. Both arrays must have the same size. The array operations are performed between the elements in corresponding locations in the arrays. For example, the array multiplication operation A.*B results in a matrix C that has the same size as A and B and has the elements $c_{ij} = a_{ij}b_{ij}$. For example, if

$$A = \begin{bmatrix} 11 & 5 \\ -9 & 4 \end{bmatrix} \qquad B = \begin{bmatrix} -7 & 8 \\ 6 & 2 \end{bmatrix}$$

then C = A.*B gives this result:

$$C = \begin{bmatrix} 11(-7) & 5(8) \\ -9(6) & 4(2) \end{bmatrix} = \begin{bmatrix} -77 & 40 \\ -54 & 8 \end{bmatrix}$$

EXAMPLE 2.3–1	Vectors and Displacement

Suppose two divers start at the surface and establish the following coordinate system: x is to the west, y is to the north, and z is down. Diver 1 swims 55 ft west, 36 ft north, and then dives 25 ft. Diver 2 dives 15 ft, then swims east 20 ft and then north 59 ft. (a) Find the distance between diver 1 and the starting point. (b) How far in each direction must diver 1 swim to reach diver 2? How far in a straight line must diver 1 swim to reach diver 2?

■ Solution

(a) Using the xyz coordinates selected, the position of diver 1 is $\mathbf{r} = 55\mathbf{i} + 36\mathbf{j} + 25\mathbf{k}$, and the position of diver 2 is $\mathbf{r} = -20\mathbf{i} + 59\mathbf{j} + 15\mathbf{k}$. (Note that diver 2 swam east, which is in the negative x direction.) The distance from the origin of a point xyz is given by $\sqrt{x^2 + y^2 + z^2}$, that is, by the magnitude of the vector pointing from the origin to the point xyz. This distance is computed in the following session.

```
>>r = [55,36,25];w = [-20,59,15];
>>dist1 = sqrt(sum(r.*r))
dist1 =
   70.3278
```

The distance is approximately 70 ft. The distance could also have been computed from norm(r).

(b) The location of diver 2 relative to diver 1 is given by the vector $\mathbf{v}$ pointing from diver 1 to diver 2. We can find this vector using vector subtraction: $\mathbf{v} = \mathbf{w} - \mathbf{r}$. Continue the above MATLAB session as follows:

```
>>v = w-r
v =
   -75      23     -10
>>dist2 = sqrt(sum(v.*v))
dist2 =
   79.0822
```

Thus to reach diver 2 by swimming along the coordinate directions, diver 1 must swim 75 ft east, 23 ft north, and 10 ft up. The straight-line distance between them is approximately 79 feet.

Vectorized Functions

The built-in MATLAB functions such as sqrt(x) and exp(x) automatically operate on array arguments to produce an array result the same size as the array argument x. Thus these functions are said to be *vectorized* functions.

Thus, when multiplying or dividing these functions, or when raising them to a power, you must use element-by-element operations if the arguments are arrays. For example, to compute $z = (e^y \sin x) \cos^2 x$, you must type z = exp(y).* sin(x).*(cos(x)).^2. Obviously, you will get an error message if the size of x is not the same as the size of y. The result z will have the same size as x and y.

Aortic Pressure Model

EXAMPLE 2.3–2

The following equation is a specific case of one model used to describe the blood pressure in the aorta during systole (the period following the closure of the heart's aortic valve). The variable t represents time in seconds, and the dimensionless variable y represents the pressure difference across the aortic valve, normalized by a constant reference pressure.

$$y(t) = e^{-8t} \sin\left(9.7t + \frac{\pi}{2} \right)$$

Plot this function for $t \geq 0$.

■ Solution
Note that if t is a vector, the MATLAB functions exp(-8*t) and sin(9.7*t+pi/2) will also be vectors the same size as t. Thus we must use element-by-element multiplication to compute $y(t)$.

We must decide on the proper spacing to use for the vector t and its upper limit. The sine function $\sin(9.7t + \pi/2)$ oscillates with a frequency of 9.7 rad/sec, which is $9.7/(2\pi)$ = 1.5 Hz. Thus its period is $1/1.5 = 2/3$ sec. The spacing of t should be a small fraction of the period in order to generate enough points to plot the curve. Thus we select a spacing of 0.003 to give approximately 200 points per period.

The amplitude of the sine wave decays with time because the sine is multiplied by the decaying exponential e^{-8t}. The exponential's initial value is $e^0 = 1$, and it will be 2 percent of its initial value at $t = 0.5$ (because $e^{-8(0.5)} = 0.02$). Thus we select the upper limit of t to be 0.5. The session is:

```
>>t = [0:0.003:0.5];
>>y = exp(-8*t).*sin(9.7*t+pi/2);
>>plot(t,y),xlabel('t (sec)'), . . .
    ylabel('Normalized Pressure Difference y(t)')
```

The plot is shown in Figure 2.3–1. Note that we do not see much of an oscillation despite the presence of a sine wave. This is because the period of the sine wave is greater than the time it takes for the exponential e^{-8t} to become essentially zero.

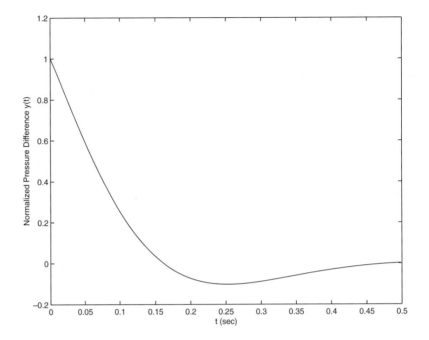

Figure 2.3–1 Aortic pressure response for Example 2.3–2.

Element-by-Element Division

The definition of element-by-element division, also called array division, is similar to the definition of array multiplication except, of course, that the elements of one array are divided by the elements of the other array. Both arrays must have the same size. The symbol for array right division is ./. For example, if

$$\mathbf{x} = [8, 12, 15] \qquad \mathbf{y} = [-2, 6, 5]$$

then $z = x./y$ gives

$$\mathbf{z} = [8/(-2), 12/6, 15/5] = [-4, 2, 3]$$

Also, if

$$\mathbf{A} = \begin{bmatrix} 24 & 20 \\ -9 & 4 \end{bmatrix} \qquad \mathbf{B} = \begin{bmatrix} -4 & 5 \\ 3 & 2 \end{bmatrix}$$

then $C = A./B$ gives

$$\mathbf{C} = \begin{bmatrix} 24/(-4) & 20/5 \\ -9/3 & 4/2 \end{bmatrix} = \begin{bmatrix} -6 & 4 \\ -3 & 2 \end{bmatrix}$$

The array left division operator (.\ ) is defined to perform element-by-element division using left division. Refer to Table 2.3–1 for examples. Note that A.\B is not equivalent to A./B.

Transportation Route Analysis

EXAMPLE 2.3–3

The following table gives data for the distance traveled along five truck routes and the corresponding time required to traverse each route. Use the data to compute the average speed required to drive each route. Find the route that has the highest average speed.

	1	2	3	4	5
Distance (mi)	560	440	490	530	370
Time (hr)	10.3	8.2	9.1	10.1	7.5

■ Solution

For example, the average speed on the first route is $560/10.3 = 54.4$ mi/hr. First we define the row vectors d and t from the distance and time data. Then, to find the average speed on each route using MATLAB, we use array division. The session is

```
>>d = [560, 440, 490, 530, 370]
>>t = [10.3, 8.2, 9.1, 10.1, 7.5]
>>speed = d./t
speed =
    54.3689    53.6585    53.8462    52.4752    49.3333
```

The results are in miles per hour. Note that MATLAB displays more significant figures than is justified by the three-significant-figure accuracy of the given data, so we should round the results to three significant figures before using them.

To find the highest average speed and the corresponding route, continue the session as follows:

```
>>[highest_speed, route] = max(speed)
highest_speed =
    54.3689
route =
    1
```

The first route has the highest speed.

If we did not need the speeds for every route, we could have solved this problem by combining two lines as follows: [highest_speed, route] = max(d./t).

Element-by-Element Exponentiation

MATLAB enables us not only to raise arrays to powers but also to raise scalars and arrays to *array* powers. To perform exponentiation on an element-by-element basis, we must use the .^ symbol. For example, if x = [3, 5, 8], then typing x.^3 produces the array $[3^3, 5^3, 8^3] = [27, 125, 512]$. If x = [0:2:6], then typing x.^2 returns the array $[0^2, 2^2, 4^2, 6^2] = [0, 4, 16, 36]$. If

$$\mathbf{A} = \begin{bmatrix} 4 & -5 \\ 2 & 3 \end{bmatrix}$$

then B = A.^3 gives this result:

$$\mathbf{B} = \begin{bmatrix} 4^3 & (-5)^3 \\ 2^3 & 3^3 \end{bmatrix} = \begin{bmatrix} 64 & -125 \\ 8 & 27 \end{bmatrix}$$

We can raise a scalar to an array power. For example, if p = [2, 4, 5], then typing 3.^p produces the array $[3^2, 3^4, 3^5]$ = [9, 81, 243]. This example illustrates a common situation in which it helps to remember that .^ is a *single* symbol; the dot in 3.^p is not a decimal point associated with the number 3. The following operations, with the value of p given here, are equivalent and give the correct answer:

```
3.^p
3.0.^p
3..^p
(3).^p
3.^[2,4,5]
```

With array exponentiation, the power may be an array if the base is a scalar or if the power's dimensions are the same as the base dimensions. For example if, x = [1,2,3] and y = [2,3,4], then y.^x gives the answer 2964. If A = [1,2; 3,4], then 2.^A gives the array [2,4;8,16].

Test Your Understanding

T2.3–1 Given the matrices

$$\mathbf{A} = \begin{bmatrix} 21 & 27 \\ -18 & 8 \end{bmatrix} \qquad \mathbf{B} = \begin{bmatrix} -7 & -3 \\ 9 & 4 \end{bmatrix}$$

find their (a) array product, (b) array right division (**A** divided by **B**), and (c) **B** raised to the third power element by element.
(Answers: (a) [−147, −81; −162, 32], (b) [−3, −9; −2, 2], and (c) [−343, −27; 729, 64].)

EXAMPLE 2.3–4   A Batch Distillation Process

Consider a system for heating a liquid benzene/toluene solution to distill a pure benzene vapor. A particular batch distillation unit is charged initially with 100 mol of a 60 percent mol benzene/40 percent mol toluene mixture. Let L (mol) be the amount of liquid remaining in the still, and let x (mol B/mol) be the benzene mole fraction in the remaining liquid. Conservation of mass for benzene and toluene can be applied to derive the following relation [Felder, 1986].

$$L = 100 \left(\frac{x}{0.6} \right)^{0.625} \left(\frac{1 - x}{0.4} \right)^{-1.625}$$

Determine what mole fraction of benzene remains when $L = 70$. Note that it is difficult to solve this equation directly for x. Use a plot of x versus L to solve the problem.

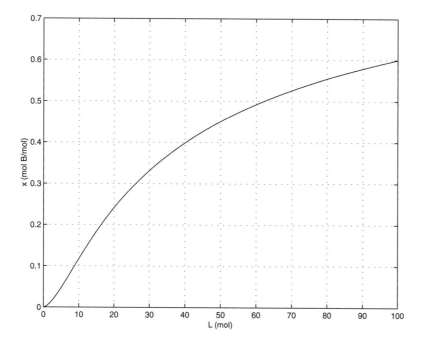

Figure 2.3–2 Plot for Example 2.3–4.

■ Solution

This equation involves both array multiplication and array exponentiation. Note that MATLAB enables us to use decimal exponents to evaluate L. It is clear that L must be in the range $0 \le L \le 100$; however, we do not know the range of x, except that $x \ge 0$. Therefore, we must make a few guesses for the range of x, using a session like the following. We find that $L > 100$ if $x > 0.6$, so we choose x = [0:0.001:0.6]. We use the ginput function to find the value of x corresponding to $L = 70$.

```
>>x = [0:0.001:0.6];
>>L = 100*(x/0.6).^(0.625).*((1-x)/0.4).^(-1.625);
>>plot(L,x),grid,xlabel('L(mol)'),ylabel('x (mol B/mol)'),
...
  [L,x] = ginput(1)
```

The plot is shown in Figure 2.3–2. The answer is $x = 0.52$ if $L = 70$. The plot shows that the remaining liquid becomes leaner in benzene as the liquid amount becomes smaller. Just before the still is empty ($L = 0$), the liquid is pure toluene.

2.4 Matrix Operations

Matrix addition and subtraction are identical to element-by-element addition and subtraction. The corresponding matrix elements are summed or subtracted.

MATRIX
OPERATIONS

However, matrix multiplication and division are not the same as element-by-element multiplication and division.

Multiplication of Vectors

Recall that vectors are simply matrices with one row or one column. Thus matrix multiplication and division procedures apply to vectors as well, and we will introduce matrix multiplication by considering the vector case first.

The *vector dot product* $\mathbf{u} \cdot \mathbf{w}$ of the vectors $\mathbf{u}$ and $\mathbf{w}$ is a scalar and can be thought of as the perpendicular projection of $\mathbf{u}$ onto $\mathbf{w}$. It can be computed from $|\mathbf{u}||\mathbf{w}| \cos \theta$, where θ is the angle between the two vectors and $|\mathbf{u}|$, $|\mathbf{w}|$ are the magnitudes of the vectors. Thus if the vectors are parallel and in the same direction, $\theta = 0$ and $\mathbf{u} \cdot \mathbf{w} = |\mathbf{u}||\mathbf{w}|$. If the vectors are perpendicular, $\theta = 90°$ and thus $\mathbf{u} \cdot \mathbf{w} = 0$. Because the unit vectors $\mathbf{i}$, $\mathbf{j}$, and $\mathbf{k}$ have unit length:

$$\mathbf{i} \cdot \mathbf{i} = \mathbf{j} \cdot \mathbf{j} = \mathbf{k} \cdot \mathbf{k} = 1 \tag{2.4–1}$$

Because the unit vectors are perpendicular:

$$\mathbf{i} \cdot \mathbf{j} = \mathbf{i} \cdot \mathbf{k} = \mathbf{j} \cdot \mathbf{k} = 0 \tag{2.4–2}$$

Thus the vector dot product can be expressed in terms of unit vectors as

$$\mathbf{u} \cdot \mathbf{w} = (u_1 \mathbf{i} + u_2 \mathbf{j} + u_3 \mathbf{k}) \cdot (w_1 \mathbf{i} + w_2 \mathbf{j} + w_3 \mathbf{k})$$

Carrying out the multiplication algebraically and using the properties given by (2.4–1) and (2.4–2), we obtain

$$\mathbf{u} \cdot \mathbf{w} = u_1 w_1 + u_2 w_2 + u_3 w_3$$

The *matrix* product of a *row* vector $\mathbf{u}$ with a *column* vector $\mathbf{w}$ is defined in the same way as the vector dot product; the result is a scalar that is the sum of the products of the corresponding vector elements; that is,

$$\begin{bmatrix} u_1 & u_2 & u_3 \end{bmatrix} \begin{bmatrix} w_1 \\ w_2 \\ w_3 \end{bmatrix} = u_1 w_1 + u_2 w_2 + u_3 w_3$$

if each vector has three elements. Thus the result of multiplying a 1×3 vector times a 3×1 vector is a 1×1 array; that is, a scalar. This definition applies to vectors having any number of elements, as long as both vectors have the same number of elements.

Thus the result of multiplying a $1 \times n$ vector times an $n \times 1$ vector is a 1×1 array, that is, a scalar.

| EXAMPLE 2.4–1 | Miles Traveled |

Table 2.4–1 gives the speed of an aircraft on each leg of a certain trip and the time spent on each leg. Compute the miles traveled on each leg and the total miles traveled.

Table 2.4–1 Aircraft speeds and times per leg

	Leg			
	1	2	3	4
Speed (mi/hr)	200	250	400	300
Time (hr)	2	5	3	4

■ **Solution**

We can define a row vector s containing the speeds and a row vector t containing the times for each leg. Thus s = [200, 250, 400, 300] and t = [2, 5, 3, 4]. To find the miles traveled on each leg, we multiply the speed by the time. To do so, we use the MATLAB symbol .*, which *specifies* the multiplication s.*t to produce the row vector whose elements are the products of the corresponding elements in s and t:

$$s.*t = [200(2), 250(5), 400(3), 300(4)] = [400, 1250, 1200, 1200]$$

This vector contains the miles traveled by the aircraft on each leg of the trip.

To find the total miles traveled, we use the matrix product, denoted by s*t'. In this definition the product is the *sum* of the individual element products; that is,

$$s*t' = [200(2) + 250(5) + 400(3) + 300(4)] = 4050$$

These two examples illustrate the difference between *array* multiplication s.*t and *matrix* multiplication s*t'.

Vector-Matrix Multiplication

Not all matrix products are scalars. To generalize the preceding multiplication to a column vector multiplied by a matrix, think of the matrix as being composed of row vectors. The scalar result of each row-column multiplication forms an element in the result, which is a column vector. For example:

$$\begin{bmatrix} 2 & 7 \\ 6 & -5 \end{bmatrix} \begin{bmatrix} 3 \\ 9 \end{bmatrix} = \begin{bmatrix} 2(3) + 7(9) \\ 6(3) - 5(9) \end{bmatrix} = \begin{bmatrix} 69 \\ -27 \end{bmatrix} \tag{2.4–3}$$

Thus the result of multiplying a 2×2 matrix times a 2×1 vector is a 2×1 array; that is, a column vector. Note that the definition of multiplication requires that the number of columns in the matrix be equal to the number of rows in the vector. In general, the product **Ax,** where **A** has p columns, is defined only if **x** has p rows. If **A** has m rows and **x** is a column vector, the result of **Ax** is a column vector with m rows.

Matrix-Matrix Multiplication

We can expand this definition of multiplication to include the product of two matrices **AB.** The number of columns in **A** must equal the number of rows in **B.** The row-column multiplications form column vectors, and these column vectors form the matrix result. The product **AB** has the same number of rows as **A** and the same number of columns as **B**. For example,

$$\begin{bmatrix} 6 & -2 \\ 10 & 3 \\ 4 & 7 \end{bmatrix}\begin{bmatrix} 9 & 8 \\ -5 & 12 \end{bmatrix} = \begin{bmatrix} (6)(9) + (-2)(-5) & (6)(8) + (-2)(12) \\ (10)(9) + (3)(-5) & (10)(8) + (3)(12) \\ (4)(9) + (7)(-5) & (4)(8) + (7)(12) \end{bmatrix}$$

$$= \begin{bmatrix} 64 & 24 \\ 75 & 116 \\ 1 & 116 \end{bmatrix} \tag{2.4-4}$$

Use the operator * to perform matrix multiplication in MATLAB. The following MATLAB session shows how to perform the matrix multiplication shown in (2.4–4).

```
>>A = [6,-2;10,3;4,7];
>>B = [9,8;-5,12];
>>A*B
```

Element-by-element multiplication is defined for the following product:

$$\begin{bmatrix} 3 & 1 & 7 \end{bmatrix}\begin{bmatrix} 4 & 6 & 5 \end{bmatrix} = \begin{bmatrix} 12 & 6 & 35 \end{bmatrix}$$

However, this product is *not* defined for *matrix* multiplication, because the first matrix has three columns, but the second matrix does not have three rows. Thus if we were to type $[3, 1, 7]*[4, 6, 5]$ in MATLAB, we would receive an error message.

The following product is defined in matrix multiplication and gives the result shown:

$$\begin{bmatrix} x_1 \\ x_2 \\ x_3 \end{bmatrix}\begin{bmatrix} y_1 & y_2 & y_3 \end{bmatrix} = \begin{bmatrix} x_1y_1 & x_1y_2 & x_1y_3 \\ x_2y_1 & x_2y_2 & x_2y_3 \\ x_3y_1 & x_3y_2 & x_3y_3 \end{bmatrix}$$

The following product is also defined:

$$\begin{bmatrix} 10 & 6 \end{bmatrix}\begin{bmatrix} 7 & 4 \\ 5 & 2 \end{bmatrix} = [10(7) + 6(5) \ 10(4) + 6(2)] = [100 \ 52]$$

Evaluating Multivariable Functions

To evaluate a function of two variables, say, $z = f(x, y)$, for the values $x = x_1, x_2, \ldots, x_m$ and $y = y_1, y_2, \ldots, y_n$, define the $m \times n$ matrices:

$$\mathbf{x} = \begin{bmatrix} x_1 & x_1 & \cdots & x_1 \\ x_2 & x_2 & \cdots & x_2 \\ \vdots & \vdots & \vdots & \vdots \\ x_m & x_m & \cdots & x_m \end{bmatrix} \qquad \mathbf{y} = \begin{bmatrix} y_1 & y_2 & \cdots & y_n \\ y_1 & y_2 & \cdots & y_n \\ \vdots & \vdots & \vdots & \vdots \\ y_1 & y_2 & \cdots & y_n \end{bmatrix}$$

When the function $z = f(x, y)$ is evaluated in MATLAB using array operations, the resulting $m \times n$ matrix **z** has the elements $z_{ij} = f(x_i, y_j)$. We can extend this technique to functions of more than two variables by using multidimensional arrays.

Height versus Velocity

EXAMPLE 2.4–2

The maximum height h achieved by an object thrown with a speed v at an angle θ to the horizontal, neglecting drag, is

$$h = \frac{v^2 \sin^2\theta}{2g}$$

Create a table showing the maximum height for the following values of v and θ:

$$v = 10, 12, 14, 16, 18, 20 \text{ m/s} \qquad \theta = 50°, 60°, 70°, 80°$$

The rows in the table should correspond to the speed values, and the columns should correspond to the angles.

■ Solution
The program is shown below.

```
g = 9.8; v = [10:2:20];
theta = [50:10:80];
h = (v'.^2)*(sind(theta).^2)/(2*g);
table = [0, theta; v', h]
```

The arrays v and theta contain the given velocities and angles. The array v is 1×6 and the array theta is 1×4. Thus the term v'.^2 is a 6×1 array, and the term sind(theta).^2 is a 1×4 array. The product of these two arrays, h, is a matrix product and is a $(6 \times 1)(1 \times 4) = (6 \times 4)$ matrix.

The array [0, theta] is 1×5 and the array [v', h] is 6×5, so the matrix table is 7×5. The following table shows the matrix table rounded to one decimal place. From this table we can see that the maximum height is 8.8 m if $v = 14$ m/s and $\theta = 70°$.

0	50	60	70	80
10	3.0	3.8	4.5	4.9
12	4.3	5.5	6.5	7.1
14	5.9	7.5	8.8	9.7
16	7.7	9.8	11.5	12.7
18	9.7	12.4	14.6	16.0
20	12.0	15.3	18.0	19.8

Test Your Understanding

T2.4–1 Use MATLAB to compute the dot product of the following vectors:

$$\mathbf{u} = 6\mathbf{i} - 8\mathbf{j} + 3\mathbf{k}$$
$$\mathbf{w} = 5\mathbf{i} + 3\mathbf{j} - 4\mathbf{k}$$

Check your answer by hand. (Answer: -6.)

T2.4–2 Use MATLAB to show that

$$
\begin{bmatrix} 7 & 4 \\ -3 & 2 \\ 5 & 9 \end{bmatrix} \begin{bmatrix} 1 & 8 \\ 7 & 6 \end{bmatrix} = \begin{bmatrix} 35 & 80 \\ 11 & -12 \\ 68 & 94 \end{bmatrix}
$$

EXAMPLE 2.4–3   Manufacturing Cost Analysis

Table 2.4–2 shows the hourly cost of four types of manufacturing processes. It also shows the number of hours required of each process to produce three different products. Use matrices and MATLAB to solve the following. (a) Determine the cost of each process to produce one unit of product 1. (b) Determine the cost to make one unit of each product. (c) Suppose we produce 10 units of product 1, 5 units of product 2, and 7 units of product 3. Compute the total cost.

■ Solution

(a) The basic principle we can use here is that cost equals the hourly cost times the number of hours required. For example, the cost of using the lathe for product 1 is ($10/h)(6 h) = $60, and so forth for the other three processes. If we define the row vector of hourly costs to be `hourly_costs` and define the row vector of hours required for product 1 to be `hours_1`, then we can compute the costs of each process for product 1 using *element-by-element* multiplication. In MATLAB the session is

```
>>hourly_cost = [10, 12, 14, 9];
>>hours_1 = [6, 2, 3, 4];
>>process_cost_1 = hourly_cost.*hours_1
process_cost_1 =
    60   24   42   36
```

These are the costs of each of the four processes to produce one unit of product 1.

(b) To compute the total cost of one unit of product 1, we can use the vectors `hourly_costs` and `hours_1` but apply *matrix* multiplication instead of element-by-element multiplication, because matrix multiplication sums the individual products. The matrix multiplication gives

Table 2.4–2 Cost and time data for manufacturing processes

| Process | Hourly cost ($) | Hours required to produce one unit | | |
		Product 1	Product 2	Product 3
Lathe	10	6	5	4
Grinding	12	2	3	1
Milling	14	3	2	5
Welding	9	4	0	3

$$[10 \ 12 \ 14 \ 9] \begin{bmatrix} 6 \\ 2 \\ 3 \\ 4 \end{bmatrix} = 10(6) + 12(2) + 14(3) + 9(4) = 162$$

We can perform similar multiplication for products 2 and 3, using the data in the table. For product 2:

$$[10 \ 12 \ 14 \ 9] \begin{bmatrix} 5 \\ 3 \\ 2 \\ 0 \end{bmatrix} = 10(5) + 12(2) + 14(3) + 9(0) = 114$$

For product 3:

$$[10 \ 12 \ 14 \ 9] \begin{bmatrix} 4 \\ 1 \\ 5 \\ 3 \end{bmatrix} = 10(4) + 12(1) + 14(5) + 9(3) = 149$$

These three operations could have been accomplished in one operation by defining a matrix whose columns are formed by the data in the last three columns of the table:

$$[10 \ 12 \ 14 \ 9] \begin{bmatrix} 6 & 5 & 4 \\ 2 & 3 & 1 \\ 3 & 2 & 5 \\ 4 & 0 & 3 \end{bmatrix} = \begin{bmatrix} 60 + 24 + 42 + 36 \\ 50 + 36 + 28 + 0 \\ 40 + 12 + 70 + 27 \end{bmatrix} = [162 \ \ 114 \ \ 149]$$

In MATLAB the session continues as follows. Remember that we must use the transpose operation to convert the row vectors into column vectors.

```
>>hours_2 = [5, 3, 2, 0];
>>hours_3 = [4, 1, 5, 3];
>>unit_cost = hourly_cost*[hours_1', hours_2', hours_3']
unit_cost =
   162    114    149
```

Thus the costs to produce one unit each of products 1, 2, and 3 is $162, $114, and $149, respectively.

(c) To find the total cost to produce 10, 5, and 7 units, respectively, we can use matrix multiplication:

$$[10 \ \ 5 \ \ 7] \begin{bmatrix} 162 \\ 114 \\ 149 \end{bmatrix} = 1620 + 570 + 1043 = 3233$$

In MATLAB the session continues as follows. Note the use of the transpose operator on the vector `unit_cost`.

```
>>units = [10, 5, 7];
>>total_cost = units*unit_cost'
total_cost =
    3233
```

The total cost is $3233.

The General Matrix Multiplication Case

We can state the general result for matrix multiplication as follows: Suppose **A** has dimension $m \times p$ and **B** has dimension $p \times q$. If **C** is the product **AB,** then **C** has dimension $m \times q$ and its elements are given by

$$c_{ij} = \sum_{k=1}^{p} a_{ik} b_{kj} \qquad (2.4\text{--}5)$$

for all $i = 1, 2, \ldots, m$ and $j = 1, 2, \ldots, q$. For the product to be defined, the matrices **A** and **B** must be *conformable;* that is, the number of *rows* in **B** must equal the number of *columns* in **A.** The product has the same number of rows as **A** and the same number of columns as **B.**

Matrix multiplication does not have the commutative property; that is, in general, $\mathbf{AB} \neq \mathbf{BA}$. Reversing the order of matrix multiplication is a common and easily made mistake.

The associative and distributive properties hold for matrix multiplication. The associative property states that

$$\mathbf{A(B + C)} = \mathbf{AB} + \mathbf{AC} \qquad (2.4\text{--}6)$$

The distributive property states that

$$\mathbf{(AB)C} = \mathbf{A(BC)} \qquad (2.4\text{--}7)$$

Applications to Cost Analysis

Project cost data stored in tables must often be analyzed in several ways. The elements in MATLAB matrices are similar to the cells in a spreadsheet, and MATLAB can perform many spreadsheet-type calculations for analyzing such tables.

EXAMPLE 2.4–4   Product Cost Analysis

Table 2.4–3 shows the costs associated with a certain product, and Table 2.4–4 shows the production volume for the four quarters of the business year. Use MATLAB to find the quarterly costs for materials, labor, and transportation; the total material, labor, and transportation costs for the year; and the total quarterly costs.

Table 2.4–3 Product costs

Product	Unit costs ($ × 10³)		
	Materials	**Labor**	**Transportation**
1	6	2	1
2	2	5	4
3	4	3	2
4	9	7	3

Table 2.4–4 Quarterly production volume

Product	Quarter 1	Quarter 2	Quarter 3	Quarter 4
1	10	12	13	15
2	8	7	6	4
3	12	10	13	9
4	6	4	11	5

■ Solution

The costs are the product of the unit cost times the production volume. Thus we define two matrices: U contains the unit costs in Table 2.4–3 in thousands of dollars, and P contains the quarterly production data in Table 2.4–4.

```
>>U = [6, 2, 1;2, 5, 4;4, 3, 2;9, 7, 3];
>>P = [10, 12, 13, 15;8, 7, 6, 4;12, 10, 13, 9;6, 4, 11, 5];
```

Note that if we multiply the first column in U times the first column in P, we obtain the total materials cost for the first quarter. Similarly, multiplying the first column in U times the *second* column in P gives the total materials cost for the *second* quarter. Also, multiplying the second column in U times the first column in P gives the total *labor* cost for the first quarter, and so on. Extending this pattern, we can see that we must multiply the *transpose* of U times P. This multiplication gives the cost matrix C.

```
>>C = U'*P
```

The result is

$$C = \begin{bmatrix} 178 & 162 & 241 & 179 \\ 138 & 117 & 172 & 112 \\ 84 & 72 & 96 & 64 \end{bmatrix}$$

Each column in **C** represents one quarter. The total first-quarter cost is the sum of the elements in the first column, the second-quarter cost is the sum of the second column, and so on. Thus because the sum command sums the columns of a matrix, the quarterly costs are obtained by typing:

```
>>Quarterly_Costs = sum(C)
```

The resulting vector, containing the quarterly costs in thousands of dollars, is [400 351 509 355]. Thus the total costs in each quarter are $400,000; $351,000; $509,000; and $355,000.

The elements in the first row of **C** are the material costs for each quarter; the elements in the second row are the labor costs, and those in the third row are the transportation costs. Thus to find the total material costs, we must sum across the first row of **C**. Similarly, the total labor and total transportation costs are the sums across the second and third rows of **C**. Because the sum command sums *columns,* we must use the transpose of **C**. Thus we type the following:

```
>>Category_Costs = sum(C')
```

The resulting vector, containing the category costs in thousands of dollars, is [760 539 316]. Thus the total material costs for the year are $760,000; the labor costs are $539,000; and the transportation costs are $316,000.

We displayed the matrix **C** only to interpret its structure. If we need not display **C,** the entire analysis would consist of only four command lines.

```
>>U = [6, 2, 1;2, 5, 4;4, 3, 2;9, 7, 3];
>>P = [10, 12, 13, 15;8, 7, 6, 4;12, 10, 13, 9;6, 4, 11, 5];
>>Quarterly_Costs = sum(U'*P)
Quarterly_Costs =
    400   351   509   355
>>Category_Costs = sum((U'*P)')
Category_Costs =
    760   539   316
```

This example illustrates the compactness of MATLAB commands.

Special Matrices

NULL MATRIX

IDENTITY MATRIX

Two exceptions to the noncommutative property are the *null matrix,* denoted by **0,** and the *identity,* or *unity, matrix,* denoted by **I.** The null matrix contains all zeros and is not the same as the empty matrix [ ], which has no elements. The identity matrix is a square matrix whose diagonal elements are all equal to one, with the remaining elements equal to zero. For example, the 2 × 2 identity matrix is

$$\mathbf{I} = \begin{bmatrix} 1 & 0 \\ 0 & 1 \end{bmatrix}$$

These matrices have the following properties:

$$\mathbf{0A} = \mathbf{A0} = \mathbf{0}$$

$$\mathbf{IA} = \mathbf{AI} = \mathbf{A}$$

MATLAB has specific commands to create several special matrices. Type help specmat to see the list of special matrix commands; also check Table 2.4–5. The identity matrix **I** can be created with the eye(n) command, where n is the desired dimension of the matrix. To create the 2 × 2 identity matrix, you type eye(2). Typing eye(size(A)) creates an identity matrix having the same dimension as the matrix **A.**

Table 2.4–5 Special matrices

Command	Description
eye(n)	Creates an $n \times n$ identity matrix.
eye(size(A))	Creates an identity matrix the same size as the matrix **A**.
ones(n)	Creates an $n \times n$ matrix of ones.
ones(m,n)	Creates an $m \times n$ array of ones.
ones(size(A))	Creates an array of ones the same size as the array **A**.
zeros(n)	Creates an $n \times n$ matrix of zeros.
zeros(m,n)	Creates an $m \times n$ array of zeros.
zeros(size(A))	Creates an array of zeros the same size as the array **A**.

Sometimes we want to initialize a matrix to have all zero elements. The `zeros` command creates a matrix of all zeros. Typing `zeros(n)` creates an $n \times n$ matrix of zeros, whereas typing `zeros(m,n)` creates an $m \times n$ matrix of zeros, as will typing `A(m,n) = 0`. Typing `zeros(size(A))` creates a matrix of all zeros having the same dimension as the matrix **A**. This type of matrix can be useful for applications in which we do not know the required dimension ahead of time. The syntax of the `ones` command is the same, except that it creates arrays filled with ones.

For example, to create and plot the function

$$f(x) = \begin{cases} 10 & 0 \leq x \leq 2 \\ 0 & 2 < x < 5 \\ -3 & 5 \leq x \leq 7 \end{cases}$$

the script file is

```
x1 = [0:0.01:2];
f1 = 10*ones(size(x1));
x2 = [2.01:0.01:4.99];
f2 = zeros(size(x2));
x3 = [5:0.01:7];
f3 = -3*ones(size(x3));
f = [f1, f2, f3];
x = [x1, x2, x3];
plot(x,f),xlabel('x'),ylabel('y')
```

(Consider what the plot would look like if the command `plot(x,f)` were replaced with the command `plot(x1,f1,x2,f2,x3,f3)`.)

Matrix Division and Linear Algebraic Equations

Matrix division uses both the right and left division operators, / and \, for various applications, a principal one being the solution of sets of linear algebraic equations. Section 2.6 covers a related topic, the matrix inverse.

You can use the left division operator (\) in MATLAB to solve sets of linear algebraic equations. For example, consider the set

$$6x + 12y + 4z = 70$$
$$7x - 2y + 3z = 5$$
$$2x + 8y - 9z = 64$$

To solve such sets in MATLAB you must create two arrays; we will call them A and B. The array A has as many rows as there are equations, and as many columns as there are variables. The rows of A must contain the coefficients of x, y, and z in that order. In this example, the first row of A must be 6, 12, 4; the second row must be 7, -2, 3, and the third row must be 2, 8, -9. The array B contains the constants on the right-hand side of the equation; it has one column and as many rows as there are equations. In this example, the first row of B is 70, the second is 5, and the third is 64. The solution is obtained by typing A\B. The session is

```
>>A = [6,12,4;7,-2,3;2,8,-9];
>>B = [70;5;64];
>>Solution = A\B
Solution =
      3
      5
     -2
```

The solution is $x = 3$, $y = 5$, and $z = -2$.

This method works fine when the equation set has a unique solution. To learn how to deal with problems having a nonunique solution (or perhaps no solution at all!), see Section 2.5.

Test Your Understanding

T2.4–3 Use MATLAB to solve the following set of equations.

$$6x - 4y + 8z = 112$$
$$-5x - 3y + 7z = 75$$
$$14x + 9y - 5z = -67$$

(Answer: $x = 2$, $y = -5$, $z = 10$.)

Matrix Exponentiation

Raising a matrix to a power is equivalent to repeatedly multiplying the matrix by itself, for example, $\mathbf{A}^2 = \mathbf{AA}$. This process requires the matrix to have the same number of rows as columns; that is, it must be a *square* matrix. MATLAB uses the symbol ^ for matrix exponentiation. To find $\mathbf{A}^2$, type A^2.

We can raise a scalar n to a matrix power **A,** if **A** is square, by typing `n^A`, but the applications for such a procedure are in advanced courses. However, raising a matrix to a matrix power—that is, $\mathbf{A^B}$—is not defined, even if **A** and **B** are square.

Special Products

Many applications in physics and engineering use the cross product and dot product—for example, calculations to compute moments and force components use these special products. If **A** and **B** are vectors with three elements, the cross product command `cross(A,B)` computes the three-element vector that is the cross product $\mathbf{A} \times \mathbf{B}$. If **A** and **B** are $3 \times n$ matrices, `cross(A,B)` returns a $3 \times n$ array whose columns are the cross products of the corresponding columns in the $3 \times n$ arrays **A** and **B.** For example, the moment **M** with respect to a reference point O due to the force **F** is given by $\mathbf{M} = \mathbf{r} \times \mathbf{F,}$ where **r** is the position vector from the point O to the point where the force **F** is applied. To find the moment in MATLAB, you type `M = cross(r,F)`.

The dot product command `dot(A,B)` computes a row vector of length n whose elements are the dot products of the corresponding columns of the $m \times n$ arrays **A** and **B.** To compute the component of the force **F** along the direction given by the vector **r,** you type `dot(F,r)`.

2.5 Matrix Methods for Linear Equations

Sets of linear algebraic equations can be expressed as a single equation, using matrix notation. This standard and compact form is useful for expressing solutions and for developing software applications with an arbitrary number of variables. In this application, a vector is taken to be a column vector unless otherwise specified.

Matrix notation enables us to represent multiple equations as a single matrix equation. For example, consider the following set.

$$2x_1 + 9x_2 = 5$$
$$3x_1 - 4x_2 = 7$$

This set can be expressed in vector-matrix form as

$$\begin{bmatrix} 2 & 9 \\ 3 & -4 \end{bmatrix} \begin{bmatrix} x_1 \\ x_2 \end{bmatrix} = \begin{bmatrix} 5 \\ 7 \end{bmatrix}$$

which can be represented in the following compact form

$$\mathbf{Ax} = \mathbf{b} \qquad (2.5\text{--}1)$$

where we have defined the following matrices and vectors:

$$\mathbf{A} = \begin{bmatrix} 2 & 9 \\ 3 & -4 \end{bmatrix} \qquad \mathbf{x} = \begin{bmatrix} x_1 \\ x_2 \end{bmatrix} \qquad \mathbf{b} = \begin{bmatrix} 5 \\ 7 \end{bmatrix}$$

In general, the set of m equations in n unknowns can be expressed in the form of (2.5–1), where $\mathbf{A}$ has m rows and n columns and $\mathbf{b}$ and $\mathbf{x}$ have one column and m rows.

Matrix Inverse

MATRIX INVERSE

The solution of the scalar equation $ax = b$ is $x = b/a$ if $a \neq 0$. The division operation of scalar algebra has an analogous operation in matrix algebra. For example, to solve the matrix equation (2.5–1) for $\mathbf{x}$, we must somehow "divide" $\mathbf{b}$ by $\mathbf{A}$. The procedure for doing this is developed from the concept of a *matrix inverse*. The inverse of a matrix $\mathbf{A}$ is denoted by $\mathbf{A}^{-1}$ and has the property that

$$\mathbf{A}^{-1}\mathbf{A} = \mathbf{A}\mathbf{A}^{-1} = \mathbf{I}$$

where $\mathbf{I}$ is the identity matrix. Using this property, we multiply both sides of (2.5–1) from the left by $\mathbf{A}^{-1}$ to obtain $\mathbf{A}^{-1}\mathbf{A}\mathbf{x} = \mathbf{A}^{-1}\mathbf{b}$. Because $\mathbf{A}^{-1}\mathbf{A}\mathbf{x} = \mathbf{I}\mathbf{x} = \mathbf{x}$, we obtain the solution

$$\mathbf{x} = \mathbf{A}^{-1}\mathbf{b} \tag{2.5–2}$$

SINGULAR MATRIX

The inverse of a matrix $\mathbf{A}$ is defined only if $\mathbf{A}$ is square and nonsingular. A matrix is *singular* if its determinant $|\mathbf{A}|$ is zero. If $\mathbf{A}$ is singular, then a unique solution to (2.5–1) does not exist. The MATLAB functions `inv(A)` and `det(A)` compute the inverse and the determinant of the matrix $\mathbf{A}$. If the `inv(A)` function is applied to a singular matrix, MATLAB will issue a warning to that effect.

An *ill-conditioned* set of equations is a set that is close to being singular. The ill-conditioned status depends on the accuracy with which the solution calculations are made. When internal numerical accuracy used by MATLAB is insufficient to obtain a solution, it prints the message warning that the matrix is close to singular, and that the results might be inaccurate.

For a 2×2 matrix $\mathbf{A}$,

$$\mathbf{A} = \begin{bmatrix} a & b \\ c & d \end{bmatrix} \qquad \mathbf{A}^{-1} = \frac{1}{ad - bc}\begin{bmatrix} d & -b \\ -c & a \end{bmatrix}$$

where $\det(\mathbf{A}) = ad - bc$. Thus $\mathbf{A}$ is singular if $ad - bc = 0$.

EXAMPLE 2.5–1   The Matrix Inverse Method

Solve the following equations, using the matrix inverse.

$$2x_1 + 9x_2 = 5$$
$$3x_1 - 4x_2 = 7$$

■ Solution

The matrix $\mathbf{A}$ and the vector $\mathbf{b}$ are

$$\mathbf{A} = \begin{bmatrix} 2 & 9 \\ 3 & -4 \end{bmatrix} \qquad \mathbf{b} = \begin{bmatrix} 5 \\ 7 \end{bmatrix}$$

The script file to solve this system is

```
% File cable.m
s34 = sqrt(34); s35 = sqrt(35); s42 = sqrt(42);
A1 = [1/s35, -3/s34, 1/s42];
A2 = [3/s35, 0, -4/s42];
A3 = [5/s35, 5/s34, 5/s42];
A = [A1; A2; A3];
b = [0; 0; 1];
rank(A)
rank([A, b])
T = A\b
```

When this file is executed by typing `cable`, we find that rank($\mathbf{A}$) = rank ([$\mathbf{A}$ $\mathbf{b}$]) = 3 and obtain the values $T_1 = 0.5071$, $T_2 = 0.2915$, and $T_3 = 0.4166$. Because $\mathbf{A}$ is 3×3 and rank($\mathbf{A}$) = 3, which is the number of unknowns, the solution is unique. Using the linearity property, we multiply these results by mg and obtain the general solution $T_1 = 0.5071mg$, $T_2 = 0.2915mg$, and $T_3 = 0.4166mg$.

Underdetermined Systems

An *underdetermined system* does not contain enough information to determine all the unknown variables, usually but not always because it has fewer equations than unknowns. Thus an infinite number of solutions can exist, with one or more of the unknowns dependent on the remaining unknowns. The left division method works for square and nonsquare $\mathbf{A}$ matrices. However, if $\mathbf{A}$ is not square, the left division method can give answers that might be misinterpreted. We will show how to interpret MATLAB results correctly.

UNDERDETER-MINED SYSTEM

When there are more equations than unknowns, the left division method will give a solution with some of the unknowns set equal to zero, but this is not the general solution. An infinite number of solutions might exist even when the number of equations equals the number of unknowns. This can occur when |$\mathbf{A}$| = 0. For such systems the left division method generates an error message warning us that the matrix $\mathbf{A}$ is singular. In such cases the *pseudoinverse method* x = `pinv(A)*b` gives one solution, the *minimum norm solution*. In cases where there are an infinite number of solutions, the `rref` function can be used to express some of the unknowns in terms of the remaining unknowns, whose values are arbitrary.

PSEUDOINVERSE METHOD

MINIMUM NORM SOLUTION

An equation set can be underdetermined even though it has as many equations as unknowns. This can happen if some of the equations are not independent. Determining by hand whether all the equations are independent might not be easy, especially if the set has many equations, but it is easily done in MATLAB.

An Underdetermined Set with Three Equations and Three Unknowns

EXAMPLE 2.5–4

Show that the following set does not have a unique solution. How many of the unknowns will be undetermined? Interpret the results given by the left division method.

$$2x_1 - 4x_2 + 5x_3 = -4$$

$$-4x_1 - 2x_2 + 3x_3 = \ \ 4$$

$$2x_1 + 6x_2 - 8x_3 = \ \ 0$$

■ Solution

A MATLAB session to check the ranks is

```
>>A = [2,-4,5;-4,-2,3;2,6,-8];
>>b = [-4;4;0];
>>rank(A)
ans =
   2
>>rank([A, b])
ans =
   2
>>x = A\b
Warning: Matrix is singular to working precision.
ans =
     NaN
     NaN
     NaN
```

Because the ranks of **A** and [**A b**] are equal, a solution exists. However, because the number of unknowns is three, and is one greater than the rank of **A,** one of the unknowns will
be undetermined. An infinite number of solutions exist, and we can solve for only two of
the unknowns in terms of the third unknown. The set is underdetermined because there
are fewer than three independent equations; the third equation can be obtained from the
first two. To see this, add the first and second equations, to obtain $-2x_1 - 6x_2 + 8x_3 = 0$,
which is equivalent to the third equation.

   Note that we could also tell that the matrix **A** is singular because its rank is less than 3.
If we use the left division method, MATLAB returns a message warning that the problem
is singular, and it does not produce an answer.

The `pinv` Function and the Euclidean Norm

The `pinv` function (which stands for "pseudoinverse") can be used to obtain a
solution of an underdetermined set. To solve the equation set **Ax** = **b** using the
`pinv` function, you type x = pinv(A)*b. The `pinv` function gives a solution

**EUCLIDEAN
NORM**

that gives the minimum value of the *Euclidean norm*, which is the magnitude of
the solution vector x. The magnitude of a vector **v** in three-dimensional space,
having components *x, y, z,* is $\sqrt{x^2 + y^2 + z^2}$. It can be computed using matrix
multiplication and the transpose as follows.

$$\sqrt{\mathbf{v}^T\mathbf{v}} = \sqrt{[x\ y\ z]^T \begin{bmatrix} x \\ y \\ z \end{bmatrix}} = \sqrt{x^2 + y^2 + z^2}$$

The generalization of this formula to an *n*-dimensional vector **v** gives the magnitude of the vector and is the Euclidean norm *N*. Thus

$$N = \sqrt{\mathbf{v}^T\mathbf{v}} \qquad (2.5\text{--}4)$$

The MATLAB function `norm(v)` computes the Euclidean norm.

An Underdetermined Set with Two Equations and Three Unknowns | EXAMPLE 2.5–5

Obtain the solution to the following set, using the left division method and the pseudoinverse method.

$$x_1 + x_2 + x_3 = 400 \qquad (2.5\text{--}5)$$

$$10x_1 + 5x_2 = 1600 \qquad (2.5\text{--}6)$$

■ Solution
These equations can be written in the matrix form $\mathbf{Ax} = \mathbf{b}$ as follows.

$$\begin{bmatrix} 1 & 1 & 1 \\ 10 & 5 & 0 \end{bmatrix} \begin{bmatrix} x_1 \\ x_2 \\ x_3 \end{bmatrix} = \begin{bmatrix} 400 \\ 1600 \end{bmatrix}$$

The MATLAB session is

```
>>A = [1,1,1;10,5,0];
>>b = [400;1600];
>>rank(A)
ans =
  2
>>rank([A, b])
ans =
  2
>>x = A\b
x =
   160.0000
   0
   240.0000
>>x = pinv(A)*b
x =
```

```
93.3333
133.3333
173.3333
```

The left division answer corresponds to $x_1 = 160$, $x_2 = 0$, and $x_3 = 240$. This illustrates how the MATLAB left division operator produces a solution with one or more variables set to zero, for underdetermined sets having more unknowns than equations.

Because the ranks of **A** and [**A b**] are both 2, a solution exists but it is not unique. Because the number of unknowns is three, and is one greater than the rank of **A,** an infinite number of solutions exist, and we can solve for only two of the unknowns in terms of the third.

The pseudoinverse solution gives $x_1 = 93.3333$, $x_2 = 133.3333$, and $x_3 = 173.3333$. This is the minimum norm solution for real values of the variables. The minimum norm solution consists of the real values of x_1, x_2, and x_3 that minimize

$$N = \sqrt{x_1^2 + x_2^2 + x_3^2}$$

To understand what MATLAB is doing, note that we can solve (2.5–5) and (2.5–6) to obtain x_1 and x_2 in terms of x_3 as $x_1 = x_3 - 80$ and $x_2 = 480 - 2x_3$. Then the Euclidean norm can be expressed as

$$N = \sqrt{(x_3 - 80)^2 + (480 - 2x_3)^2 + x_3^2} = \sqrt{6x_3^2 - 2080x_3 + 236{,}800}$$

The real value of x_3 that minimizes N can be found by plotting N versus x_3, or by using calculus. The answer is $x_3 = 173.3333$, the same as the minimum norm solution given by the pseudoinverse method.

Where there are an infinite number of solutions, we must decide whether the solutions given by the left division and the pseudoinverse methods are useful for applications. This must be done in the context of the specific application.

Test Your Understanding

T2.5–4 Find two solutions to the following set.

$$x_1 + 3x_2 + 2x_3 = 2$$

$$x_1 + x_2 + x_3 = 4$$

(Answer: Minimum norm solution: $x_1 = 4.33$, $x_2 = -1.67$, $x_3 = 1.34$. Left division solution: $x_1 = 5$, $x_2 = -1$, $x_3 = 0$.)

REDUCED ROW ECHELON FORM

The Reduced Row Echelon Form

We can express some of the unknowns in an underdetermined set as functions of the remaining unknowns. In Example 2.5–5, we wrote the solutions for two of the unknowns in terms of the third: $x_1 = x_3 - 80$ and $x_2 = 480 - 2x_3$. These two equations are equivalent to

$$x_1 - x_3 = -80 \qquad x_2 + 2x_3 = 480$$

In matrix form these are

$$\begin{bmatrix} 1 & 0 & -1 \\ 0 & 1 & 2 \end{bmatrix} \begin{bmatrix} x_1 \\ x_2 \\ x_3 \end{bmatrix} = \begin{bmatrix} -80 \\ 480 \end{bmatrix}.$$

The augmented matrix [**A** **b**] for the above set is

$$\begin{bmatrix} 1 & 0 & -1 & -80 \\ 0 & 1 & 2 & 480 \end{bmatrix}$$

Note that the first two columns form a 2×2 identity matrix. This indicates that the corresponding equations can be solved directly for x_1 and x_2 in terms of x_3.

We can always reduce an underdetermined set to such a form by multiplying the set's equations by suitable factors and adding the resulting equations to eliminate an unknown variable. The MATLAB `rref` function provides a procedure for reducing an equation set to this form, which is called the *reduced row echelon form*. Its syntax is `rref([A b])`. Its output is the augmented matrix [**C** **d**] that corresponds to the equation set **Cx** = **d.** This set is in reduced row echelon form.

Three Equations in Three Unknowns, Continued

EXAMPLE 2.5–6

The following underdetermined equation set was analyzed in Example 2.5–4. There it was shown that an infinite number of solutions exist. Use the `rref` function to obtain the solutions.

$$2x_1 - 4x_2 + 5x_3 = -4$$

$$-4x_1 - 2x_2 + 3x_3 = 4$$

$$2x_1 + 6x_2 - 8x_3 = 0$$

■ **Solution**
The MATLAB session is

```
>>A = [2,-4,5;-4,-2,3;2,6,-8];
>>b = [-4;4;0];
>>rref([A, b])
ans =
   1    0    -0.1    -1.2000
   0    1    -1.3     0.4000
   0    0     0       0
```

The answer corresponds to the augmented matrix [**C d**], where

$$[\mathbf{C} \ \ \mathbf{d}] = \begin{bmatrix} 1 & 0 & -0.1 & -1.2 \\ 0 & 1 & -1.3 & 0.4 \\ 0 & 0 & 0 & 0 \end{bmatrix}$$

This matrix corresponds to the matrix equation $\mathbf{Cx} = \mathbf{d}$, or

$$x_1 + 0x_2 - 0.1x_3 = -1.2$$
$$0x_1 + x_2 - 1.3x_3 = \quad 0.4$$
$$0x_1 + 0x_2 - 0x_3 = \quad 0$$

These can be easily solved for x_1 and x_2 in terms of x_3 as follows: $x_1 = 0.1x_3 - 1.2, x_2 = 1.3x_3 + 0.4$. This is the general solution to the problem, where x_3 is taken to be the arbitrary variable.

Supplementing Underdetermined Systems

In underdetermined cases we might be able to include additional information, objectives, or constraints to find a unique solution.

EXAMPLE 2.5–7

Traffic Engineering

A traffic engineer wants to know if measurements of traffic flow entering and leaving a road network are sufficient to predict the traffic flow on each street in the network. For example, consider the network of one-way streets shown in Figure 2.5–2. The numbers shown are the measured traffic flows in vehicles per hour. Assume that no vehicles park anywhere within the network. If possible, calculate the traffic flows f_1, f_2, f_3, and f_4. If this is not possible, suggest how to obtain the necessary information.

■ **Solution**

The flow *into* intersection 1 must equal the flow *out* of the intersection. Thus gives

$$100 + 200 = f_1 + f_4$$

Similarly, for the other three intersections, we have

$$f_1 + f_2 = 300 + 200$$
$$600 + 400 = f_2 + f_3$$
$$f_3 + f_4 = 300 + 500$$

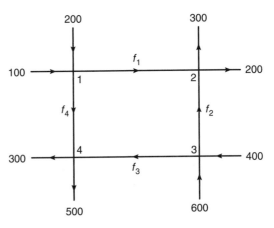

Figure 2.5–2 A network of one-way streets.

Putting these in the matrix form $\mathbf{Ax} = \mathbf{b}$, we obtain

$$
\mathbf{A} = \begin{bmatrix} 1 & 0 & 0 & 1 \\ 1 & 1 & 0 & 0 \\ 0 & 1 & 1 & 0 \\ 0 & 0 & 1 & 1 \end{bmatrix} \qquad \mathbf{b} = \begin{bmatrix} 300 \\ 500 \\ 1000 \\ 800 \end{bmatrix} \qquad \mathbf{x} = \begin{bmatrix} f_1 \\ f_2 \\ f_3 \\ f_4 \end{bmatrix}
$$

First, check the ranks of $\mathbf{A}$ and $[\mathbf{A}\ \mathbf{b}]$, using the MATLAB `rank` function. Both have a rank of 3, which is one less than the number of unknowns, so we can determine three of the unknowns in terms of the fourth. Thus we cannot determine all the traffic flows based on the given measurements.

Using the `rref([A b])` function produces the reduced augmented matrix

$$
\begin{bmatrix} 1 & 0 & 0 & 1 & 300 \\ 0 & 1 & 0 & -1 & 200 \\ 0 & 0 & 1 & 1 & 800 \\ 0 & 0 & 0 & 0 & 0 \end{bmatrix}
$$

which corresponds to the reduced system

$$ f_1 + f_4 = 300 $$

$$ f_2 - f_4 = 200 $$

$$ f_3 + f_4 = 800 $$

These can be solved easily as follows: $f_1 = 300 - f_4$, $f_2 = 200 + f_4$, and $f_3 = 800 - f_4$. If we could measure the flow on one of the internal roads, say f_4, then we could compute the other flows. So we recommend that the engineer arrange to have this additional measurement made.

Test Your Understanding

T2.5–5 Use the `rref`, `pinv`, and the left division methods to solve the following set.

$$
\begin{aligned}
3x_1 + 5x_2 + 6x_3 &= 6 \\
8x_1 - x_2 + 2x_3 &= 1 \\
5x_1 - 6x_2 - 4x_3 &= -5
\end{aligned}
$$

(Answer: There are an infinite number of solutions. The result obtained with the `rref` function is $x_1 = 0.2558 - 0.3721x_3$, $x_2 = 1.0465 - 0.9767x_3$, x_3 arbitrary. The `pinv` function gives $x_1 = 0.0571$, $x_2 = 0.5249$, $x_3 = 0.5340$. The left division method generates an error message.)

T2.5–6 Use the `rref`, `pinv`, and the left division methods to solve the following set.

$$3x_1 + 5x_2 + 6x_3 = 4$$

$$x_1 - 2x_2 - 3x_3 = 10$$

(Answer: There are an infinite number of solutions. The result obtained with the `rref` function is $x_1 = 0.2727x_3 + 5.2727$, $x_2 = -1.3636x_3 - 2.2626$, x_3 arbitrary. The solution obtained with left division is $x_1 = 4.8000$, $x_2 = 0$, $x_3 = -1.7333$. The one obtained with the pseudoinverse method is $x_1 = 4.8394$, $x_2 = -0.1972$, $x_3 = -1.5887$.)

Overdetermined Systems

OVERDETER-MINED SYSTEM

An *overdetermined system* is a set of equations that has more independent equations than unknowns. Some overdetermined systems have exact solutions, and they can be obtained with the left division method $x = A\backslash b$. For other overdetermined systems, no exact solution exists; in some of these cases, the left division method does not yield an answer, while in other cases the left division method gives an answer that satisfies the equation set only in a "least squares" sense. We will show what this means in the next example. When MATLAB gives an answer to an overdetermined set, it does not tell us whether the answer is the exact solution. We must determine this information ourselves, and we will now show how to do this.

EXAMPLE 2.5–8

The Least Squares Method

Suppose we have the following three data points, and we want to find the straight line $y = c_1 x + c_2$ that best fits the data in some sense.

x	y
0	2
5	6
10	11

(a) Find the coefficients c_1 and c_2 using the least squares criterion. (b) Find the coefficients by using the left division method to solve the three equations (one for each data point) for the two unknowns c_1 and c_2. Compare with the answer from part (a).

■ Solution

(a) Because two points define a straight line, unless we are extremely lucky, our three data points will not lie on the same straight line. A common criterion for obtaining the straight line that best fits the data is the *least squares* criterion. According to this criterion, the line that minimizes J, the sum of the squares of the vertical differences between the line and the data points, is the "best" fit. Here J is

LEAST SQUARES METHOD

$$J = \sum_{i=1}^{i=3}(c_1 x_i + c_2 - y_i)^2 = (0c_1 + c_2 - 2)^2 + (5c_1 + c_2 - 6)^2 + (10c_1 + c_2 - 11)^2$$

If you are familiar with calculus, you know that the values of c_1 and c_2 that minimize J are found by setting the partial derivatives $\partial J/\partial c_1$ and $\partial J/\partial c_2$ equal to zero.

$$\frac{\partial J}{\partial c_1} = 250c_1 + 30c_2 - 280 = 0$$

$$\frac{\partial J}{\partial c_2} = 30c_1 + 6c_2 - 38 = 0$$

The solution is $c_1 = 0.9$ and $c_2 = 11/6$. The best straight line in the least squares sense is $y = 0.9x + 11/6$.

(b) Evaluating the equation $y = c_1x + c_2$ at each data point gives the following three equations, which are overdetermined, because there are more equations than unknowns.

$$0c_1 + c_2 = 2 \tag{2.5--7}$$

$$5c_1 + c_2 = 6 \tag{2.5--8}$$

$$10c_1 + c_2 = 11 \tag{2.5--9}$$

These equations can be written in the matrix form $\mathbf{Ax} = \mathbf{b}$ as follows.

$$\mathbf{Ax} = \begin{bmatrix} 0 & 1 \\ 5 & 0 \\ 10 & 1 \end{bmatrix} \begin{bmatrix} c_1 \\ c_2 \end{bmatrix} = \begin{bmatrix} 2 \\ 6 \\ 11 \end{bmatrix} = \mathbf{b}$$

where

$$[\mathbf{A\ b}] = \begin{bmatrix} 0 & 1 & 2 \\ 5 & 1 & 6 \\ 10 & 1 & 11 \end{bmatrix}$$

To use left division, the MATLAB session is

```
>>A = [0,1;5,1;10,1];
>>b = [2;6;11];
>>rank(A)
ans =
  2
>>rank([A, b])
ans =
  3
>>x = A\b
x =
    0.9000
    1.8333
>>A*x
```

```
ans =
    1.833
    6.333
   10.8333
```

This result for **x** agrees with the least squares solution obtained previously: $c_1 = 0.9$, $c_2 = 11/6 = 1.8333$. The rank of **A** is 2, but the rank of [**A b**] is 3, so no exact solution exists for c_1 and c_2. Note that A*x gives the y values generated by the line $y = 0.9x + 1.8333$ at the x data values: $x = 0, 5, 10$. These are different from the right-hand sides of the original three equations (2.5–7) through (2.5–9). This is not unexpected, because the least squares solution is not an exact solution of the equations.

Some overdetermined systems have an exact solution. The left division method sometimes gives an answer for overdetermined systems, but it does not indicate whether the answer is the exact solution. We need to check the ranks of **A** and [**A b**] to know if the answer is the exact solution. The next example illustrates this situation.

EXAMPLE 2.5–9	An Overdetermined Set

Solve the following equations and discuss the solution for two cases: $c = 9$ and $c = 10$.

$$x_1 + x_2 = 1$$

$$x_1 + 2x_2 = 3$$

$$x_1 + 5x_2 = c$$

■ **Solution**
The coefficient matrix and the augmented matrix for this problem are

$$\mathbf{A} = \begin{bmatrix} 1 & 1 \\ 1 & 2 \\ 1 & 5 \end{bmatrix} \qquad [\mathbf{A\ b}] = \begin{bmatrix} 1 & 1 & 1 \\ 1 & 2 & 3 \\ 1 & 5 & c \end{bmatrix}$$

Making the computations in MATLAB, we find that for $c = 9$, rank (**A**) = rank([**A b**]) = 2. Thus the system has a solution, and because the number of unknowns (two) equals the rank of **A,** there is a unique solution. The left division method A\b gives this solution, which is $x_1 = -1$ and $x_2 = 2$.
   For $c = 10$ we find that rank(**A**) = 2, but rank([**A b**]) = 3. Because rank(**A**) ≠ rank([**A b**]), there is no solution. However, the left division method **A\b** gives $x_1 = -1.3846$ and $x_2 = 2.2692$, which is *not* an exact solution! This can be verified by substituting these values into the original equation set. This answer is the solution to the equation set in a least squares sense. That is, these values are the values of x_1 and x_2 that minimize J, the sum of the squares of the differences between the equations' left- and right-hand sides.

$$J = (x_1 + x_2 - 1)^2 + (x_1 + 2x_2 - 3)^2 + (x_1 + 5x_2 - 10)^2$$

Table 2.5–1 Matrix functions and commands for solving linear equations

Function	Description
det(A)	Computes the determinant of the array **A**.
inv(A)	Computes the inverse of the matrix **A**.
pinv(A)	Computes the pseudoinverse of the matrix **A**.
rank(A)	Computes the rank of the matrix **A**.
rref([A b])	Computes the reduced row echelon form corresponding to the augmented matrix [**A b**].
x = inv(A)*b	Solves the matrix equation **Ax** = **b** using the matrix inverse.
x = A\b	Solves the matrix equation **Ax** = **b** using left division.

To interpret MATLAB answers correctly for an overdetermined system, first check the ranks of **A** and [**A b**] to see if an exact solution exists; if one does not exist, then we know that the left division answer is a least squares solution. In Chapter 4 we develop a general-purpose program that checks the ranks and solves a general set of linear equations.

Table 2.5–1 summarizes this section's functions and commands.

Test Your Understanding

T2.5–7 Solve the following set.

$$x_1 - 3x_2 = \quad 2$$
$$3x_1 + 5x_2 = \quad 7$$
$$70x_1 - 28x_2 = 153$$

(Answer: There is a unique solution: $x_1 = 2.2143$, $x_2 = 0.0714$, which is given by the left division method.)

T2.5–8 Show why there is no solution to the following set.

$$x_1 - 3x_2 = \quad 2$$
$$3x_1 + 5x_2 = \quad 7$$
$$5x_1 - 2x_2 = -4$$

2.6 Polynomial Operations Using Arrays

MATLAB has some convenient tools for working with polynomials. Type help polyfun for more information on this category of commands. We will use the following notation to describe a polynomial:

$$f(x) = a_1 x^n + a_2 x^{n-1} + a_3 x^{n-2} + \cdots + a_{n-1} x^2 + a_n x + a_{n+1}$$

We can describe a polynomial in MATLAB with a row vector whose elements are the polynomial's coefficients, *starting with the coefficient of the highest power of x.* This vector is $[a_1, a_2, a_3, \ldots, a_{n-1}, a_n, a_{n+1}]$. For example, the vector [4,-8,7,-5] represents the polynomial $4x^3 - 8x^2 + 7x - 5$.

Polynomial roots can be found with the roots(a) function, where a is the array containing the polynomial coefficients. For example, to obtain the roots of $x^3 + 12x^2 + 45x + 50 = 0$, you type y = roots([1,12,45,50]). The answer (y) is a *column* array containing the values $-2, -5, -5$.

The poly(r) function computes the coefficients of the polynomial whose roots are specified by the array r. The result is a *row* array that contains the polynomial's coefficients. For example, to find the polynomial whose roots are 1 and $3 \pm 5i$, the session is

```
>>p = poly([1,3+5i, 3-5i])
p =
    1    -7    40    -34
```

Thus the polynomial is $x^3 - 7x^2 + 40x - 34$.

Polynomial Addition and Subtraction

To add two polynomials, add the arrays that describe their coefficients. If the polynomials are of different degrees, add zeros to the coefficient array of the lower-degree polynomial. For example, consider

$$f(x) = 9x^3 - 5x^2 + 3x + 7$$

whose coefficient array is f = [9,-5,3,7] and

$$g(x) = 6x^2 - x + 2$$

whose coefficient array is g = [6,-1,2]. The degree of $g(x)$ is one less that of $f(x)$. Therefore, to add $f(x)$ and $g(x)$, we append one zero to g to "fool" MATLAB into thinking $g(x)$ is a third-degree polynomial. That is, we type g = [0 g] to obtain [0,6,-1,2] for g. This vector represents $g(x) = 0x^3 + 6x^2 - x + 2$. To add the polynomials, type h = f+g. The result is h = [9,1,2,9], which corresponds to $h(x) = 9x^3 + x^2 + 2x + 9$. Subtraction is done in a similar way.

Polynomial Multiplication and Division

To multiply a polynomial by a scalar, simply multiply the coefficient array by that scalar. For example, $5h(x)$ is represented by [45,5,10,45].

Multiplication and division of polynomials are easily done with MATLAB. Use the conv function (it stands for "convolve") to multiply polynomials and use the deconv function (deconv stands for "deconvolve") to perform synthetic division. Table 2.6–1 summarizes these functions, as well as the poly, polyval, and roots functions.

Table 2.6–1 Polynomial functions

Command	Description
conv(a,b)	Computes the product of the two polynomials described by the coefficient arrays a and b. The two polynomials need not be the same degree. The result is the coefficient array of the product polynomial.
[q,r] = deconv(num,den)	Computes the result of dividing a numerator polynomial, whose coefficient array is num, by a denominator polynomial represented by the coefficient array den. The quotient polynomial is given by the coefficient array q, and the remainder polynomial is given by the coefficient array r.
poly(r)	Computes the coefficients of the polynomial whose roots are specified by the vector r. The result is a *row* vector that contains the polynomial's coefficients arranged in descending order of power.
polyval(a,x)	Evaluates a polynomial at specified values of its independent variable x, which can be a matrix or a vector. The polynomial's coefficients of descending powers are stored in the array a. The result is the same size as x.
roots(a)	Computes the roots of a polynomial specified by the coefficient array a. The result is a *column* vector that contains the polynomial's roots.

The product of the polynomials $f(x)$ and $g(x)$ is

$$f(x)g(x) = (9x^3 - 5x^2 + 3x + 7)(6x^2 - x + 2)$$
$$= 54x^5 - 39x^4 + 41x^3 + 29x^2 - x + 14$$

Dividing $f(x)$ by $g(x)$ using synthetic division gives a quotient of

$$\frac{f(x)}{g(x)} = \frac{9x^3 - 5x^2 + 3x + 7}{6x^2 - x + 2} = 1.5x - 0.5833$$

with a remainder of $-0.5833x + 8.1667$. Here is the MATLAB session to perform these operations.

```
>>f = [9,-5,3,7];
>>g = [6,-1,2];
>>product = conv(f,g)
product =
    54    -39    41    29    -1    14
>>[quotient, remainder] = deconv(f,g)
quotient =
    1.5    -0.5833
remainder =
    0    0    -0.5833    8.1667
```

The conv and deconv functions do not require that the polynomials have the same degree, so we did not have to fool MATLAB as we did when adding the polynomials.

Plotting Polynomials

The polyval(a,x) function evaluates a polynomial at specified values of its independent variable x, which can be a matrix or a vector. The polynomial's coefficient array is a. The result is the same size as x. For example, to evaluate the polynomial $f(x) = 9x^3 - 5x^2 + 3x + 7$ at the points $x = 0, 2, 4, \ldots, 10$, type

```
>>f = polyval([9,-5,3,7],[0:2:10]);
```

The resulting vector f contains six values that correspond to $f(0), f(2), f(4), \ldots, f(10)$.

The `polyval` function is very useful for plotting polynomials. To do this you should define an array that contains many values of the independent variable x in order to obtain a smooth plot. For example, to plot the polynomial $f(x) = 9x^3 - 5x^2 + 3x + 7$ for $-2 \leq x \leq 5$, you type

```
>>polyval([9,-5,3,7], [-2:0.01:5]);
>>plot (x,f),xlabel('x'),ylabel('f(x)'),grid
```

Polynomial derivatives and integrals are covered in Chapter 7.

Test Your Understanding

T2.6–1 Use MATLAB to obtain the roots of

$$x^3 + 13x^2 + 52x + 6 = 0$$

Use the `poly` function to confirm your answer.

T2.6–2 Use MATLAB to confirm that

$$(20x^3 - 7x^2 + 5x + 10)(4x^2 + 12x - 3)$$
$$= 80x^5 + 212x^4 - 124x^3 + 121x^2 + 105x - 30$$

T2.6–3 Use MATLAB to confirm that

$$\frac{12x^3 + 5x^2 - 2x + 3}{3x^2 - 7x + 4} = 4x + 11$$

with a remainder of $59x - 41$.

T2.6–4 Use MATLAB to confirm that

$$\frac{6x^3 + 4x^2 - 5}{12x^3 - 7x^2 + 3x + 9} = 0.7108$$

when $x = 2$.

T2.6–5 Plot the polynomial

$$y = x^3 + 13x^2 + 52x + 6$$

over the range $-7 \leq x \leq 1$.

2.7 Cell Arrays

CELL ARRAY

The *cell array* is an array in which each element is a *bin,* or *cell,* which can contain an array. You can store different classes of arrays in a cell array, and you can group data sets that are related but have different dimensions. You access cell arrays using the same indexing operations used with ordinary arrays.

This is the only section in the text that uses cell arrays. Coverage of this section is therefore optional. Some more advanced MATLAB applications, such as those found in some of the toolboxes, do use cell arrays.

Creating Cell Arrays

You can create a cell array by using assignment statements or by using the `cell` function. You can assign data to the cells by using either *cell indexing* or *content indexing*. To use cell indexing, enclose in parentheses the cell subscripts on the left side of the assignment statement and use the standard array notation. Enclose the cell contents on the right side of the assignment statement in braces { }.

CELL INDEXING

CONTENT INDEXING

An Environmental Database

EXAMPLE 2.7–1

Suppose you want to create a 2 × 2 cell array A, whose cells contain the location, the date, the air temperature (measured at 8 A.M., 12 noon, and 5 P.M.), and the water temperatures measured at the same time in three different points in a pond. The cell array looks like the following.

Walden Pond	June 13, 1997
[60 72 65]	$\begin{bmatrix} 55 & 57 & 56 \\ 54 & 56 & 55 \\ 52 & 55 & 53 \end{bmatrix}$

■ Solution

You can create this array by typing the following either in interactive mode or in a script file and running it.

```
A(1,1) = {'Walden Pond'};
A(1,2) = {'June 13, 1997'};
A(2,1) = {[60,72,65]};
A(2,2) = {[55,57,56;54,56,55;52,55,53]};
```

If you do not yet have contents for a particular cell, you can type a pair of empty braces { } to denote an empty cell, just as a pair of empty brackets [ ] denotes an empty numeric array. This notation creates the cell but does not store any contents in it.

To use content indexing, enclose in braces the cell subscripts on the left side using the standard array notation. Then specify the cell contents on the right side of the assignment operator. For example:

```
A{1,1} = 'Walden Pond';
A{1,2} = 'June 13, 1997';
A{2,1} = [60,72,65];
A{2,2} = [55,57,56;54,56,55;52,55,53];
```

Type A at the command line. You will see

```
A  =
    'Walden Pond'     'June 13, 1997'
     [1x3 double]      [3x3 double]
```

You can use the `celldisp` function to display the full contents. For example, typing `celldisp(A)` displays

```
A{1,1} =
   Walden Pond
A{2,1} =
   60   72   65
   .
   .
   .
   etc.
```

The `cellplot` function produces a graphical display of the cell array's contents in the form of a grid. Type `cellplot(A)` to see this display for the cell array A. Use commas or spaces with braces to indicate columns of cells and use semicolons to indicate rows of cells (just as with numeric arrays). For example, typing

```
B = {[2,4], [6,-9;3,5]; [7;2], 10};
```

creates the following 2 × 2 cell array:

$$\begin{array}{|c|c|} \hline [2 \quad 4] & \begin{bmatrix} 6 & -9 \\ 3 & 5 \end{bmatrix} \\ \hline [7 \quad 2] & 10 \\ \hline \end{array}$$

You can preallocate empty cell arrays of a specified size by using the `cell` function. For example, type C = `cell(3,5)` to create the 3 × 5 cell array C and fill it with empty matrices. Once the array has been defined in this way, you can use assignment statements to enter the contents of the cells. For example, type C(2,4) = {[6,-3,7]} to put the 1 × 3 array in cell (2,4) and type C(1,5) = {1:10} to put the numbers from 1 to 10 in cell (1,5). Type C(3,4) = {'30 mph'} to put the string in cell (3,4).

Accessing Cell Arrays

You can access the contents of a cell array by using either cell indexing or content indexing. To use cell indexing to place the contents of cell (3,4) of the array C in the new variable `Speed`, type Speed = C(3,4). To place the contents of the cells in rows 1 to 3, columns 2 to 5 in the new cell array D, type D = C(1:3,2:5). The new cell array D will have three rows, four columns, and 12 arrays. To use content indexing to access some or all of the contents in a *single*

5. Type this matrix in MATLAB and use MATLAB to answer the following
   questions:

$$\mathbf{A} = \begin{bmatrix} 3 & 7 & -4 & 12 \\ -5 & 9 & 10 & 2 \\ 6 & 13 & 8 & 11 \\ 15 & 5 & 4 & 1 \end{bmatrix}$$

   a. Create a vector **v** consisting of the elements in the second column of **A**.
   b. Create a vector **w** consisting of the elements in the second row of **A**.

6. Type this matrix in MATLAB and use MATLAB to answer the following
   questions:

$$\mathbf{A} = \begin{bmatrix} 3 & 7 & -4 & 12 \\ -5 & 9 & 10 & 2 \\ 6 & 13 & 8 & 11 \\ 15 & 5 & 4 & 1 \end{bmatrix}$$

   a. Create a 4 × 3 array **B** consisting of all elements in the second
   through fourth columns of **A**.
   b. Create a 3 × 4 array **C** consisting of all elements in the second
   through fourth rows of **A**.
   c. Create a 2 × 3 array **D** consisting of all elements in the first two rows
   and the last three columns of **A**.

7.* Compute the length and absolute value of the following vectors:
   a. $\mathbf{x} = [2, 4, 7]$
   b. $\mathbf{y} = [2, -4, 7]$
   c. $\mathbf{z} = [5 + 3i, -3 + 4i, 2 - 7i]$

8. Given the matrix

$$\mathbf{A} = \begin{bmatrix} 3 & 7 & -4 & 12 \\ -5 & 9 & 10 & 2 \\ 6 & 13 & 8 & 11 \\ 15 & 5 & 4 & 1 \end{bmatrix}$$

   a. Find the maximum and minimum values in each column.
   b. Find the maximum and minimum values in each row.

9. Given the matrix

$$\mathbf{A} = \begin{bmatrix} 3 & 7 & -4 & 12 \\ -5 & 9 & 10 & 2 \\ 6 & 13 & 8 & 11 \\ 15 & 5 & 4 & 1 \end{bmatrix}$$

a. Sort each column and store the result in an array **B**.
b. Sort each row and store the result in an array **C**.
c. Add each column and store the result in an array **D**.
d. Add each row and store the result in an array **E**.

10. Consider the following arrays.

$$A = \begin{bmatrix} 1 & 4 & 2 \\ 2 & 4 & 100 \\ 7 & 9 & 7 \\ 3 & \pi & 42 \end{bmatrix} \qquad B = \ln(A)$$

Write MATLAB expressions to do the following.
a. Select just the second row of **B**.
b. Evaluate the sum of the second row of **B**.
c. Multiply the second column of **B** and the first column of **A** element-by-element.
d. Evaluate the maximum value in the vector resulting from element-by-element multiplication of the second column of **B** with the first column of **A**.
e. Use element-by-element division to divide the first row of **A** by the first three elements of the third column of **B**. Evaluate the sum of the elements of the resulting vector.

Section 2.2

11.* a. Create a three-dimensional array **D** whose three "layers" are these matrices:

$$A = \begin{bmatrix} 3 & -2 & 1 \\ 6 & 8 & -5 \\ 7 & 9 & 10 \end{bmatrix} \qquad B = \begin{bmatrix} 6 & 9 & -4 \\ 7 & 5 & 3 \\ -8 & 2 & 1 \end{bmatrix} \qquad C = \begin{bmatrix} -7 & -5 & 2 \\ 10 & 6 & 1 \\ 3 & -9 & 8 \end{bmatrix}$$

b. Use MATLAB to find the largest element in each layer of **D** and the largest element in **D**.

Section 2.3

12.* Given the matrices

$$A = \begin{bmatrix} -7 & 16 \\ 4 & 9 \end{bmatrix} \qquad B = \begin{bmatrix} 6 & -5 \\ 12 & -2 \end{bmatrix} \qquad C = \begin{bmatrix} -3 & -9 \\ 6 & 8 \end{bmatrix}$$

Use MATLAB to:
a. Find **A** + **B** + **C**.
b. Find **A** − **B** + **C**.
c. Verify the associative law

$$(A + B) + C = A + (B + C)$$

Section 2.4

25.* Use MATLAB to find the products **AB** and **BA** for the following matrices:

$$A = \begin{bmatrix} 11 & 5 \\ -9 & -4 \end{bmatrix} \qquad B = \begin{bmatrix} -7 & -8 \\ 6 & 2 \end{bmatrix}$$

26. Given the matrices

$$A = \begin{bmatrix} 3 & -2 & 1 \\ 6 & 8 & -5 \\ 7 & 9 & 10 \end{bmatrix} \qquad B = \begin{bmatrix} 6 & 9 & -4 \\ 7 & 5 & 3 \\ -8 & 2 & 1 \end{bmatrix} \qquad C = \begin{bmatrix} -7 & -5 & 2 \\ 10 & 6 & 1 \\ 3 & -9 & 8 \end{bmatrix}$$

Use MATLAB to:

a. Verify the associative property

$$A(B + C) = AB + AC$$

b. Verify the distributive property

$$(AB)C = A(BC)$$

27. The following tables show the costs associated with a certain product and the production volume for the four quarters of the business year. Use MAT-LAB to find (*a*) the quarterly costs for materials, labor, and transportation; (*b*) the total material, labor, and transportation costs for the year; and (*c*) the total quarterly costs.

	Unit product costs ($\$ \times 10^3$)		
Product	**Materials**	**Labor**	**Transportation**
1	7	3	2
2	3	1	3
3	9	4	5
4	2	5	4
5	6	2	1

	Quarterly production volume			
Product	**Quarter 1**	**Quarter 2**	**Quarter 3**	**Quarter 4**
1	16	14	10	12
2	12	15	11	13
3	8	9	7	11
4	14	13	15	17
5	13	16	12	18

28.* Aluminum alloys are made by adding other elements to aluminum to improve its properties, such as hardness or tensile strength. The following table shows the composition of five commonly used alloys, which are known by their alloy numbers (2024, 6061, and so on) [Kutz, 1999]. Obtain a matrix algorithm to compute the amounts of raw materials needed to produce a given amount of each alloy. Use MATLAB to determine how much raw material of each type is needed to produce 1000 tons of each alloy.

	Composition of aluminum alloys				
Alloy	**%Cu**	**%Mg**	**%Mn**	**%Si**	**%Zn**
2024	4.4	1.5	0.6	0	0
6061	0	1	0	0.6	0
7005	0	1.4	0	0	4.5
7075	1.6	2.5	0	0	5.6
356.0	0	0.3	0	7	0

29. Redo Example 2.4–3 as a script file to allow the user to examine the effects of labor costs. Allow the user to input the four labor costs in the following table. When you run the file, it should display the quarterly costs and the category costs. Run the file for the case where the unit labor costs are $3000, $7000, $4000, and $8000, respectively.

Product costs

	Unit costs ($ \times 10^3$)		
Product	**Materials**	**Labor**	**Transportation**
1	6	2	1
2	2	5	4
3	4	3	2
4	9	7	3

Quarterly production volume

Product	**Quarter 1**	**Quarter 2**	**Quarter 3**	**Quarter 4**
1	10	12	13	15
2	8	7	6	4
3	12	10	13	9
4	6	4	11	5

30. Vectors with three elements can represent position, velocity, and acceleration. A mass of 5 kg, which is 3 m away from the x-axis, starts at $x = 2$ m and moves with a speed of 10 m/s parallel to the y-axis. Its velocity is thus described by $\mathbf{v} = [0, 10, 0]$, and its position is described

by $\mathbf{r} = [2, 10t + 3, 0]$. Its angular momentum vector $\mathbf{L}$ is found from $\mathbf{L} = m(\mathbf{r} \times \mathbf{v})$, where m is the mass. Use MATLAB to:

a. Compute a matrix $\mathbf{P}$ whose 11 rows are the values of the position vector $\mathbf{r}$ evaluated at the times $t = 0, 0.5, 1, 1.5, \ldots 5$ s.
b. What is the location of the mass when $t = 5$ s?
c. Compute the angular momentum vector $\mathbf{L}$. What is its direction?

31.* The *scalar triple product* computes the magnitude M of the moment of a force vector $\mathbf{F}$ about a specified line. It is $M = (\mathbf{r} \times \mathbf{F}) \cdot \mathbf{n}$, where $\mathbf{r}$ is the position vector from the line to the point of application of the force and $\mathbf{n}$ is a unit vector in the direction of the line.

Use MATLAB to compute the magnitude M for the case where $\mathbf{F} = [10, -5, 4]$ N, $\mathbf{r} = [-3, 7, 2]$ m, and $\mathbf{n} = [6, 8, -7]$.

32. Verify the identity

$$\mathbf{A} \times (\mathbf{B} \times \mathbf{C}) = \mathbf{B}\,(\mathbf{A} \cdot \mathbf{C}) - \mathbf{C}(\mathbf{A} \cdot \mathbf{B})$$

for the vectors $\mathbf{A} = 5\mathbf{i} - 3\mathbf{j} + 7\mathbf{k}$, $\mathbf{B} = -6\mathbf{i} + 4\mathbf{j} + 3\mathbf{k}$, and $\mathbf{C} = 2\mathbf{i} + 8\mathbf{j} - 9\mathbf{k}$.

33. The area of a parallelogram can be computed from $|\mathbf{A} \times \mathbf{B}|$, where $\mathbf{A}$ and $\mathbf{B}$ define two sides of the parallelogram (see Figure P33). Compute the area of a parallelogram defined by $\mathbf{A} = 7\mathbf{i}$ and $\mathbf{B} = \mathbf{i} + 3\mathbf{j}$.

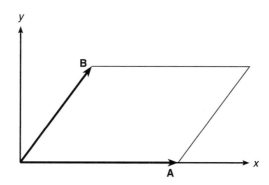

Figure P33

34. The volume of a parallelepiped can be computed from $|\mathbf{A} \cdot (\mathbf{B} \times \mathbf{C})|$, where $\mathbf{A}$, $\mathbf{B}$, and $\mathbf{C}$ define three sides of the parallelepiped (see Figure P34). Compute the volume of a parallelepiped defined by $\mathbf{A} = 6\mathbf{i}$, $\mathbf{B} = 2\mathbf{i} + 4\mathbf{j}$, and $\mathbf{C} = 3\mathbf{i} - 2\mathbf{k}$.

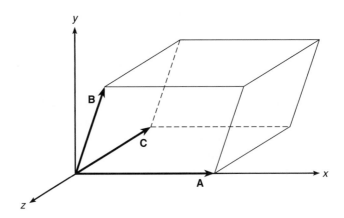

Figure P34

Section 2.5

35. Solve the following problems using matrix inversion. Check your solutions by computing $\mathbf{A}^{-1}\mathbf{A}$.

a. $2x + y = 5$
   $3x - 9y = 2$

b. $-8x - 5y = 4$
   $-2x + 7y = 10$

c. $12x - 5y = 11$
   $-3x + 4y + 7x_3 = -3$
   $6x + 2y + 3x_3 = 22$

d. $6x - 3y + 4x_3 = 41$
   $12x + 5y - 7x_3 = -26$
   $-5x + 2y + 6x_3 = 14$

36.* *a.* Solve the following matrix equation for the matrix $\mathbf{C}$.

$$\mathbf{A}(\mathbf{BC} + \mathbf{A}) = \mathbf{B}$$

b. Evaluate the solution obtained in part *a* for the case

$$\mathbf{A} = \begin{bmatrix} 3 & 9 \\ -2 & 4 \end{bmatrix} \quad \mathbf{B} = \begin{bmatrix} 2 & -3 \\ 7 & 6 \end{bmatrix}$$

37. Use MATLAB to solve the following problems.

a. $-2x + y = -5$
   $-2x + y = 3$

b. $-2x + y = 3$
   $-8x + 4y = 12$

c. $-2x + y = -5$
   $-2x + y = -5.00001$

d. $x_1 + 5x_2 - x_3 + 6x_4 = 19$
   $2x_1 - x_2 + x_3 - 2x_4 = 7$
   $-x_1 + 4x_2 - x_3 + 3x_4 = 20$
   $3x_1 - 7x_2 - 2x_3 + x_4 = -75$

38. The circuit shown in Figure P38 has five resistances and one applied voltage. Kirchhoff's voltage law applied to each loop in the circuit shown gives

$$v - R_2 i_2 - R_4 i_4 = 0$$

$$-R_2 i_2 + R_1 i_1 + R_3 i_3 = 0$$

$$-R_4 i_4 - R_3 i_3 + R_5 i_5 = 0$$

Conservation of charge applied at each node in the circuit gives

$$i_6 = i_1 + i_2$$

$$i_2 + i_3 = i_4$$

$$i_1 = i_3 + i_5$$

$$i_4 + i_5 = i_6$$

a. Write a MATLAB script file that uses given values of the applied voltage v and the values of the five resistances, and solves for the six currents.

b. Use the program developed in part a to find the currents for the case where $R_1 = 1$, $R_2 = 5$, $R_3 = 2$, $R_4 = 10$, $R_5 = 5$ kΩ, and $v = 100$ V. (1 kΩ = 1000 Ω.)

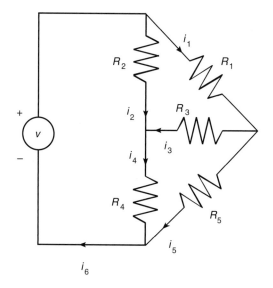

Figure P38

39.* *a.* Use MATLAB to solve the following equations for x, y, and z as functions of the parameter c.

$$x - 5y - 2z = 11c$$

$$6x + 3y + z = 13c$$

$$7x + 3y - 5z = 10c$$

b. Plot the solutions for x, y, and z versus c on the same plot, for $-10 \leq c \leq 10$.

40. Fluid flows in pipe networks can be analyzed in a manner similar to that used for electric resistance networks. Figure P40 shows a network with three pipes. The volume flow rates in the pipes are q_1, q_2, and q_3. The pressures at the pipe ends are p_a, p_b, and p_c. The pressure at the junction is p_1. Under certain conditions, the pressure-flow rate relation in a pipe has the same form as the voltage-current relation in a resistor. Thus, for the three pipes, we have

$$q_1 = \frac{1}{R_1}(p_a - p_1)$$

$$q_2 = \frac{1}{R_2}(p_1 - p_b)$$

$$q_3 = \frac{1}{R_3}(p_1 - p_c)$$

where the R_i are the pipe resistances. From conservation of mass, $q_1 = q_2 + q_3$.

a. Set up these equations in a matrix form $\mathbf{Ax} = \mathbf{b}$ suitable for solving for the three flow rates q_1, q_2, and q_3, and the pressure p_1, given the values of pressures p_a, p_b, and p_c and the values of resistances R_1, R_2, and R_3. Find the expressions for $\mathbf{A}$ and $\mathbf{b}$.

b. Use MATLAB to solve the matrix equations obtained in part a for the case where $p_a = 4320$ lb/ft^2, $p_b = 3600$ lb/ft^2, and $p_c = 2880$ lb/ft^2. These correspond to 30, 25, and 20 psi, respectively (1 psi = 1 lb/in^2, and atmospheric pressure is 14.7 psi). Use the resistance values $R_1 = 10,000$, $R_2 = 14,000$ lb sec/ft^5. These values correspond to fuel oil flowing through pipes 2 ft long, with 2- and 1.4-in diameters, respectively. The units of the answers are ft^3/sec for the flow rates and lb/ft^2 for pressure.

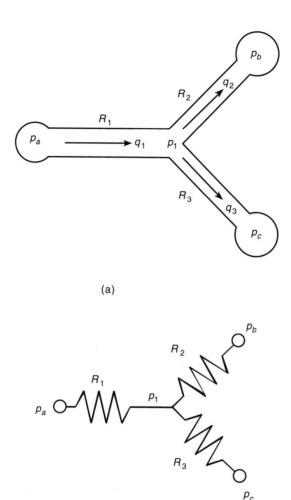

(a)

(b)

Figure P40

41. Figure P41 illustrates a robot arm that has two "links" connected by two "joints"—a shoulder or base joint and an elbow joint. There is a motor at each joint. The joint angles are θ_1 and θ_2. The (x, y) coordinates of the hand at the end of the arm are given by

$$x = L_1 \cos\theta_1 + L_2 \cos(\theta_1 + \theta_2)$$
$$y = L_1 \sin\theta_1 + L_2 \sin(\theta_1 + \theta_2)$$

where L_1 and L_2 are the lengths of the links.

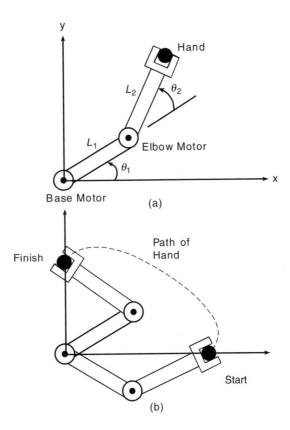

Figure P41

Polynomials are used for controlling the motion of robots. If we start the
arm from rest with zero velocity and acceleration, the following
polynomials are used to generate commands to be sent to the joint
motor controllers.

$$\theta_1(t) = \theta_1(0) + a_1 t^3 + a_2 t^4 + a_3 t^5$$

$$\theta_2(t) = \theta_2(0) + b_1 t^3 + b_2 t^4 + b_3 t^5$$

where $\theta_1(0)$ and $\theta_2(0)$ are the starting values at time $t = 0$. The angles
$\theta_1(t_f)$ and $\theta_2(t_f)$ are the joint angles corresponding to the desired destination of the arm at time t_f. The values of $\theta_1(0)$, $\theta_2(0)$, $\theta_1(t_f)$, and $\theta_2(t_f)$ can
be found from trigonometry, if the starting and ending (x, y) coordinates
of the hand are specified.

a. Set up a matrix equation to be solved for the coefficients a_1, a_2, and a_3,
    given values for $\theta_1(0)$, $\theta_1(t_f)$, and t_f. Obtain a similar equation for the
    coefficients b_1, b_2, and b_3.

b. Use MATLAB to solve for the polynomial coefficients given the values $t_f = 2$ secs, $\theta_1(0) = -19°$, $\theta_2(0) = 44°$, $\theta_1(t_f) = 43°$, and $\theta_2(t_f) = 151°$. (These values correspond to a starting hand location of $x = 6.5$, $y = 0$ ft, and a destination location of $x = 0$, $y = 2$ ft, for $L_1 = 4$ and $L_2 = 3$ ft.)

c. Use the results of part b to plot the path of the hand.

42.* Engineers must be able to predict the rate of heat loss through a building wall to determine the heating system requirements. They do this by using the concept of *thermal resistance R*, which relates the heat flow rate q through a material to the temperature difference ΔT across the material: $q = \Delta T/R$. This relation is like the voltage-current relation for an electrical resistor: $i = v/R$. So the heat flow rate plays the role of electric current, and the temperature difference plays the role of the voltage difference . The SI unit for q is the *watt* (W), which is 1 Joule/s.

The wall shown in Figure P42 consists of four layers: an inner layer of plaster/lathe 10 mm thick, a layer of fiberglass insulation 125 mm thick, a layer of wood 60 mm thick, and an outer layer of brick 50 mm thick. If we assume that the inner and outer temperatures T_i and T_o have remained constant for some time, then the heat energy stored in the layers is constant, and thus the heat flow rate through each layer is the same. Applying conservation of energy gives the following equations.

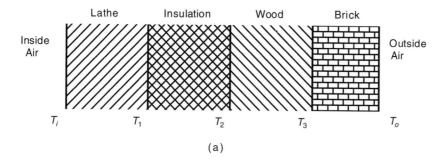

(a)

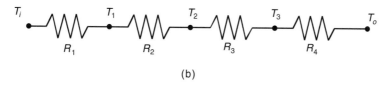

(b)

Figure P42

$$q = \frac{1}{R_1}(T_i - T_1) = \frac{1}{R_2}(T_1 - T_2) = \frac{1}{R_3}(T_2 - T_3) = \frac{1}{R_4}(T_3 - T_o)$$

The thermal resistance of a solid material is given by $R = D/k$, where D is the material thickness and k is the material's *thermal conductivity*. For the given materials, the resistances for a wall area of 1 m^2 are $R_1 = 0.036$, $R_2 = 4.01$, $R_3 = 0.408$, and $R_4 = 0.038$ K/W.

Suppose that $T_i = 20°C$ and $T_o = -10°C$. Find the other three temperatures and the heat loss rate q in watts. Compute the heat loss rate if the wall's area is 10 m^2.

43. The concept of thermal resistance described in the previous problem can be used to find the temperature distribution in the flat square plate shown in Figure P43(a). The plate's edges are insulated so that no heat can escape, except at two points where the edge temperature is heated to T_a

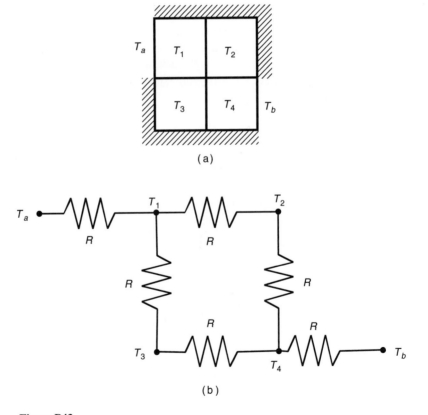

(a)

(b)

Figure P43

and T_b, respectively. The temperature varies through the plate, so no single point can describe the plate's temperature. One way to estimate the temperature distribution is to imagine that the plate consists of four subsquares and to compute the temperature in each subsquare. Let R be the thermal resistance of the material between the centers of adjacent subsquares. Then we can think of the problem as a network of electrical resistors, as shown in part (b) of the figure. Let q_{ij} be the heat flow rate between the points whose temperatures are T_i and T_j. If T_a and T_b remain constant for some time, then the heat energy stored in each subsquare is constant also, and the heat flow rate between each subsquare is constant. Under these conditions, conservation of energy says that the heat flow into a subsquare equals the heat flow out. Applying this principle to each subsquare gives the following equations.

$$q_{a1} = q_{12} + q_{13}$$

$$q_{12} = q_{24}$$

$$q_{13} = q_{34}$$

$$q_{34} + q_{24} = q_{4b}$$

Substituting $q = (T_i - T_j)/R$, we find that R can be canceled out of every equation, and they can be rearranged as follows:

$$T_1 = \frac{1}{3}(T_a + T_2 + T_3)$$

$$T_2 = \frac{1}{2}(T_1 + T_4)$$

$$T_3 = \frac{1}{2}(T_1 + T_4)$$

$$T_4 = \frac{1}{3}(T_2 + T_3 + T_5)$$

These equations tell us that the temperature of each subsquare is the average of the temperatures in the adjacent subsquares!

Solve these equations for the case where $T_a = 150°C$ and $T_b = 20°C$.

44. Use the averaging principle developed in Problem 43 to find the temperature distribution of the plate shown in Figure P44, using the 3×3 grid, and the given values $T_a = 150°C$ and $T_b = 20°C$.

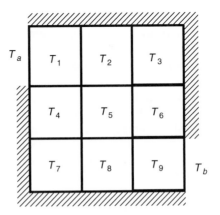

Figure P44

45.* Solve the following equations:

$$7x + 9y - 9z = 22$$
$$3x + 2y - 4z = 12$$
$$x + 5y - z = -2$$

46. The following table shows how many hours in process reactors A and B are required to produce 1 ton each of chemical products 1, 2, and 3. The two reactors are available for 35 and 40 hours per week, respectively.

Hours	Product 1	Product 2	Product 3
Reactor A	6	2	10
Reactor B	3	5	2

Let x, y, and z be the number of tons each of products 1, 2, and 3 that can be produced in one week.

   a. Use the data in the table to write two equations in terms of x, y, and z. Determine whether a unique solution exists. If not, use MATLAB to find the relations between x, y, and z.

   b. Note that negative values x, y, and z have no meaning here. Find the allowable ranges for x, y, and z.

   c. Suppose the profits for each product are $200, $300, and $100 for products 1, 2, and 3, respectively. Find the values of x, y, and z to maximize the profit.

   d. Suppose the profits for each product are $200, $500, and $100 for products 1, 2, and 3, respectively. Find the values of x, y, and z to maximize the profit.

47. See Figure P47. Assume that no vehicles stop within the network. A traffic engineer wants to know if the traffic flows $f_1, f_2, \ldots, f_7$ (in vehicles per hour) can be computed given the measured flows shown in the figure. If not, then determine how many more traffic sensors need to be installed, and obtain the expressions for the other traffic flows in terms of the measured quantities.

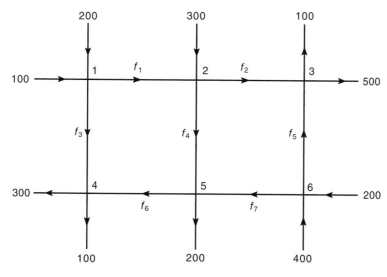

Figure P47

48.* Use MATLAB to solve the following problem:

$$x - 3y = 2$$
$$x + 5y = 18$$
$$4x - 6y = 20$$

49.* Use MATLAB to solve the following problem:

$$x - 3y = 2$$
$$x + 5y = 18$$
$$4x - 6y = 10$$

50. *a.* Use MATLAB to find the coefficients of the quadratic polynomial $y = ax^2 + bx + c$ that passes through the three points: $(x, y) = (1, 4)$, $(4, 73)$, $(5, 120)$.

   b. Use MATLAB to find the coefficients of the cubic polynomial $y = ax^3 + bx^2 + cx + d$ that passes through the three points given in part *a*.

Section 2.6

51. Use MATLAB to plot the polynomials $y = 3x^4 - 6x^3 + 8x^2 + 4x + 90$ and $z = 3x^3 + 5x^2 - 8x + 70$ over the interval $-3 \le x \le 3$. Properly label the plot and each curve. The variables y and z represent current in milliamps; the variable x represents voltage in volts.

52. Use MATLAB to plot the polynomial $y = 3x^4 - 5x^3 - 28x^2 - 5x + 200$ on the interval $-1 \le x \le 1$. Put a grid on the plot and use the `ginput` function to determine the coordinates of the peak of the curve.

53. Use MATLAB to find the following product:

$$(10x^3 - 9x^2 - 6x + 12)(5x^3 - 4x^2 - 12x + 8)$$

54.* Use MATLAB to find the quotient and remainder of

$$\frac{14x^3 - 6x^2 + 3x + 9}{5x^2 + 7x - 4}$$

55.* Use MATLAB to evaluate

$$\frac{8x^3 - 9x^2 - 7}{10x^3 + 5x^2 - 3x - 7}$$

at $x = 5$.

56. The *ideal gas law* provides one way to estimate the pressures and volumes of a gas in a container. The law is

$$P = \frac{RT}{\hat{V}}$$

More accurate estimates can be made with the *van der Waals* equation

$$P = \frac{RT}{\hat{V} - b} - \frac{a}{\hat{V}^2}$$

where the term b is a correction for the volume of the molecules and the term $a/\hat{V}^2$ is a correction for molecular attractions. The values of a and b depend on the type of gas. The gas constant is R, the *absolute* temperature is T, and the gas specific volume is $\hat{V}$. If 1 mol of an ideal gas were confined to a volume of 22.41 L at 0°C (273.2 K), it would exert a pressure of 1 atm. In these units, $R = 0.08206$.

For chlorine (Cl_2), $a = 6.49$ and $b = 0.0562$. Compare the specific volume estimates $\hat{V}$ given by the ideal gas law and the van der Waals equation for 1 mol of Cl_2 at 300 K and a pressure of 0.95 atm.

57. Aircraft A is flying east at 320 mi/hr, while aircraft B is flying south at 160 mi/hr. At 1:00 P.M. the aircraft are located as shown in Figure P57.

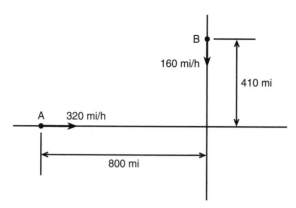

Figure P57

a. Obtain the expression for the distance *D* between the aircraft as a function of time. Plot *D* versus time until *D* reaches its minimum value.

b. Use the `roots` function to compute the time when the aircraft are first within 30 mi of each other.

58. The function

$$y = \frac{3x^2 - 12x + 20}{x^2 - 7x + 10}$$

approaches ∞ as $x \rightarrow 2$ and as $x \rightarrow 5$. Plot this function over the range $0 \leq x \leq 7$. Choose an appropriate range for the *y*-axis.

59. The following formulas are commonly used by engineers to predict the lift and drag of an airfoil:

$$L = \frac{1}{2}\rho C_L S V^2$$

$$D = \frac{1}{2}\rho C_D S V^2$$

where *L* and *D* are the lift and drag forces, *V* is the airspeed, *S* is the wing span, ρ is the air density, and C_L and C_D are the *lift* and *drag* coefficients. Both C_L and C_D depend on α, the angle of attack, the angle between the relative air velocity and the airfoil's chord line.

Wind tunnel experiments for a particular airfoil have resulted in the following formulas.

$$C_L = 4.47 \times 10^{-5}\, \alpha^3 + 1.15 \times 10^{-3}\alpha^2 + 6.66 \times 10^{-2}\alpha + 1.02 \times 10^{-1}$$

$$C_D = 5.75 \times 10^{-6}\, \alpha^3 + 5.09 \times 10^{-4}\alpha^2 + 1.81 \;\times\; 10^{-4}\alpha + 1.25 \times 10^{-2}$$

where α is in degrees.

Plot the lift and drag of this airfoil versus V for $0 \le V \le 150$ mi/hr (you must convert V to ft /sec; there is 5280 ft/mi). Use the values $\rho = 0.002378$ slug/ft^3 (air density at sea level), $\alpha = 10°$, and $S = 36$ ft. The resulting values of L and D will be in pounds.

60. The lift-to-drag ratio is an indication of the effectiveness of an airfoil. Referring to Problem 59, the equations for lift and drag are

$$L = \frac{1}{2}\, \rho C_L S V^2$$

$$D = \frac{1}{2}\, \rho C_D S V^2$$

where, for a particular airfoil, the lift and drag coefficients versus angle of attack α are given by

$$C_L = 4.47 \times 10^{-5}\alpha^3 + 1.15 \;\times\; 10^{-3}\alpha^2 + 6.66 \times 10^{-2}\alpha + 1.02 \times 10^{-1}$$

$$C_D = 5.75 \times 10^{-6}\alpha^3 + 5.09 \times 10^{-4}\alpha^2 + 1.81 \times 10^{-4}\alpha + 1.25 \times 10^{-2}$$

Using the first two equations, we see that the lift-to-drag ratio is given simply by the ratio C_L/C_D.

$$\frac{L}{D} = \frac{\frac{1}{2}\rho C_L S V^2}{\frac{1}{2}\rho C_D S V^2} = \frac{C_L}{C_D}$$

Plot L/D versus α for $-2° \le \alpha \le 22°$. Determine the angle of attack that maximizes L/D.

Section 2.7

61. *a.* Use both cell indexing and content indexing to create the following 2 $\times$ 2 cell array.

Motor 28C	Test ID 6
$\begin{bmatrix} 3 & 9 \\ 7 & 2 \end{bmatrix}$	[6 5 1]

b. What are the contents of the (1,1) element in the (2,1) cell in this array?

62. The capacitance of two parallel conductors of length L and radius r, separated by a distance d in air, is given by

$$C = \frac{\pi \epsilon L}{\ln\left(\dfrac{d-r}{r}\right)}$$

where ϵ is the permittitivity of air ($\epsilon = 8.854 \times 10^{-12}$ F/m). Create a cell array of capacitance values versus d, L, and r for $d = 0.003, 0.004, 0.005,$ and 0.01 m; $L = 1, 2, 3$ m; and $r = 0.001, 0.002, 0.003$ m. Use MATLAB to determine the capacitance value for $d = 0.005$, $L = 2$, and $r = 0.001$.

Section 2.8

63. *a.* Create a structure array that contains the conversion factors for converting units of mass, force, and distance between the metric SI system and the British Engineering System.

b. Use your array to compute the following:

- The number of meters in 24 ft.
- The number of feet in 65 m.
- The number of pounds equivalent to 18 N.
- The number of newtons equivalent to 5 lb.
- The number of kilograms in 6 slugs.
- The number of slugs in 15 kg.

64. Create a structure array that contains the following information fields concerning the road bridges in a town: bridge location, maximum load (tons), year built, year due for maintenance. Then enter the following data into the array:

Location	Max. load	Year built	Due for maintenance
Smith St.	80	1928	1997
Hope Ave.	90	1950	1999
Clark St.	85	1933	1998
North Rd.	100	1960	1998

65. Edit the structure array created in Problem 64 to change the maintenance data for the Clark St. bridge from 1998 to 2000.

66. Add the following bridge to the structure array created in Problem 64.

Location	Max. load	Year built	Due for maintenance
Shore Rd.	85	1997	2002

3

CHAPTER

Functions and Files

MATLAB has many built-in functions, including trigonometric, logarithmic, and hyperbolic functions, as well as functions for processing arrays. These functions are summarized in Section 3.1. In addition, you can define your own functions with a *function* file, and you can use them just as conveniently as the built-in functions. We explain this technique in Section 3.2. Section 3.3 covers additional topics in function programming, including function handles, anonymous functions, subfunctions, and nested functions. Another type of file that is useful in MATLAB is the data file. Importing and exporting such files is covered in Section 3.4.

Sections 3.1 and 3.2 contain essential topics and must be covered. The material in Section 3.3 is useful for creating large programs. The material in Section 3.4 is useful for readers who must work with large data sets.

3.1 Elementary Mathematical Functions

You can use the `lookfor` command to find functions that are relevant to your application. For example, type `lookfor imaginary` to get a list of the functions that deal with imaginary numbers. You will see listed:

```
imag  Complex imaginary part
i     Imaginary unit
j     Imaginary unit
```

Note that `imaginary` is not a MATLAB function, but the word is found in the help descriptions of the MATLAB function `imag` and the special symbols `i` and `j`.

Their names and brief descriptions are displayed when you type `lookfor imaginary`. If you know the correct spelling of a MATLAB function—for example, `disp`—you can type `help disp` to obtain a description of the function.

Some of the functions, like `sqrt` and `sin`, are built-in. These are stored as image files and are not M-files. They are part of the MATLAB core so they are very efficient, but the computational details are not readily accessible. Other functions, like `sind`, are implemented in M-files. You can see the code and even modify it, although this is not recommended.

Exponential and Logarithmic Functions

Table 3.1–1 summarizes some of the common elementary functions. An example is the square root function `sqrt`. To compute $\sqrt{9}$, you type `sqrt(9)` at the command line. When you press **Enter,** you see the result `ans = 3`. You can use functions with variables. For example, consider the session:

```
>>x = -9; y = sqrt(x)
y =
   0 + 3.0000i
```

Note that the `sqrt` function returns the positive root only.

One of the strengths of MATLAB is that it will treat a variable as an array automatically. For example, to compute the square roots of 5, 7, and 15, type

```
>>x = [5,7,15]; y = sqrt(x)
y =
   2.2361     2.6358     3.8730
```

Table 3.1–1 Some common mathematical functions

Exponential	
`exp(x)`	Exponential; e^x.
`sqrt(x)`	Square root; $\sqrt{x}$.
Logarithmic	
`log(x)`	Natural logarithm; ln x.
`log10(x)`	Common (base 10) logarithm; log $x = \log_{10} x$.
Complex	
`abs(x)`	Absolute value; x.
`angle(x)`	Angle of a complex number x.
`conj(x)`	Complex conjugate.
`imag(x)`	Imaginary part of a complex number x.
`real(x)`	Real part of a complex number x.
Numeric	
`ceil(x)`	Round to the nearest integer toward ∞.
`fix(x)`	Round to the nearest integer toward zero.
`floor(x)`	Round to the nearest integer toward $-\infty$.
`round(x)`	Round toward the nearest integer.
`sign(x)`	Signum function: $+1$ if $x > 0$; 0 if $x = 0$; -1 if $x < 0$.

The square root function operates on every element in the array x.

Similarly, we can type exp(2) to obtain $e^2 = 7.3891$, where e is the base of the natural logarithms. Typing exp(1) gives 2.7183, which is e. Note that in mathematics text, ln x denotes the *natural* logarithm, where $x = e^y$ implies that

$$\ln x = \ln(e^y) = y \ln e = y$$

because ln $e = 1$. However, this notation has not been carried over into MATLAB, which uses log(x) to represent ln x.

The *common* (base 10) logarithm is denoted in text by log x or $\log_{10} x$. It is defined by the relation $x = 10^y$; that is,

$$\log_{10} x = \log_{10} 10^y = y \log_{10} 10 = y$$

because $\log_{10} 10 = 1$. The MATLAB common logarithm function is log10(x). A common mistake is to type log(x), instead of log10(x).

Another common error is to forget to use the array multiplication operator .*. Note that in the MATLAB expression y = exp(x).*log(x), we need to use the operator .* if x is an array because both exp(x) and log(x) will be arrays.

Complex Number Functions

Chapter 1 explained how MATLAB easily handles complex number arithmetic. In the *rectangular* representation the number $a + ib$ represents a point in the xy plane. The number's real part a is the x coordinate of the point, and the imaginary part b is the y coordinate. The *polar* representation uses the distance M of the point from the origin, which is the length of the hypotenuse, and the angle θ the hypotenuse makes with the *positive real* axis. The pair (M, θ) is simply the polar coordinates of the point. From the Pythagorean theorem, the length of the hypotenuse is given by $M = \sqrt{a^2 + b^2}$ which is called the *magnitude* of the number. The angle θ can be found from the trigonometry of the right triangle. It is $\theta = \arctan (b/a)$.

Adding and subtracting complex numbers by hand is easy when they are in the rectangular representation. However, the polar representation facilitates multiplication and division of complex numbers by hand. We must enter complex numbers in MATLAB using the rectangular form, and its answers will be given in that form. We can obtain the rectangular representation from the polar representation as follows:

$$a = M \cos \theta \qquad b = M \sin \theta$$

The MATLAB abs(x) and angle(x) functions calculate the magnitude M and angle θ of the complex number x. The functions real(x) and imag(x) return the real and imaginary parts of x. The function conj(x) computes the complex conjugate of x.

The magnitude of the product z of two complex numbers x and y is equal to the product of their magnitudes: $|z| = |x||y|$. The angle of the product is equal to the sum of the angles: $\angle z = \angle x + \angle y$. These facts are demonstrated below.

```
>>x = -3 + 4i; y = 6 - 8i;
>>mag_x = abs(x)
mag_x =
    5.0000
>>mag_y = abs(y)
mag_y =
    10.0000
>>mag_product = abs(x*y)
    50.0000
>>angle_x = angle(x)
angle_x =
    2.2143
>>angle_y = angle(y)
angle_y =
    -0.9273
>>sum_angles = angle_x + angle_y
sum_angles =
    1.2870
>>angle_product = angle(x*y)
angle_product =
    1.2870
```

Similarly, for division, if $z = x/y$, then $|z| = |x|/|y|$ and $\angle z = \angle x - \angle y$.

Note that when x is a vector of *real* values, abs(x) does not give the geometric length of the vector. This length is given by norm(x). If x is a complex number representing a geometric vector, then abs(x) gives its geometric length.

Numeric Functions

The round function rounds to the nearest integer. If y=[2.3,2.6,3.9], typing round(y) gives the results 2, 3, 4. The fix function truncates to the nearest integer toward zero. Typing fix(y) gives the results 2, 2, 3. The ceil function (which stands for "ceiling") rounds to the nearest integer toward ∞. Typing ceil(y) produces the answers 3, 3, 4.

Suppose z = [-2.6,-2.3,5.7]. The floor function rounds to the nearest integer toward $-\infty$. Typing floor(z) produces the result $-3, -3, 5$. Typing fix(z) produces the answer $-2, -2, 5$. The abs function computes the absolute value. Thus abs(z) produces 2.6, 2.3, 5.7.

Test Your Understanding

T3.1–1 For several values of x and y, confirm that $\ln (xy) = \ln x + \ln y$.

T3.1–2 Find the magnitude, angle, real part, and imaginary part of the number $\sqrt{2 + 6i}$.

Function Arguments

FUNCTION
ARGUMENT

When writing mathematics in text, we use parentheses ( ), brackets [ ], and braces { } to improve the readability of expressions, and we have much latitude over their use. For example, we can write sin 2 in text, but MATLAB requires parentheses surrounding the 2 (which is called the *function argument* or *parameter*). Thus to evaluate sin 2 in MATLAB, we type `sin(2)`. The MATLAB function name must be followed by a pair of parentheses that surround the argument. To express in text the sine of the second element of the array x, we would type `sin[x(2)]`. However, in MATLAB you cannot use brackets or braces in this way, and you must type `sin(x(2))`.

You can include expressions and other functions as arguments. For example, to evaluate $\sin(x^2 + 5)$, you type `sin(x.^2 + 5)`. To evaluate $\sin(\sqrt{x} + 1)$, you type `sin(sqrt(x)+1)`. Be sure to check the order of precedence and the number and placement of parentheses when typing such expressions. Every left-facing parenthesis requires a right-facing mate. However, this condition does not guarantee that the expression is correct!

Another common mistake involves expressions like $\sin^2 x$, which means $(\sin x)^2$. In MATLAB we write this expression as `(sin(x))^2`, *not* as `sin^2(x)`, `sin^2x`, or `sin(x^2)`!

Trigonometric Functions

Other commonly used functions are `cos(x)`, `tan(x)`, `sec(x)`, and `csc(x)`, which return cos x, tan x, sec x, and csc x, respectively. Table 3.1–2 lists the

Table 3.1–2 Trigonometric functions

Trigonometric*	
`cos(x)`	Cosine; cos x.
`cot(x)`	Cotangent; cot x.
`csc(x)`	Cosecant; csc x.
`sec(x)`	Secant; sec x.
`sin(x)`	Sine; sin x.
`tan(x)`	Tangent; tan x.
Inverse trigonometric[†]	
`acos(x)`	Inverse cosine; arccos $x = \cos^{-1} x$.
`acot(x)`	Inverse cotangent; arccot $x = \cot^{-1} x$.
`acsc(x)`	Inverse cosecant; arccsc $x = \csc^{-1} x$.
`asec(x)`	Inverse secant; arcsec $x = \sec^{-1} x$.
`asin(x)`	Inverse sine; arcsin $x = \sin^{-1} x$.
`atan(x)`	Inverse tangent; arctan $x = \tan^{-1} x$.
`atan2(y,x)`	Four-quadrant inverse tangent.

*These functions accept x in radians.
[†]These functions return a value in radians.

MATLAB trigonometric functions that operate in radian mode. Thus `sin(5)` computes the sine of 5 rad, not the sine of 5°. Similarly, the inverse trigonometric functions return an answer in radians. The functions that operate in degree mode have the letter d appended to their names. To compute the inverse sine, type `asin(x)`. For example, `asin(0.5)` returns the answer: 0.5236 rad. *Note*: In MATLAB `sin(x)^(-1)` does not give $\sin^{-1}(x)$; it gives $1/\sin(x)$!

MATLAB has two inverse tangent functions. The function `atan(x)` computes arctan x—the arctangent or inverse tangent—and returns an angle between $-\pi/2$ and $\pi/2$. Another correct answer is the angle that lies in the opposite quadrant. The user must be able to choose the correct answer. For example, `atan(1)` returns the answer 0.7854 rad, which corresponds to 45°. Thus tan 45° = 1. However, tan(45° + 180°) = tan 225° = 1 also. Thus arctan(1) = 225° is also correct.

MATLAB provides the `atan2(y,x)` function to determine the arctangent unambiguously, where x and y are the coordinates of a point. The angle computed by `atan2(y,x)` is the angle between the positive real axis and line from the origin (0, 0) to the point (x, y). For example, the point $x = 1, y = -1$ corresponds to $-45°$ or -0.7854 rad, and the point $x = -1, y = 1$ corresponds to 135° or 2.3562 rad. Typing `atan2(-1,1)` returns -0.7854, while typing `atan2(1,-1)` returns 2.3562. The `atan2(y,x)` function is an example of a function that has two arguments. The order of the arguments is important for such functions. At present there is no `atan2d` function.

Test Your Understanding

T3.1–3 For several values of x, confirm that $e^{ix} = \cos x + i \sin x$.

T3.1–4 For several values of x in the range $0 \leq x \leq 2\pi$, confirm that $\sin^{-1} x + \cos^{-1} x = \pi/2$.

T3.1–5 For several values of x in the range $0 \leq x \leq 2\pi$, confirm that $\tan(2x) = 2 \tan x/(1 - \tan^2 x)$.

Hyperbolic Functions

The *hyperbolic functions* are the solutions of some common problems in engineering analysis. For example, the *catenary* curve, which describes the shape of a hanging cable supported at both ends, can be expressed in terms of the hyperbolic cosine, cosh x, which is defined as

$$\cosh x = \frac{e^x + e^{-x}}{2}$$

The hyperbolic sine, sinh x, is defined as

$$\sinh x = \frac{e^x - e^{-x}}{2}$$

The inverse hyperbolic sine, $\sinh^{-1} x$, is the value y that satisfies sinh $y = x$.

Table 3.1–3 Hyperbolic functions

Hyperbolic	
cosh(x)	Hyperbolic cosine; $\cosh x = (e^x + e^{-x})/2$.
coth(x)	Hyperbolic cotangent; $\cosh x/\sinh x$.
csch(x)	Hyperbolic cosecant; $1/\sinh x$.
sech(x)	Hyperbolic secant; $1/\cosh x$.
sinh(x)	Hyperbolic sine; $\sinh x = (e^x - e^{-x})/2$.
tanh(x)	Hyperbolic tangent; $\sinh x/\cosh x$.
Inverse hyperbolic	
acosh(x)	Inverse hyperbolic cosine;
acoth(x)	Inverse hyperbolic cotangent;
acsch(x)	Inverse hyperbolic cosecant;
asech(x)	Inverse hyperbolic secant;
asinh(x)	Inverse hyperbolic sine;
atanh(x)	Inverse hyperbolic tangent;

Several other hyperbolic functions have been defined. Table 3.1–3 lists these hyperbolic functions and the MATLAB commands to obtain them.

Test Your Understanding

T3.1–6 For several values of x in the range $0 \le x \le 5$, confirm that $\sin(ix) = i \sinh x$.

T3.1–7 For several values of x in the range $-10 \le x \le 10$, confirm that $\sinh^{-1} x = \ln\left(x + \sqrt{x^2 + 1}\right)$.

3.2 User-Defined Functions

FUNCTION FILE

LOCAL VARIABLE

FUNCTION DEFINITION LINE

Another type of M-file is a *function file*. Unlike a script file, all the variables in a function file are *local,* which means their values are available only within the function. Function files are useful when you need to repeat a set of commands several times. They are the building blocks of larger programs.

To create a function file, open the Editor/Debugger as described in Chapter 1. The first line in a function file must begin with a *function definition line* that has a list of inputs and outputs. This line distinguishes a function M-file from a script M-file. Its syntax is as follows:

```
function [output variables] = function_name(input variables)
```

The output variables are those variables whose values are computed by the function, using the given values of the input variables. Note that the output variables are enclosed in *square brackets* (which are optional if there is only one output), while the input variables must be enclosed with *parentheses*. The function_name should be the same as the file name in which it is saved (with the .m extension). That is, if we name a function drop, it should be saved in the file drop.m. The function is "called" by typing its name (for example, drop) at the command line. The word function in the function definition line must be *lowercase*. Before naming a function, you can use the exist function to see if another function has the same name.

Some Simple Function Examples

Functions operate on variables within their own workspace (called *local variables*), which is separate from the workspace you access at the MATLAB command prompt. Consider the following user-defined function fun.

```
function z = fun(x,y)
u = 3*x;
z = u + 6*y.^2;
```

Note the use of the array exponentiation operator (.^). This enables the function to accept y as an array. Now consider what happens when you call this function in various ways in the Command window. Call the function with its output argument:

```
>>x = 3; y = 7;
>>z = fun (x,y)
z =
   303
```

or

```
>>z = fun(3,7)
z =
   303
```

The function uses $x = 3$ and $y = 7$ to compute z.

Call the function without its output argument and try to access its value. You see an error message.

```
>>fun(3,7)
ans =
    303
>>z
???   Undefined function or variable 'z'.
```

Assign the output argument to another variable:

```
>>q = fun(3,7)
q =
   303
```

You can suppress the output by putting a semicolon after the function call. For example, if you type q = fun(3,7); the value of q will be computed but not displayed.

The variables x and y are *local* to the function fun, so unless you pass their values by naming them x and y, their values will not be available in the workspace outside the function. The variable u is also local to the function. For example,

```
>>x = 3; y = 7; q = fun(x,y);
>>u
??? Undefined function or variable 'u'.
```

Compare this to

```
>>q = fun(3,7);
>>x
??? Undefined function or variable 'x'.
>>y
??? Undefined function or variable 'y'.
```

Only the order of the arguments is important, not the names of the arguments:

```
>>a = 7;b = 3;
>>z = fun(b,a)     % This is equivalent to z = fun(3,7)
z =
    303
```

You can use arrays as input arguments:

```
>>r = fun([2:4],[7:9])
r =
    300     393     498
```

A function may have more than one output. These are enclosed in square brackets. For example, the function circle computes the area A and circumference C of a circle, given its radius as an input argument.

```
function [A, C] = circle(r)
A = pi*r.^2;
C = 2*pi*r;
```

The function is called as follows, if $r = 4$.

```
>>[A, C] = circle(4)
A =
    50.2655
C =
    25.1327
```

A function may have no input arguments and no output list. For example, the following user-defined function show_date computes and stores the date in the variable today, and displays the value of today.

```
function show_date
today = date
```

Variations in the Function Line

The following examples show permissible variations in the format of the function line. The differences depend on whether there is no output, a single output, or multiple outputs.

Function definition line	File name
1. `function [area_square] = square(side);`	`square.m`
2. `function area_square = square(side);`	`square.m`
3. `function [volume_box] = box(height,width,length);`	`box.m`
4. `function [area_circle,circumf] = circle(radius);`	`circle.m`
5. `function sqplot(side);`	`sqplot.m`

Example 1 is a function with one input and one output. The square brackets are optional when there is only one output (see example 2). Example 3 has one output and three inputs. Example 4 has two outputs and one input. Example 5 has no output variable (for example, a function that generates a plot). In such cases the equal sign may be omitted.

Comment lines starting with the % sign can be placed anywhere in the function file. However, if you use `help` to obtain information about the function, MATLAB displays all comment lines immediately following the function definition line up to the first blank line or first executable line. The first comment line can be accessed by the `lookfor` command.

We can call both built-in and user-defined functions either with the output variables explicitly specified, as in examples 1 through 4, or without any output variables specified. For example, we can call the function `square` as `square (side)` if we are not interested in its output variable `area_square`. (The function might perform some other operation that we want to occur, such as producing a plot.) Note that if we omit the semicolon at the end of the function call statement, the first variable in the output variable list will be displayed using the default variable name `ans`.

Variations in Function Calls

The following function, called `drop`, computes a falling object's velocity and distance dropped. The input variables are the acceleration g, the initial velocity v_0, and the elapsed time t. Note that we must use the element-by-element operations for any operations involving function inputs that are arrays. Here we anticipate that `t` will be an array, so we use the element-by-element operator (`.^`).

```
function [dist,vel] = drop(g,v0,t);
% Computes the distance traveled and the
% velocity of a dropped object, as functions
% of g, the initial velocity v0, and the time t.
```

```
vel = g*t + v0;
dist = 0.5*g*t.^2 + v0*t;
```

The following examples show various ways to call the function drop:

1. The variable names used in the function definition may, but need not, be used when the function is called:

```
a = 32.2;
initial_speed = 10;
time = 5;
[feet_dropped,speed] = drop(a,initial_speed,time)
```

2. The input variables need not be assigned values outside the function prior to the function call:

```
[feet_dropped,speed] = drop(32.2,10,5)
```

3. The inputs and outputs may be arrays:

```
[feet_dropped,speed]=drop(32.2,10,[0:1:5])
```

This function call produces the arrays feet_dropped and speed, each with six values corresponding to the six values of time in the array [0:1:5].

Local Variables

The names of the input variables given in the function definition line are local to that function. This means that other variable names can be used when you call the function. All variables inside a function are erased after the function finishes executing, except when the same variable names appear in the output variable list used in the function call.

For example, when using the drop function in a program, we can assign a value to the variable dist before the function call, and its value will be unchanged after the call because its name was not used in the output list of the call statement (the variable feet_dropped was used in the place of dist). This is what is meant by the function's variables being "local" to the function. This feature allows us to write generally useful functions using variables of our choice, without being concerned that the calling program uses the same variable names for other calculations. This means that our function files are "portable," and need not be rewritten every time they are used in a different program.

You might find the M-file Debugger to be useful for locating errors in function files. Runtime errors in functions are more difficult to locate because the function's local workspace is lost when the error forces a return to the MATLAB base workspace. The Debugger provides access to the function workspace, and allows you to change values. It also enables you to execute lines one at a time and to set *breakpoints,* which are specific locations in the file where execution is temporarily halted. The applications in this text will probably not require use of the Debugger, which is useful mainly for very large programs. For more information, see Chapter 4 of this text and also Chapter 4 of [Palm, 2005].

Global Variables

The `global` command declares certain variables global, and therefore their values are available to the basic workspace and to other functions that declare these variables global. The syntax to declare the variables `A`, `X`, and `Q` is `global A X Q`. Use a space, not a comma, to separate the variables. Any assignment to those variables, in any function or in the base workspace, is available to all the other functions declaring them global. If the global variable doesn't exist the first time you issue the `global` statement, it will be initialized to the empty matrix. If a variable with the same name as the global variable already exists in the current workspace, MATLAB issues a warning and changes the value of that variable to match the global. In a user-defined function, make the global command the first executable line. Place the same command in the calling program. It is customary, but not required, to capitalize the names of global variables and to use long names, to make them easily recognizable.

GLOBAL VARIABLES

The decision to declare a variable global is not always clear-cut. It is recommended to avoid using global variables. This can often be done by using anonymous and nested functions, as discussed in Section 3.3.

Function Handles

A *function handle* is a way to reference a given function. First introduced in MATLAB 6.0, function handles have become widely used and frequently appear in examples throughout the MATLAB documentation. You can create a function handle to any function by using the @ sign before the function name. You can then give the handle a name, if you wish, and you can use the handle to reference the function.

FUNCTION HANDLE

For example, consider the following user-defined function, which computes $y = x + 2e^{-x} - 3$.

```
function y = f1(x)
y = x + 2*exp(-x) - 3;
```

To create a handle to this function and name the handle `fh1`, you type `fh1 = @f1`.

Function Functions

Some MATLAB functions act on functions. These commands are called *function functions*. If the function acted upon is not a simple function, it is more convenient to define the function in an M-file. You can pass the function to the calling function by using a function handle.

Finding the Zeros of a Function You can use the `fzero` function to find the zero of a function of a single variable, which is denoted by `x`. Its basic syntax is

```
fzero(@function, x0)
```

where `@function` is a function handle and `x0` is a user-supplied guess for the zero. The `fzero` function returns a value of `x` that is near `x0`. It identifies only points where the function crosses the *x*-axis, not points where the function just touches the axis. For example, fzero (@cos,2) returns the value *x* = 1.5708. As another example, *y* = *x*² is a parabola that touches the *x*-axis at *x* = 0. Because the function never crosses the *x*-axis, however, no zero will be found.

The function fzero(@function,x0) tries to find a zero of function near x0, if x0 is a scalar. The value returned by fzero is near a point where function changes sign, or NaN if the search fails. In this case, the search terminates when the search interval is expanded until an Inf, NaN, or a complex value is found (fzero cannot find complex zeros). If x0 is a *vector* of length two, fzero assumes that x0 is an interval where the sign of function(x0(1)) differs from the sign of function(x0(2)). An error occurs if this is not true. Calling fzero with such an interval guarantees that fzero will return a value near a point where function changes sign. Plotting the function first is a good way to get a value for the vector x0. If the function is not continuous, fzero might return values that are discontinuous points instead of zeros. For example, x = fzero(@tan,1) returns x = 1.5708, a discontinuous point in tan(*x*).

Functions can have more than one zero, so it helps to plot the function first and then use fzero to obtain an answer that is more accurate than the answer read off the plot. Figure 3.2–1 shows the plot of the function, *y* = *x* + 2*e*⁻ˣ − 3, which has two zeros, one near *x* = −0.5 and one near *x* = 3. Using the function file f1 created earlier, to find the zero near *x* = −0.5, type x = fzero(@f1,-0.5).

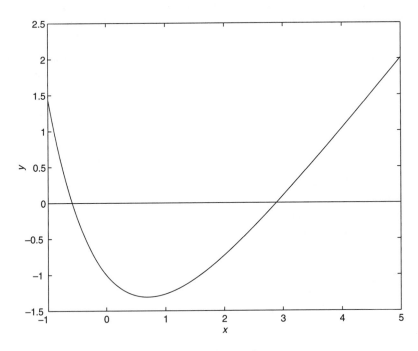

Figure 3.2–1 Plot of the function $y = x + 2e^{-x} - 3$.

The answer is $x = -0.5831$. To find the zero near $x = 3$, type `x = fzero (@f1,3)`. The answer is $x = 2.8887$.

The syntax `fzero (@f1, -0.5)` is preferred to the older syntax `fzero ('f1', -0.5)`.

Minimizing a Function of One Variable   The `fminbnd` function finds the minimum of a function of a single variable, which is denoted by x. Its basic syntax is

```
fminbnd(@function, x1, x2)
```

where `@function` is a function handle. The `fminbnd` function returns a value of x that minimizes the function in the interval `x1` $\leq$ `x` $\leq$ `x2`. For example, `fminbnd(@cos,0,4)` returns the value $x = 3.1416$.

However, to use this function to find the minimum of more-complicated functions, it is more convenient to define the function in a function file. For example, if $y = 1 - xe^{-x}$, define the following function file:

```
function y = f2(x)
y = 1-x.*exp(-x);
```

To find the value of x that gives a minimum of y for $0 \leq x \leq 5$, type `x = fminbnd(@f2,0,5)`. The answer is $x = 1$. To find the minimum value of y, type `y = f2(x)`. The result is `y = 0.6321`.

Whenever we use a minimization technique, we should check to make sure that the solution is a true minimum. For example, consider the following polynomial:
$$y = 0.025x^5 - 0.0625x^4 - 0.333x^3 + x^2$$

Its plot is shown in Figure 3.2–2. The function has two minimum points in the interval $-1 < x < 4$. The minimum near $x = 3$ is called a *relative* or *local* minimum because it forms a valley whose lowest point is higher than the minimum at $x = 0$. The minimum at $x = 0$ is the true minimum and is also called the *global* minimum. First create the function file

```
function y = f3(x)
y = polyval ([0.025, -0.0625, -0.333, 1, 0, 0], x);
```

To specify the interval $-1 \leq x \leq 4$, type `x = fminbnd (@f3, -1, 4)`. MATLAB gives the answer `x = 2.0438e-006`, which is essentially 0, the true minimum point. If we specify the interval $0.1 \leq x \leq 2.5$, MATLAB gives the answer `x = 0.1001`, which corresponds to the minimum value of y on the interval $0.1 \leq x \leq 2.5$. Thus we will miss the true minimum point if our specified interval does not include it.

Also, `fminbnd` can give misleading answers. If we specify the interval $1 \leq x \leq 4$, MATLAB (R 2007 a) gives the answer `x = 2.8236`, which corresponds to the "valley" shown in the plot, but which is not the minimum point on the interval $1 \leq x \leq 4$. On this interval the minimum point is at the boundary $x = 1$. The `fminbnd` procedure looks for a minimum point corresponding to a zero slope. In practice, the best use of the `fminbnd` function is to determine precisely the

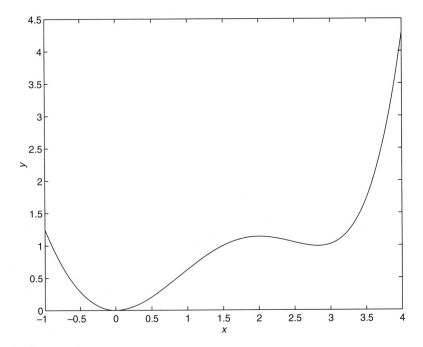

Figure 3.2–2 Plot of the function $y = 0.025x^5 - 0.0625x^4 - 0.333x^3 + x^2$.

location of a minimum point whose approximate location was found by other means, such as by plotting the function.

Minimizing a Function of Several Variables   To find the minimum of a function of more than one variable, use the `fminsearch` function. Its basic syntax is

```
fminsearch(@function, x0)
```

where `@function` is a function handle. The vector `x0` is a guess that must be supplied by the user. For example, to use the function $f = xe^{-x^2-y^2}$, first define it in an M-file, using the vector x whose elements are `x(1)` `=` x and `x(2)` `=` y.

```
function f = f4(x)
f = x(1).*exp(-x(1).^2-x(2).^2);
```

Suppose we guess that the minimum is near $x = y = 0$. The session is

```
>>fminsearch(@f4, [0, 0])
ans =
   -0.7071   0.000
```

Thus the minimum occurs at $x = -0.7071$, $y = 0$.

The `fminsearch` function can often handle discontinuities, particularly if they do not occur near the solution. The `fminsearch` function might give local solutions only, and it minimizes over the real numbers only; that is, x

Table 3.2–1 Minimization and root-finding functions

Function	Description
fminbnd(@function,x1,x2)	Returns a value of x in the interval x1 $\leq$ x $\leq$ x2 that corresponds to a minimum of the single-variable function described by the handle @function.
fminsearch(@function,x0)	Uses the starting vector x0 to find a minimum of the multivariable function described by the handle @function.
fzero(@function,x0)	Uses the starting value x0 to find a zero of the single-variable function described by the handle @function.

must consist of real variables only and the function must return real numbers only. When x has complex variables, they must be split into real and imaginary parts.

Table 3.2–1 summarizes the basic syntax of the fminbnd, fminsearch, and fzero commands.

These functions have extended syntax not described here. With these forms you can specify the accuracy required for the solution, as well as the number of steps to use before stopping. Use the help facility to find out more about these functions.

Optimization of an Irrigation Channel

EXAMPLE 3.2–1

Figure 3.2–3 shows the cross section of an irrigation channel. A preliminary analysis has shown that the cross-sectional area of the channel should be 100 ft^2 to carry the desired water-flow rate. To minimize the cost of concrete used to line the channel, we want to minimize the length of the channel's perimeter. Find the values of d, b, and θ that minimize this length.

■ Solution

The perimeter length L can be written in terms of the base b, depth d, and angle θ as follows:

$$L = b + \frac{2d}{\sin \theta}$$

The area of the trapezoidal cross section is

$$100 = db + \frac{d^2}{\tan \theta}$$

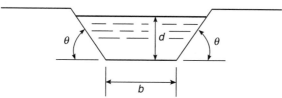

Figure 3.2–3 Cross section of an irrigation channel.

The variables to be selected are b, d, and θ. We can reduce the number of variables by solving the latter equation for b to obtain

$$b = \frac{1}{d}\left(100 - \frac{d^2}{\tan\theta}\right)$$

Substitute this expression into the equation for L. The result is

$$L = \frac{100}{d} - \frac{d}{\tan\theta} + \frac{2d}{\sin\theta}$$

We must now find the values of d and θ to minimize L.

First define the function file for the perimeter length. Let the vector **x** be $[d\ \theta]$.

```
function L = channel(x)
L = 100./x(1) - x(1)./tan(x(2)) + 2*x(1)./sin(x(2));
```

Then use the `fminsearch` function. Using a guess of $d = 20$ and $\theta = 1$ rad, the session is

```
>>x = fminsearch(@channel,[20,1])
x =
    7.5984    1.0472
```

Thus the minimum perimeter length is obtained with $d = 7.5984$ ft and $\theta = 1.0472$ rad, or $\theta = 60°$. Using a different guess, $d = 1$, $\theta = 0.1$, produces the same answer. The value of the base b corresponding to these values is $b = 8.7738$.

However, using the guess $d = 20$, $\theta = 0.1$ produces the physically meaningless result $d = -781$, $\theta = 3.1416$. The guess $d = 1$, $\theta = 1.5$ produces the physically meaningless result $d = 3.6058$, $\theta = -3.1416$.

The equation for L is a function of the two variables d and θ, and it forms a surface when L is plotted versus d and θ on a three-dimensional coordinate system. This surface might have multiple peaks, multiple valleys, and "mountain passes" called saddle points that can fool a minimization technique. Different initial guesses for the solution vector can cause the minimization technique to find different valleys and thus report different results. We can use the surface-plotting functions covered in Chapter 5 to look for multiple valleys, or we can use a large number of initial values for d and θ, say, over the physically realistic ranges $0 < d < 30$ and $0 < \theta < \pi/2$. If all the physically meaningful answers are identical, then we can be reasonably sure that we have found the minimum.

Test Your Understanding

T3.2–1 The equation $e^{-0.2x}\sin(x + 2) = 0.1$ has three solutions in the interval $0 < x < 10$. Find these three solutions.

T3.2–2 The function $y = 1 + e^{-0.2x}\sin(x + 2)$ has two minimum points in the interval $0 < x < 10$. Find the values of x and y at each minimum.

T3.2–3 Find the depth d and angle θ to minimize the perimeter length of the channel shown in Figure 3.2–3 to provide an area of 200 ft^2. (Answer: $d = 10.7457$ ft, $\theta = 60°$.)

3.3 Additional Function Topics

In addition to function handles, *anonymous functions, subfunctions,* and *nested functions* are some of the newer features of MATLAB. This section covers the basic features of these new types of functions.

Methods for Calling Functions

There are four ways to invoke, or "call," a function into action. These are:

1. As a character string identifying the appropriate function M-file,
2. As a function handle,
3. As an "inline" function object, or
4. As a string expression.

Examples of these ways follow for the `fzero` function used with the user-defined function `fun1`, which computes $y = x^2 - 4$.

1. As a character string identifying the appropriate function M-file, which is

   ```
   function y = fun1(x)
   y = x.^2-4;
   ```

   The function may be called as follows, to compute the zero over the range $0 \le x \le 3$:

   ```
   >>x = fzero('fun1',[0, 3])
   ```

2. As a function handle to an existing function M-file:

   ```
   >>x = fzero(@fun1,[0, 3])
   ```

3. As an "inline" function object:

   ```
   >>fun1 = 'x.^2-4';
   >>fun_inline = inline(fun1);
   >>x = fzero(fun_inline,[0, 3])
   ```

4. As a string expression:

   ```
   >>fun1 = 'x.^2-4';
   >>x = fzero(fun1,[0, 3])
   ```

   or as

   ```
   >>x = fzero('x.^2-4',[0, 3])
   ```

Method 2 was not available prior to MATLAB 6.0, and it is now preferred over method 1. The third method is not discussed in this text because it is a slower method than the first two. The third and fourth methods are equivalent because they both utilize the `inline` function; the only difference is that with the fourth method MATLAB determines that the first argument of `fzero` is a string variable and calls `inline` to convert the string variable to an inline function object. The function handle method (method 2) is the fastest method, followed by method 1.

In addition to speed improvement, another advantage of using a function handle is that it provides access to subfunctions, which are normally not visible outside of their defining M-file. This is discussed later in this section.

Types of Functions

At this point it is helpful to review the types of functions provided for in MATLAB. MATLAB provides built-in functions, such as `clear`, `sin`, and `plot`, which are not M-files, and also some functions that are M-files, such as the function `mean`. In addition, the following types of *user-defined* functions can be created in MATLAB.

PRIMARY FUNCTION

■ The *primary function* is the first function in an M-file and typically contains the main program. Following the primary function in the same file can be any number of subfunctions, which can serve as subroutines to the primary function. Usually the primary function is the only function in an M-file that you can call from the MATLAB command line or from another M-file function. You invoke this function using the name of the M-file in which it is defined. We normally use the same name for the function and its file, but if the function name differs from the file name, you must use the file name to invoke the function.

ANONYMOUS FUNCTIONS

■ *Anonymous functions* enable you to create a simple function without needing to create an M-file for it. You can construct an anonymous function either at the MATLAB command line or from within another function or script. Thus, anonymous functions provide a quick way of making a function from any MATLAB expression without the need to create, name, and save a file.

SUBFUNCTIONS

■ *Subfunctions* are placed in the primary function and are called by the primary function. You can use multiple functions within a single primary function M-file.

NESTED FUNCTIONS

■ *Nested functions* are functions defined within another function. They can help to improve the readability of your program and also give you more flexible access to variables in the M-file. The difference between nested functions and subfunctions is that subfunctions normally cannot be accessed outside of their primary function file.

■ *Overloaded* functions are functions that respond differently to different types of input arguments. They are similar to overloaded functions in any object-oriented language. For example, an overloaded function can be created to treat integer inputs differently than inputs of class double.

PRIVATE FUNCTION

■ *Private functions* enable you to restrict access to a function. They can be called only from an M-file function in the parent directory.

Anonymous Functions

Anonymous functions enable you to create a simple function without needing to create an M-file for it. You can construct an anonymous function either at the MATLAB command line or from within another function or script. The syntax for creating an anonymous function from an expression is

```
fhandle = @ (arglist) expr
```

where `arglist` is a comma-separated list of input arguments to be passed to the function, and `expr` is any single, valid MATLAB expression. This syntax creates the function handle `fhandle`, which enables you to invoke the function. Note that this syntax is different from that used to create other function handles, `fhandle = @functionname`. The handle is also useful for passing the anonymous function in a call to some other function in the same way as any other function handle.

For example, to create a simple function called `sq` to calculate the square of a number, type

```
sq = @(x) x.^2;
```

To improve readability, you may enclose the expression in parentheses, as `sq = @(x) (x.^2);`. To execute the function, type the name of the function handle, followed by any input arguments enclosed in parentheses. For example,

```
>>sq(5)
ans =
     25
>>sq([5,7])
ans =
     25     49
```

You might think that this particular anonymous function will not save you any work because typing `sq([5,7])` requires nine keystrokes, one more than is required to type `[5,7].^2`. Here, however, the anonymous function protects you from forgetting to type the period `(.)` required for array exponentiation. Anonymous functions are useful, however, for more complicated functions involving numerous keystrokes.

You can pass the handle of an anonymous function to other functions. For example, to find the minimum of the polynomial $4x^2 - 50x + 5$ over the interval $[-10, 10]$, you type

```
>>poly1 = @(x) 4*x.^2 - 50*x + 5;
>>fminbnd(poly1, -10, 10)
ans =
     6.2500
```

If you are not going to use that polynomial again, you can omit the handle definition line and type instead

```
>>fminbnd(@(x) 4*x.^2 - 50*x + 5, -10, 10)
```

Multiple-Input Arguments   You can create anonymous functions having more than one input. For example, to define the function $\sqrt{x^2 + y^2}$, type

```
>>sqrtsum = @(x,y) sqrt(x.^2 + y.^2);
```

Then

```
>>sqrtsum(3, 4)
ans =
     5
```

As another example, consider the function defining a plane, $z = Ax + By$. The scalar variables A and B must be assigned values before you create the function handle. For example,

```
>>A = 6; B = 4:
>>plane = @(x,y) A*x + B*y;
>>z = plane(2,8)
z =
    44
```

No-Input Arguments   To construct a handle for an anonymous function that has no input arguments, use empty parentheses for the input argument list, as shown by the following: d = @() date;.
  Use empty parentheses when invoking the function, as follows:

```
>>d()
ans =
     01-Mar-2007
```

You must include the parentheses. If you do not, MATLAB just identifies the handle; it does not execute the function.

Calling One Function within Another   One anonymous function can call another to implement function composition. Consider the function $5 \sin(x^3)$. It is composed of the functions $g(y) = 5 \sin(y)$ and $f(x) = x^3$. In the following session the function whose handle is h calls the functions whose handles are f and g to compute $5 \sin(2^3)$.

```
>>f = @(x) x.^3;
>>g = @(x) 5*sin(x);
>>h = @(x) g(f(x));
>>h(2)
ans =
     4.9468
```

To preserve an anonymous function from one MATLAB session to the next, save the function handle to a MAT-file. For example, to save the function associated with the handle h, type `save anon.mat h`. To recover it in a later session, type `load anon.mat h`.

Variables and Anonymous Functions Variables can appear in anonymous functions in two ways:

■ As variables specified in the argument list, as for example `f = @(x) x.^3;`, and

■ As variables specified in the body of the expression, as for example with the variables A and B in `plane = @(x,y) A*x + B*y`. In this case, when the function is created MATLAB captures the values of these variables and retains those values for the lifetime of the function handle. In this example, if the values of A or B are changed after the handle is created, their values associated with the handle do not change. This feature has both advantages and disadvantages, so you must keep it in mind.

Subfunctions

A function M-file may contain more than one user-defined function. The first defined function in the file is called the *primary function,* whose name is the same as the M-file name. All other functions in the file are called *subfunctions.* Subfunctions are normally "visible" only to the primary function and other subfunctions in the same file; that is, they normally cannot be called by programs or functions outside the file. However, this limitation can be removed with the use of function handles, as we will see later in this section.

Create the primary function first with a function definition line and its defining code, and name the file with this function name as usual. Then create each subfunction with its own function definition line and defining code. The order of the subfunctions does not matter, but function names must be unique within the M-file.

The order in which MATLAB checks for functions is very important. When a function is called from within an M-file, MATLAB first checks to see if the function is a built-in function such as `sin`. If not, it checks to see if it is a *subfunction* in the file, then checks to see if it is a *private* function (which is a function M-file residing in the `private` subdirectory of the calling function). Then MATLAB checks for a standard M-file on your search path. Thus, because MATLAB checks for a subfunction before checking for private and standard M-file functions, you may use subfunctions with the same name as another existing M-file. This feature allows you to name subfunctions without being concerned about whether another function exists with the *same name,* so you need not choose long function names to avoid conflict. This feature also protects you from using another function unintentionally.

Note that you may even supercede a MATLAB M-function in this way. The following example shows how the MATLAB M-function `mean` can be superceded by our own definition of the mean, one which gives the root-mean-square value. The function `mean` is a subfunction. The function `subfun_demo` is the primary function.

```
function y = subfun_demo(a)
y = a - mean(a);
%
function w = mean(x)
w = sqrt(sum(x.^2))/length(x);
```

A sample session follows.

```
>>y = subfn_demo([4, -4])
y =
    1.1716    -6.8284
```

If we had used the MATLAB M-function mean, we would have obtained a different answer; that is,

```
>>a=[4,-4];
>>b = a - mean(a)
b =
    4     -4
```

Thus the use of subfunctions enables you to reduce the number of files that define your functions. For example, if it were not for the subfunction mean in the previous example, we would have had to define a separate M-file for our mean function and give it a different name so as not to confuse it with the MATLAB function of the same name.

Subfunctions are normally visible only to the primary function and other sub-functions in the same file. However, we can use a function handle to allow access to the subfunction from outside the M-file, as the following example shows. Create the following M-file with the primary function fn_demo1 (range) and the subfunction testfun(x) to compute the zeros of the function $(x^2 - 4) \cos x$ over the range specified in the input variable range. Note the use of a function handle in the second line.

```
function yzero = fn_demo1(range)
fun = @testfun;
[yzero,value] = fzero(fun,range);
%
function y = testfun(x)
y = (x.^2-4).*cos(x);
```

A test session gives the following results.

```
>>yzero = fn_demo1([3, 6])
yzero =
    4.7124
```

So the zero of $(x^2 - 4) \cos x$ over $3 \le x \le 6$ occurs at $x = 4.7124$.

Nested Functions

With MATLAB 7 you can now place the definitions of one or more functions within another function. Functions so defined are said to be *nested* within the main function. You can also nest functions within other nested functions. Like any M-file function, a nested function contains the usual components of an M-file function. You must, however, always terminate a nested function with an end statement. In fact, if an M-file contains at least one nested function, you must terminate *all* functions, including subfunctions, in the file with an end statement, whether or not they contain nested functions.

The following example constructs a function handle for a nested function p(x) and then passes the handle to the MATLAB function fminbnd to find the minimum point on the parabola. The parabola function constructs and returns a function handle f for the nested function p that evaluates the parabola $ax^2 + bx + c$. This handle gets passed to fminbnd.

```
function f = parabola(a, b, c)
f = @p;
    % Nested function
    function y = p(x)
      y = polyval ([a,b,c],x);
    end
end
```

In the Command window type

```
>>f = parabola(4, -50, 5);
>>fminbnd(f, -10, 10)
ans =
    6.2500
```

Note than the function p(x) can see the variables a, b, and c in the calling function's workspace.

Contrast this approach to that required using global variables. First create the function p(x).

```
function y = p(x)
global a b c
y = polyval ([a, b, c], x);
```

Then, in the command window, type

```
>>global a b c
>>a = 4; b = -50; c = 5;
>> fminbnd (@p, -10, 10)
```

Nested functions might seem to be the same as subfunctions, but they are not. Nested functions have two unique properties:

1. A nested function can access the workspaces of all functions inside of which it is nested. So for example, a variable that has a value assigned to it

by the primary function can be read or overwritten by a function nested at any level within the main function. In addition, a variable assigned in a nested function can be read or overwritten by any of the functions containing that function.

2. If you construct a function handle for a nested function, the handle not only stores the information needed to access the nested function; it also stores the values of all variables shared between the nested function and those functions that contain it. This means that these variables persist in memory between calls made by means of the function handle.

Consider the following representation of some functions named A, B, . . ., E.

```
function A(x, y)      % The primary function
B(x, y);
D(y);

    function B(x, y)      % Nested in A
    C(x);
    D(y);

        function C(x)      % Nested in B
        D(x);
        end    % This terminates C
    end      % This terminates B

    function D(x)      % Nested in A
    E(x);

        function E      % Nested in D
        . . .
        end    % This terminates E
    end        % This terminates D
end      % This terminates A
```

You call a nested function in several ways.

1. You can call it from the level immediately above it. (In the previous code, function A can call B or D, but not C or E.)
2. You can call it from a function nested at the same level within the same parent function. (Function B can call D, and D can call B.)
3. You can call it from a function at any lower level. (Function C can call B or D, but not E.)
4. If you construct a function handle for a nested function, you can call the nested function from any MATLAB function that has access to the handle.

You can call a subfunction from any nested function in the same M-file.

Private Functions

Private functions reside in subdirectories with the special name `private`, and they are visible only to functions in the parent directory. Assume the directory `rsmith` is on the MATLAB search path. A subdirectory of `rsmith` called `private` may contain functions that only the functions in `rsmith` can call. Because private functions are invisible outside the parent directory `rsmith`, they can use the same names as functions in other directories. This is useful if the main directory used by several individuals including R. Smith, but R. Smith wants to create a personal version of a particular function while retaining the original in the main directory. Because MATLAB looks for private functions before standard M-file functions, it will find a private function named, say `cylinder.m`, before a nonprivate M-file named `cylinder.m`.

Primary functions and subfunctions can be implemented as private functions. Create a private directory by creating a subdirectory called `private` using the standard procedure for creating a directory or a folder on your computer, but do not place the private directory on your path.

3.4 Working with Data Files

An ASCII data file may have one or more lines of text, called the header, at the beginning. These might be comments that describe what the data represents, the date it was created, and who created the data, for example. One or more lines of data, arranged in rows and columns, follow the header. The numbers in each row might be separated by spaces or by commas.

If it is inconvenient to edit the data file, the MATLAB environment provides many ways to bring data created by other applications into the MATLAB workspace, a process called *importing data,* and to package workspace variables so that they can be exported to other applications.

If the file has a header or the data is separated by commas, MATLAB will produce an error message. To correct this situation, first load the data file into a text editor, remove the header, and replace the commas with spaces. To retrieve this data into MATLAB, type `load filename`. If the file has m lines with n values in each line, the data will be assigned to an $m \times n$ matrix having the same name as the file with the extension stripped off. Your data file can have any extension except `.mat`, so that MATLAB will not try to load the file as a workspace file.

Importing Spreadsheet Files

Some spreadsheet programs store data in the `.wk1` format. You can use the command `M = wk1read('filename')` to import this data into MATLAB and store it in the matrix M. The command `A = xlsread('filename')` imports the Microsoft Excel workbook file `filename.xls` into the array A. The command `[A, B] = xlsread('filename')` imports all numeric data into the array A and all text data into the cell array B.

The Import Wizard

You can use the Import Wizard to import many types of ASCII data formats, including data on the clipboard. The Import Wizard presents a series of dialog boxes in which you specify the name of the file, the delimiter used in the file, and the variables that you want to import.

Do the following to import this sample tab-delimited, ASCII data file, `testdata.txt`:

```
1     2     3     4     5;
17    12    8     15    25;
```

1. Activate the Import Wizard either by typing `uiimport` or by selecting the **Import Data** option on the MATLAB Desktop **File** menu. The Import Wizard displays a dialog box that asks you to specify the name of the file you want to import.

2. The Import Wizard processes the contents of the file and displays tabs identifying the variables it recognizes in the file, and displays a portion of the data in a grid, similar to a spreadsheet. The Import Wizard uses the space character as the default delimiter. After you click **Next,** the Import Wizard attempts to identify the delimiter (see Figure 3.4–1).

3. In the next dialog box, the Import Wizard displays a list of the variables it found in the file. It also displays the contents of the first variable in the list. In this example there is only one variable, named `testdata`.

4. Choose the variables you want to import by clicking the check box next to their names. By default, all variables are checked for import. After selecting the variables you want to import, click the **Finish** button to import the data into the MATLAB workspace.

To import data from the clipboard, select **Paste Special** from the **Edit** menu. Then proceed with step 2. The default variable name is `A_pastespecial`.

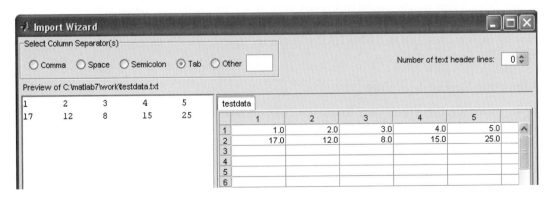

Figure 3.4–1 The first screen in the Import Wizard.

Exporting ASCII Data Files

You might want to export a MATLAB matrix as an ASCII data file where the rows and columns are represented as space-delimited, numeric values. To export a MATLAB matrix as a delimited ASCII data file, you can use either the `save` command, specifying the `-ASCII` qualifier, or the `dlmwrite` function. The `save` command is easy to use; however, the `dlmwrite` function provides more flexibility, allowing you to specify any character as a delimiter and to export subsets of an array by specifying a range of values.

Suppose you have created the array `A = [1 2 3 4; 5 6 7 8]` in MATLAB. To export the array using the `save` command, type the following in the Command window.

```
>>save my_data.out A -ASCII
```

By default, `save` uses spaces as delimiters, but you can use tabs instead of spaces by specifying the `-tab` qualifier.

3.5 Summary

In Section 3.1 we introduced just some of the most commonly used mathematical functions. You should now be able to use the MATLAB help to find other functions you need. If necessary, you can create your own functions, using the methods of Section 3.2. This section also covered function handles and their use with function functions.

Anonymous functions, subfunctions, and nested functions extend the capabilities of MATLAB. These topics were treated in Section 3.3. In addition to function files, data files are useful for many applications. Section 3.4 shows how to import and export such files in MATLAB.

Key Terms with Page References

Anonymous functions,   138
Function argument,   124
Function definition line,   126
Function file,   126
Function handle,   131
Global variables,   131

Local variable,   126
Nested functions,   138
Primary function,   138
Private function,   138
Subfunctions,   138

Problems

You can find the answers to problems marked with an asterisk at the end of the text.

Section 3.1

1.* Suppose that $y = -3 + ix$. For $x = 0$, 1, and 2, use MATLAB to compute the following expressions. Hand check the answers.

    a. $|y|$ *b.* $\sqrt{y}$

    c. $(-5-7i)y$ *d.* $\dfrac{y}{6-3i}$

2.* Let $x = -5 - 8i$ and $y = 10 - 5i$. Use MATLAB to compute the following expressions. Hand check the answers.

   a. The magnitude and angle of xy.
   b. The magnitude and angle of $\frac{x}{y}$.

3.* Use MATLAB to find the angles corresponding to the following coordinates. Hand check the answers.

   a. $(x, y) = (5, 8)$ b. $(x, y) = (-5, 8)$
   c. $(x, y) = (5, -8)$ d. $(x, y) = (-5, -8)$

4. For several values of x, use MATLAB to confirm that $\sinh x = (e^x - e^{-x})/2$.

5. For several values of x, use MATLAB to confirm that $\cosh^{-1} x = \ln (x + \sqrt{x^2 \geq 1})$, $-\infty < x < \infty$.

6. The capacitance of two parallel conductors of length L and radius r, separated by a distance d in air, is given by

$$C = \frac{\pi \epsilon L}{\ln \left(\dfrac{d - r}{r} \right)}$$

   where ϵ is the permittivity of air ($\epsilon = 8.854 \times 10^{-12}$ F/m).

   Write a script file that accepts user input for d, L, and r, and computes and displays C. Test the file with the values: $L = 1$ m, $r = 0.001$ m, and $d = 0.004$ m.

7.* When a belt is wrapped around a cylinder, the relation between the belt forces on each side of the cylinder is

$$F_1 = F_2 e^{\mu \beta}$$

   where β is the angle of wrap of the belt and μ is the friction coefficient. Write a script file that first prompts a user to specify β, μ, and F_2 and then computes the force F_1. Test your program with the values $\beta = 130°$, $\mu = 0.3$, and $F_2 = 100$ N. (Hint: Be careful with β!)

Section 3.2

8. The output of the MATLAB `atan2` function is in radians. Write a function called `atan2d` that produces an output in degrees.

9. Write a function that accepts temperature in degrees F and computes the corresponding value in degrees C. The relation between the two is

$$T \, °C = \frac{5}{9} (T \, °F - 32)$$

   Be sure to test your function.

10.* An object thrown vertically with a speed v_0 reaches a height h at time t, where

$$h = v_0 t - \frac{1}{2}gt^2$$

Write and test a function that computes the time t required to reach a specified height h, for a given value of v_0. The function's inputs should be h, v_0, and g. Test your function for the case where $h = 100$ m, $v_0 = 50$ m/s, and $g = 9.81$ m/s². Interpret both answers.

11. A water tank consists of a cylindrical part of radius r and height h, and a hemispherical top. The tank is to be constructed to hold 500 m³ when filled. The surface area of the cylindrical part is $2\pi rh$, and its volume is $\pi r^2 h$. The surface area of the hemispherical top is given by $2\pi r^2$, and its volume is given by $2\pi r^3/3$. The cost to construct the cylindrical part of the tank is \$300 per square meter of surface area; the hemispherical part costs \$400 per square meter. Use the `fminbnd` function to compute the radius that results in the least cost. Compute the corresponding height h.

12. A fence around a field is shaped as shown in Figure P12. It consists of a rectangle of length L and width W, and a right triangle that is symmetrical about the central horizontal axis of the rectangle. Suppose the width W is known (in meters), and the enclosed area A is known (in square meters). Write a user-defined function file with W and A as inputs. The outputs are the length L required so that the enclosed area is A, and the total length of fence required. Test your function for the values $W = 6$ m and $A = 80$ m².

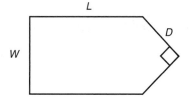

Figure P12

13. A fenced enclosure consists of a rectangle of length L and width $2R$, and a semicircle of radius R, as shown in Figure P13. The enclosure is to be built to have an area A of 1600 ft². The cost of the fence is \$40 per foot for the curved portion, and \$30 per foot for the straight sides. Use the `fminbnd` function to determine with a resolution of 0.01 ft the values of R and L required to minimize the total cost of the fence. Also compute the minimum cost.

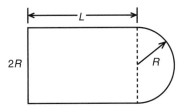

Figure P13

14. Using estimates of rainfall, evaporation, and water consumption, the town engineer developed the following model of the water volume in the reservoir as a function of time.

$$V(t) = 10^9 + 10^8(1 - e^{-t/100}) - rt$$

where V is the water volume in liters, t is time in days, and r is the town's consumption rate in liters/day. Write two user-defined functions. The first function should define the function $V(t)$ for use with the `fzero` function. The second function should use `fzero` to compute how long it will take for the water volume to decrease to x percent of its initial value of 10^9 L. The inputs to the second function should be x and r. Test your functions for the case where $x = 50$ percent and $r = 10^7$ L/day.

15. The volume V and paper surface area A of a conical paper cup are given by

$$V = \frac{1}{3}\pi r^2 h \quad A = \pi r \sqrt{r^2 + h^2}$$

where r is the radius of the base of the cone and h is the height of the cone.

a. By eliminating h, obtain the expression for A as a function of r and V.

b. Create a user-defined function that accepts R as the only argument and computes A for a given value of V. Declare V to be global within the function.

c. For $V = 10$ in.^{3}, use the function with the `fminbnd` function to compute the value of r that minimizes the area A. What is the corresponding value of the height h? Investigate the sensitivity of the solution by plotting V versus r. How much can R vary about its optimal value before the area increases 10 percent above its minimum value?

16. A torus is shaped like a doughnut. If its inner radius is a and its outer radius is b, its volume and surface area are given by

$$V = \frac{1}{4}\pi^2(a + b)(b - a)^2 \quad A = \pi^2(b^2 - a^2)$$

a. Create a user-defined function that computes V and A from the arguments a and b.

b. Suppose that the outer radius is constrained to be 2 in. greater than the inner radius. Write a script file that uses your function to plot A and V versus a for $0.25 \leq a \leq 4$ in.

17. Suppose it is known that the graph of the function $y = ax^3 + bx^2 + cx + d$ passes through four given points (x_i, y_i), $i = 1, 2, 3, 4$. Write a user-defined function that accepts these four points as input and computes the coefficients a, b, c, and d. The function should solve four linear equations in terms of the four unknowns a, b, c, and d. Test your function for the case where $(x_i, y_i) = (-2, -20)$, $(0, 4)$, $(2, 68)$, and $(4, 508)$, whose answer is $a = 7$, $b = 5$, $c = -6$, and $d = 4$.

Section 3.3

18. Create an anonymous function for $10e^{-2x}$ and use it to plot the function over the range $0 \leq x \leq 2$.

19. Create an anonymous function for $20x^2 - 200x + 3$ and use it
    a. to plot the function to determine the approximate location of its minimum, and
    b. with the `fminbnd` function to precisely determine the location of the minimum.

20. Create four anonymous functions to represent the function $6e^{3\cos x^2}$, which is composed of the functions $h(z) = 6e^z$, $g(y) = 3\cos y$, and $f(x) = x^2$. Use the anonymous functions to plot $6e^{3\cos x^2}$ over the range $0 \leq x \leq 4$.

21. Use a primary function with a subfunction to compute the zeros of the function $3x^3 - 12x^2 - 33x + 90$ over the range $-10 \leq x \leq 10$.

22. Create a primary function that uses a function handle with a nested function to compute the minimum of the function $20x^2 - 200x + 3$ over the range $0 \leq x \leq 10$.

Section 3.4

23. Use a text editor to create a file containing the following data. Then use the `load` function to load the file into MATLAB, and use the `mean` function to compute the mean value of each column.

55	42	98
51	39	95
63	43	94
58	45	90

24. Enter and save the data given in Problem 23 in a spreadsheet. Then import the spreadsheet file into the MATLAB variable A. Use MATLAB to compute the sum of each column.

25. Use a text editor to create a file from the data given in Problem 23, but separate each number with a semicolon. Then use the Import Wizard to load and save the data in the MATLAB variable A.

CHAPTER 4

Decision-Making Programs

OUTLINE

The MATLAB interactive mode is very useful for simple problems, but more complex problems require a script file. Such a file can be called a *computer program*, and writing such a file is called *programming*. The usefulness of MATLAB is greatly increased by the use of decision-making functions in its programs. These functions enable you to write programs whose operations depend on the results of calculations made by the program. Sections 4.1 through 4.3 deal with these functions.

MATLAB can also repeat calculations a specified number of times or until some condition is satisfied. This feature enables engineers to solve problems of great complexity or requiring numerous calculations. These "loop" structures are covered in Section 4.4.

The `switch` structure enhances the MATLAB decision-making capabilities. This topic is covered in Section 4.5. Use of the MATLAB Editor/Debugger for debugging programs is covered in Section 4.6.

4.1 Relational Operators and Logical Variables

RELATIONAL OPERATOR

MATLAB has six *relational operators* to make comparisons between arrays. These operators are shown in Table 4.1–1. Note that the *equal to* operator consists of

Table 4.1–1 Relational operators

Relational operator	Meaning
<	Less than.
<=	Less than or equal to.
>	Greater than.
>=	Greater than or equal to.
==	Equal to.
~=	Not equal to.

two = signs, not a single = sign as you might expect. The single = sign is the *assignment,* or *replacement,* operator in MATLAB.

The result of a comparison using the relational operators is either 0 (if the comparison is *false*), or 1 (if the comparison is *true*), and the result can be used as a variable. For example, if x = 2 and y = 5, typing z = x < y returns the value z = 1 and typing u = x==y returns the value u = 0. To make the statements more readable, we can group the logical operations using parentheses. For example, z = (x < y) and u = (x==y).

When used to compare arrays, the relational operators compare the arrays on an element-by-element basis. The arrays being compared must have the same dimension. The only exception occurs when we compare an array to a scalar. In that case all the elements of the array are compared to the scalar. For example, suppose that x = [6,3,9] and y = [14,2,9]. The following MATLAB session shows some examples.

```
>>z = (x < y)
z =
   1   0   0
>>z = (x ~= y)
z =
   1   1   0
>>z = (x > 8)
z =
   0   0   1
```

The relational operators can be used for array addressing. For example, with x = [6,3,9] and y = [14,2,9], typing z = x(x<y) finds all the elements in x that are less than the corresponding elements in y. The result is z = 6.

The arithmetic operators +, −, *, /, and \ have precedence over the relational operators. Thus the statement z = 5 > 2 + 7 is equivalent to z = 5 > (2+7) and returns the result z = 0. We can use parentheses to change the order of precedence; for example, z = (5 > 2) + 7 evaluates to z = 8.

The relational operators have equal precedence among themselves, and MATLAB evaluates them in order from left to right. Thus the statement

```
z = 5 > 3 ~= 1
```

is equivalent to

```
z = (5>3) ~= 1
```

Both statements return the result z = 0.

With relational operators that consist of more than one character, such as ==
or >=, be careful not to put a space between the characters.

The logical Class

When the relational operators are used, such as x = (5 > 2), they create a logical variable, in this case, x. Prior to MATLAB 6.5 logical was an attribute of
any numeric data type. Now logical is a first-class data type and a MATLAB
class, and so logical is now equivalent to other first-class types such as character and cell arrays. Logical variables may have only the values 1 (true) and 0
(false).

Just because an array contains only 0s and 1s, however, it is not necessarily
a logical array. For example, in the following session k and w appear the same,
but k is a logical array and w is a numeric array, and thus an error message is issued. The logical array k can be used to select those elements of x whose absolute
values are greater than 1. These elements are then stored in the array z.

```
>>x = [-2:2]
x =
    -2   -1   0   1   2
>>k = (abs(x)>1)
k =
    1   0   0   0   1
>>z = x(k)
z =
    -2   2
>>w = [1,0,0,0,1];
>>v = x(w)
??? Subscript indices must either be real positive ...
    integers or logicals.
```

The logical Function

Logical arrays can be created with the relational and logical operators and with
the logical function. The logical function returns an array that can be used
for logical indexing and logical tests. Typing B = logical(A), where A is a
numeric array, returns the logical array B. So to correct the error in the previous
session, you may type instead w = logical([1,0,0,0,1]) before typing
v = x(w).

When a finite, real value other than 1 or 0 is assigned to a logical variable,
the value is converted to logical 1 and a warning message is issued. For example,

when you type y = logical(9), y will be assigned the value logical 1 and a warning will be issued. You may use the double function to convert a logical array to an array of class double. For example, x = (5>3); y = double(x);. Some arithmetic operations convert a logical array to a double array. For example, if we add zero to each element of B by typing B = B + 0, the array B will be converted to a numeric (double) array. However, not all MATLAB functions can accept logical variables as input arguments. For example, typing

```
>>x = ([2, 3] > [1, 6]);
>>y = sin(x);
```

in Release 2007a will generate an error message, whereas y = sind(x) gives the correct answer. The reason is that in sind.m the operation x = x - 90 * round(x/90) converts x from class logical to class double. This, however, is not an important issue because it hardly makes sense to use the sin function with a logical argument. The help text for sin.m warns that the argument must be in radians; that is, a real number.

Accessing Arrays Using Logical Arrays

When a logical array is used to address another array, it extracts from that array the elements in the locations where the logical array has 1s. So typing A(B), where B is a logical array of the same size as A, returns the values of A at the indices where B is 1.

Given A = [5,6,7;8,9,10;11,12,13] and B = logical (eye(3)), we can extract the diagonal elements of A by typing C = A(B) to obtain C = [5;9;13]. Specifying array subscripts with logical arrays extracts the elements that correspond to the true (1) elements in the logical array.

Note, however, that using the *numeric* array eye(3), as C = A(eye(3)), results in an error message because the elements of eye(3) do not correspond to locations in A. If the numeric array values correspond to valid locations, you may use a numeric array to extract the elements. For example, to extract the diagonal elements of A with a numeric array, type C = A([1,5,9]).

MATLAB data types are preserved when indexed assignment is used. So now that logical is a MATLAB data type, if A is a logical array, for example A = logical(eye(4)), then typing A(3,4) = 1 does not change A to a double array. However, typing A(3,4) = 5 will set A(3,4) to logical 1 and cause a warning to be issued.

4.2 Logical Operators and Functions

LOGICAL
OPERATOR

MATLAB has five *logical operators,* which are sometimes called *Boolean* operators (see Table 4.2–1). These operators perform element-by-element operations. With the exception of the NOT operator (~), they have a lower precedence than the arithmetic and relational operators (see Table 4.2–2). The NOT symbol is called the *tilde.*

Table 4.2–1 Logical operators

Operator	Name	Definition
~	NOT	~A returns an array the same dimension as A; the new array has ones where A is zero and zeros where A is nonzero.
&	AND	A & B returns an array the same dimension as A and B; the new array has ones where both A and B have nonzero elements and zeros where either A or B is zero.
|	OR	A | B returns an array the same dimension as A and B; the new array has ones where at least one element in A or B is nonzero and zeros where A and B are both zero.
&&	Short-Circuit AND	Operator for scalar logical expressions. A && B returns true if both A and B evaluate to true, and false if they do not.
||	Short-Circuit OR	Operator for scalar logical expressions. A || B returns true if either A or B or both evaluate to true, and false if they do not.

Table 4.2–2 Order of precedence for operator types

Precedence	Operator type
First	Parentheses; evaluated starting with the innermost pair.
Second	Arithmetic operators and logical NOT (~); evaluated from left to right.
Third	Relational operators; evaluated from left to right.
Fourth	Logical AND.
Fifth	Logical OR.

The NOT operation ~A returns an array of the same dimension as A; the new array has ones where A is zero and zeros where A is nonzero. If A is logical, then ~A replaces ones with zeros and zeros with ones. For example, if x = [0,3,9] and y = [14,-2,9], then z = ~x returns the array z = [1,0,0] and the statement u = ~x > y returns the result u = [0,1,0]. This expression is equivalent to u = (~x) > y, whereas v = ~(x > y) gives the result v = [1,0,1]. This expression is equivalent to v = (x <= y).

The & and \| operators compare two arrays of the same dimension. The only exception, as with the relational operators, is that an array can be compared to a scalar. The AND operation A&B returns ones where both A and B have nonzero elements and zeros where any element of A or B is zero. The expression z = 0&3 returns z = 0; z = 2&3 returns z = 1; z = 0&0 returns z = 0, and z = [5,-3,0,0]&[2,4,0,5] returns z = [1,1,0,0]. Because of operator precedence, z = 1&2+3 is equivalent to z = 1&(2+3), which returns z = 1. Similarly, z = 5<6&1 is equivalent to z = (5<6)&1, which returns z = 1.

Let x = [6,3,9] and y = [14,2,9] and let a = [4,3,12]. The expression

z = (x>y) & a

gives $z = [0,1,0]$, and

$z = (x>y) \& (x>a)$

returns the result $z = [0,0,0]$. This is equivalent to

$z = x>y\&x>a$

which is much less readable.

    Be careful when using the logical operators with inequalities. For example, note that $\sim(x > y)$ is equivalent to $x <= y$. It is *not* equivalent to $x < y$. As another example, the relation $5 < x < 10$ must be written as

$(5 < x) \& (x < 10)$

in MATLAB.

    The OR operation $A|B$ returns ones where at least one of A and B has nonzero elements and zeros where both A and B are zero. The expression $z = 0|3$ returns $z = 1$; the expression $z = 0|0$ returns $z = 0$; and

$z = [5,-3,0,0]|[2,4,0,5]$

returns $z = [1,1,0,1]$. Because of operator precedence,

$z = 3<5|4==7$

is equivalent to

$z = (3<5)|(4==7)$

which returns $z = 1$. Similarly, $z = 1|0\&1$ is equivalent to $z = (1|0)\&1$, which returns $z = 1$, while $z = 1|0\&0$ is equivalent to $z = 1|(0\&0)$, which returns $z = 1$.

    Because of the precedence of the NOT operator, the statement

$z = \sim3==7|4==6$

returns the result $z = 0$, which is equivalent to

$z = ((\sim3)==7)|(4==6)$

    The exclusive OR function $xor(A,B)$ returns zeros where A and B are either both nonzero or both zero, and ones where either A or B is nonzero, *but not both.* The function is defined in terms of the AND, OR, and NOT operators as follows.

```
function z = xor(A,B)
z = (A|B) & ~(A&B);
```

The expression

$z = xor([3,0,6],[5,0,0])$

returns $z = [0,0,1]$, whereas

$z = [3,0,6]|[5,0,0]$

returns $z = [1,0,1]$.

Table 4.2–3 Truth table

x	y	~x	x|y	x&y	xor(x,y)
true	true	false	true	true	false
true	false	false	true	false	true
false	true	true	true	false	true
false	false	true	false	false	false

Table 4.2–3 is a so-called *truth table* that defines the operations of the logical operators and the function xor. Until you acquire more experience with the logical operators, you should use this table to check your statements. Remember that *true* is equivalent to logical 1, and *false* is equivalent to logical 0. We can test the truth table by building its numerical equivalent as follows. Let x and y represent the first two columns of the truth table in terms of ones and zeros.

TRUTH TABLE

The following MATLAB session generates the truth table in terms of ones and zeros.

```
>>x = [1,1,0,0]';
>>y = [1,0,1,0]';
>>Truth_Table = [x,y,~x,x|y,x&y,xor(x,y)]
Truth_Table =
   1  1  0  1  1  0
   1  0  0  1  0  1
   0  1  1  1  0  1
   0  0  1  0  0  0
```

Starting with MATLAB 6, the AND operator (&) was given a higher precedence than the OR operator (|). This was not true in earlier versions of MATLAB, so if you are using code created in an earlier version, you should make the necessary changes before using it in MATLAB 6 or higher. For example, now the statement y = 1|5&0 is evaluated as y = 1|(5&0), yielding the result y = 1, whereas in MATLAB 5.3 and earlier, the statement would have been evaluated as y = (1|5)&0, yielding the result y = 0. To avoid potential problems due to precedence, it is important to use parentheses in statements containing arithmetic, relational, or logical operators, even where parentheses are optional. MATLAB now provides a feature to enable the system to produce either an error message or a warning for any expression containing & and | that would be evaluated differently than in earlier versions. If you do not use this feature, MATLAB will issue a warning as the default. To activate the error feature, type feature('OrAndError',1). To reinstate the default, type feature('OrAndError',0).

Short-Circuit Operators

The following operators perform AND and OR operations on logical expressions containing *scalar* values only. They are called short-circuit operators because they evaluate their second operand only when the result is not fully determined by the first operand. They are defined as follows in terms of the two logical variables A and B.

A&&B Returns true (logical 1) if both A and B evaluate to true, and false (logical 0) if they do not.

A||B Returns true (logical 1) if either A or B, or both, evaluate to true, and false (logical 0) if they do not.

Thus in the statement A&&B, if A equals logical zero, then the entire expression will evaluate to false, regardless of the value of B, and therefore there is no need to evaluate B.

For A||B, if A is true, regardless of the value of B, the statement will evaluate to true.

Table 4.2–4 lists several useful logical functions.

Table 4.2–4 Logical functions

Logical function	Definition
all(x)	Returns a scalar, which is 1 if all the elements in the vector x are nonzero and 0 otherwise.
all(A)	Returns a row vector having the same number of columns as the matrix A and containing ones and zeros, depending on whether or not the corresponding column of A has all nonzero elements.
any(x)	Returns a scalar, which is 1 if any of the elements in the vector x is nonzero and 0 otherwise.
any(A)	Returns a row vector having the same number of columns as A and containing ones and zeros, depending on whether or not the corresponding column of the matrix A contains any nonzero elements.
find(A)	Computes an array containing the indices of the nonzero elements of the array A.
[u,v,w] = find(A)	Computes the arrays u and v containing the row and column indices of the nonzero elements of the array A and computes the array w containing the values of the nonzero elements. The array w may be omitted.
finite(A)	Returns an array of the same dimension as A with ones where the elements of A are finite and zeros elsewhere.
ischar(A)	Returns a 1 if A is a character array and 0 otherwise.
isempty(A)	Returns a 1 if A is an empty array and 0 otherwise.
isinf(A)	Returns an array of the same dimension as A, with ones where A has 'inf' and zeros elsewhere.
isnan(A)	Returns an array of the same dimension as A with ones where A has 'NaN' and zeros elsewhere. ('NaN' stands for "not a number," which means an undefined result.)
isnumeric(A)	Returns a 1 if A is a numeric array and 0 otherwise.
isreal(A)	Returns a 1 if A has no elements with imaginary parts and 0 otherwise.
logical(A)	Converts the elements of the array A into logical values.
xor(A,B)	Returns an array the same dimension as A and B; the new array has ones where either A or B is nonzero, but not both, and zeros where A and B are either both nonzero or both zero.

Logical Operators and the `find` Function

The `find` function is very useful for creating decision-making programs, especially when combined with the relational or logical operators. The function `find(x)` computes an array containing the indices of the nonzero elements of the array x. For example, consider the session

```
>>x = [-2, 0, 4];
>>y = find(x)
y =
    1    3
```

The resulting array y = [1, 3] indicates that the first and third elements of x are nonzero. Note that the `find` function returns the *indices*, not the *values*. In the following session, note the difference between the result obtained by x(x<y) and the result obtained by `find(x<y)`.

```
>>x = [6, 3, 9, 11];y = [14, 2, 9, 13];
>>values = x(x<y)
values =
        6    11
>>how_many = length (values)
how_many =
        2
>>indices = find(x<y)
indices =
        1    4
```

Thus two values in the array x are less than the *corresponding* values in the array y. They are the first and fourth values, 6 and 11. To find out how many, we could also have typed `length(indices)`.

The `find` function is also useful when combined with the logical operators. For example, consider the session

```
>>x = [5, -3, 0, 0, 8]; y = [2, 4, 0, 5, 7];
>>z = find(x&y)
z =
    1    2    5
```

The resulting array z = [1, 2, 5] indicates that the first, second, and fifth elements of both x and y are nonzero. Note that the `find` function returns the *indices*, and not the *values*. In the following session, note the difference between the result obtained by y(x&y) and the result obtained by `find(x&y)` above.

```
>>x = [5, -3, 0, 0, 8];y = [2, 4, 0, 5, 7];
>>values = y(x&y)
values =
    2    4    7
```

```
>>how_many = length(values)
how_many =
   3
```

Thus there are three nonzero values in the array x that correspond to nonzero values in the array x. They are the first, second, and fifth values, which are 2, 4, and 7.

In the above examples, there were only a few numbers in the arrays x and y, and thus we could have obtained the answers by visual inspection. However, these MATLAB methods are very useful either where there is so much data that visual inspection would be very time-consuming, or where the values are generated internally in a program.

Test Your Understanding

T4.2–1 If x = [5,-3,18,4] and y = [-9,13,7,4], what will be the result of the following operations? Use MATLAB to check your answer.

    a.   z = ~y > x
    b.   z = x&y
    c.   z = x|y
    d.   z = xor(x,y)

T4.2–2 Suppose that x = [-9, -6, 0, 2, 5] and y = [-10, -6 2, 4, 6]. What is the result of the following operations? Determine the answers by hand, and then use MATLAB to check your answers.

    a.   z = (x < y)
    b.   z = (x > y)
    c.   z = (x ~= y)
    d.   z = (x == y)
    e.   z = (x > 2)

T4.2–3 Suppose that x = [-4, -1, 0, 2, 10] and y = [-5, -2, 2, 5, 9]. Use MATLAB to find the values and the indices of the elements in x that are greater than the corresponding elements in y.

EXAMPLE 4.2–1	Height and Speed of a Projectile

The height and speed of a projectile (such as a thrown ball) launched with a speed of v_0 at an angle A to the horizontal are given by

$$h(t) = v_0 t \sin A - 0.5gt^2$$

$$v(t) = \sqrt{v_0^2 - 2v_0 gt \sin A + g^2 t^2}$$

where g is the acceleration due to gravity. The projectile will strike the ground when $h(t) = 0$, which gives the time to hit, $t_{hit} = 2(v_0/g)\sin A$. Suppose that $A = 40°$, $v_0 = 20$ m/s, and $g = 9.81$ m/s^2. Use the MATLAB relational and logical operators to find the times when the height is no less than 6 m and the speed is simultaneously no greater than 16 m/s. In addition, discuss another approach to obtaining a solution.

■ Solution

The key to solving this problem with relational and logical operators is to use the `find` command to determine the times at which the logical expression `(h >= 6)&(v <= 16)` is true. First we must generate the vectors h and v corresponding to times t_1 and t_2 between $0 \leq t \leq t_{hit}$, using a spacing for time t that is small enough to achieve sufficient accuracy for our purposes. We will choose a spacing of $t_{hit}/100$, which provides 101 values of time. The program follows. When computing the times t_1 and t_2, we must subtract 1 from `u(1)` and from `length(u)` because the first element in the array t corresponds to $t = 0$ (that is, `t(1)` is 0).

```
% Set the values for initial speed, gravity, and angle.
v0 = 20; g = 9.81; A = 40*pi/180;
% Compute the time to hit.
t_hit = 2*v0*sin(A)/g;
% Compute the arrays containing time, height, and speed.
t = [0:t_hit/100:t_hit];
h = v0*t*sin(A) - 0.5*g*t.^2;
v = sqrt(v0^2 - 2*v0*g*sin(A)*t + g^2*t.^2);
% Determine when the height is no less than 6,
% and the speed is no greater than 16.
u = find(h>=6&v<=16);
% Compute the corresponding times.
t_1 = (u(1)-1)*(t_hit/100)
t_2 = u(length(u)-1)*(t_hit/100)
```

The results are $t_1 = 0.8649$ and $t_2 = 1.7560$. Between these two times $h \geq 6$ m and $v \leq 16$ m/s.

We could have solved this problem by plotting $h(t)$ and $v(t)$, but the accuracy of the results would be limited by our ability to pick points off the graph; in addition, if we had to solve many such problems, the graphical method would be more time-consuming.

Test Your Understanding

T4.2–4 Consider the problem given in Example 4.2–1. Use relational and logical operators to find the times for which either the projectile's height is less than 4 m or the speed is greater than 17 m/s. Plot $h(t)$ and $v(t)$ to confirm your answer.

4.3 Conditional Statements

The MATLAB *conditional statements* enable us to write programs that make decisions. Conditional statements contain one or more of the `if`, `else`, and `elseif` statements. The `end` statement denotes the end of a conditional statement.

CONDITIONAL STATEMENTS

The `if` Statement

The `if` statement's basic form is

if *logical expression*
    statements
end

Every `if` statement must have an accompanying `end` statement. The `end` statement marks the end of the *statements* that are to be executed if the *logical expression* is true. A space is required between the `if` and the *logical expression,* which may be a scalar, a vector, or a matrix.

For example, suppose that x is a scalar and that we want to compute $y = \sqrt{x}$ only if $x \geq 0$. In English, we could specify this procedure as follows: If x is greater than or equal to zero, compute y from $y = \sqrt{x}$. The following `if` statement implements this procedure in MATLAB assuming x already has a scalar value.

```
if x >= 0
    y = sqrt(x)
end
```

If `x` is negative, the program takes no action. The *logical expression* here is `x >= 0`, and the *statement* is the single line `y = sqrt(x)`.

The `if` structure may be written on a single line; for example:

```
if x >= 0, y = sqrt(x), end
```

However, this form is less readable than the previous form. The usual practice is to indent the *statements* to clarify which statements belong to the `if` and its corresponding `end` and thereby improve readability.

The *logical expression* may be a compound expression; the *statements* may be a single command or a series of commands separated by commas or semicolons or on separate lines. For example if x and y have scalar values:

```
z = 0;w = 0;
if (x > 0)&(y > 0)
  z = sqrt(x) + sqrt(y)
  w = log(x) - 3*log(y)
end
```

The values of z and w are computed only if both `x` and `y` are positive. Otherwise, z and w retain their values of zero.

We may "nest" `if` statements, as shown by the following example.

if *logical expression 1*
    statement group 1
    if *logical expression 2*
        statement group 2
    end
end

Note that each `if` statement has an accompanying `end` statement.

The `else` Statement

When more than one action can occur as a result of a decision, we can use the `else` and `elseif` statements along with the `if` statement. The basic structure for the use of the `else` statement is

```
if  logical expression
      statement group 1
else
         statement group 2
end
```

For example, suppose that $y = \sqrt{x}$ for $x \geq 0$ and that $y = e^x - 1$ for $x < 0$. The following statements will calculate y, assuming that x already has a scalar value.

```
if x >= 0
    y = sqrt(x)
else
    y = exp(x) - 1
end
```

When the test, `if` *logical expression,* is performed, where the logical expression may be an *array,* the test returns a value of true only if *all* the elements of the logical expression are true! For example, if we fail to recognize how the test works, the following statements do not perform the way we might expect.

```
x = [4,-9,25];
if x < 0
    disp('Some of the elements of x are negative.')
else
    y = sqrt(x)
end
```

When this program is run it gives the result

```
y =
   2     0 + 3.000i      5
```

The program does not test each element in x in sequence. Instead it tests the truth of the vector relation x < 0. The test `if x < 0` returns a false value because it generates the vector [0,1,0]. Compare the preceding program with the following program.

```
x = [4,-9,25];
if x >= 0
    y = sqrt(x)
else
```

```
        disp('Some of the elements of x are negative.')
end
```

When executed, it produces the following result: `Some of the elements of x are negative.` The test `if x < 0` is false, and the test `if x >= 0` also returns a false value because `x >= 0` returns the vector `[1,0,1]`.

We sometimes must choose between a program that is concise, but perhaps more difficult to understand, and one that uses more statements than is necessary. For example, the statements

if *logical expression 1*
    if *logical expression 2*
       statements
    end
end

can be replaced with the more concise program

if *logical expression 1* & *logical expression 2*
    statements
end

The `elseif` Statement

The general form of the `if` statement is

if *logical expression 1*
    statement group 1
elseif *logical expression 2*
    statement group 2
else
    statement group 3
end

The `else` and `elseif` statements may be omitted if not required. However, if both are used, the `else` statement must come after the `elseif` statement to take care of all conditions that might be unaccounted for.

For example, suppose that $y = \ln x$ if $x \geq 5$ and that $y = \sqrt{x}$ if $0 \leq x < 5$. The following statements will compute y if x has a scalar value.

```
if x >= 5
    y = log(x)
else
    if x >= 0
        y = sqrt(x)
    end
end
```

If $x = -2$, for example, no action will be taken. If we use an `elseif`, we need fewer statements. For example:

```
if x >= 5
   y = log(x)
elseif x >= 0
   y = sqrt(x)
end
```

Note that the `elseif` statement does not require a separate `end` statement.

The `else` statement can be used with `elseif` to create detailed decision-making programs. For example, suppose that $y = \ln x$ for $x > 10$, $y = \sqrt{x}$ for $0 \le x \le 10$, and $y = e^x - 1$ for $x < 0$. The following statements will compute y if x already has a scalar value.

```
if x > 10
   y = log(x)
elseif x >= 0
   y = sqrt(x)
else
   y = exp(x) - 1
end
```

Decision structures may be *nested;* that is, one structure can contain another structure, which in turn can contain another, and so on.

Test Your Understanding

T4.3–1 Given a number x and the quadrant q ($q = 1, 2, 3, 4$), write a program to compute $\sin^{-1}(x)$ in degrees, taking into account the quadrant. The program should display an error message if $|x| > 1$.

Checking the Number of Input and Output Arguments

Sometimes you will want to have a function act differently depending on how many inputs it has. You can use the function `nargin`, which stands for "number of input arguments." Within the function you can use conditional statements to direct the flow of the computation depending on how many input arguments there are. For example, suppose you want to compute the square root of the input if there is only one, but compute the square root of the average if there are two inputs. The following function does this.

```
function z = sqrtfun(x, y)
if (nargin == 1)
   z = sqrt(x);
elseif (nargin == 2)
   z = sqrt((x + y)/2);
end
```

The `nargout` function can be used to determine the number of output arguments.

Strings and Conditional Statements

A string is a variable that contains characters. Strings are useful for creating input prompts and messages and for storing and operating on data such as names and addresses. To create a string variable, enclose the characters in single quotes. For example, the string variable `name` is created as follows:

```
>>name = 'Leslie Student';
```

The following string, `number`

```
>>number = '123';
```

is *not* the same as the variable `number` created by typing `number = 123`.

Strings are stored as row vectors in which each column represents a character. For example, the variable `name` has 1 row and 14 columns (each blank space occupies one column). We can access any column the way we access any other vector. For example, the letter S in the name Leslie Student occupies the eighth column in the vector `name`. It can be accessed by typing `name(8)`.

One of the most important applications for strings is to create input prompts and output messages. The following prompt program uses the `isempty(x)` function, which returns a 1 if the array `x` is empty and 0 otherwise. It also uses the `input` function, whose syntax is

```
x = input('prompt', 'string')
```

This function displays the string *prompt* on the screen, waits for input from the keyboard, and returns the entered value in the string variable `x`. The function returns an empty array if you press the **Enter** key without typing anything.

The following prompt program is a script file that allows the user to answer Yes by typing either Y or y or by pressing the **Enter** key. Any other response is treated as a No answer.

```
response = input('Do you want to continue? Y/N [Y]: ','s');
    if (isempty(response))|(response == 'Y')|(response == 'y')
        response = 'Y'
    else
        response = 'N'
    end
```

Many more string functions are available in MATLAB. Type `help strfun` to obtain information on these.

Solving Sets of Equations

In Section 2.5 we saw that the set of linear algebraic equations $\mathbf{Ax} = \mathbf{b}$ with m equations and n unknowns has solutions if and only if (1) rank[$\mathbf{A}$] = rank[$\mathbf{A}$ $\mathbf{b}$]. Let r = rank[$\mathbf{A}$]. If condition (1) is satisfied and if $r = n$, then the solution is unique. If

Table 4.3–1 Pseudocode for the linear equation solver

If the rank of **A** equals the rank of [**A b**], then
    determine whether the rank of **A** equals the number of unknowns. If so, there is a unique solution, which can be computed using left division. Display the results and stop.
    Otherwise, there is an infinite number of solutions, which can be found from the augmented matrix. Display the results and stop.
Otherwise (if the rank of **A** does not equal the rank of [**A b**]), then there are no solutions. Display this message and stop.

---

condition (1) is satisfied but $r < n$, an infinite number of solutions exists; in addition, r unknown variables can be expressed as linear combinations of the other $n - r$ unknown variables, whose values are arbitrary. In this case we can use the `rref` command to find the relations between the variables. *Pseudocode* uses natural language and mathematical expressions to construct statements that look like computer statements but without detailed syntax. The pseudocode in Table 4.3–1 can be used to outline an equation solver program before writing it.

A *flowchart* can be used to describe the possible paths that a program's computations can take, depending on how the conditional statements are executed. A condensed flowchart appears in Figure 4.3–1. From this chart or the pseudocode,

FLOWCHART

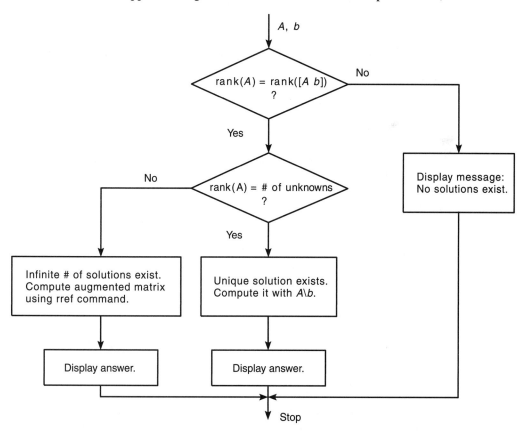

Figure 4.3–1 Flowchart illustrating a program to solve linear equations.

Table 4.3–2 MATLAB program to solve linear equations

```
% Script file lineq.m
% Solves the set Ax = b, given A and b.
% Check the ranks of A and [A b].
if rank(A) == rank([A b])
    % The ranks are equal.
    size_A = size (A);
    % Does the rank of A equal the number of unknowns?
    if rank(A) == size_A(2)
        % Yes. Rank of A equals the number of unknowns.
        disp ('There is a unique solution, which is:')
        x = A\b % Solve using left division.
    else
        % Rank of A does not equal the number of unknowns.
        disp('There is an infinite number of solutions.')
        disp('The augmented matrix of the reduced system is:')
        rref([A b]) % Compute the augmented matrix.
    end
else
    % The ranks of A and [A b] are not equal.
    disp ('There are no solutions.')
end
```

we can develop the script file shown in Table 4.3–2. The program uses the given arrays A and b to check the rank conditions, the left-division method to obtain the solution, if one exists, and the `rref` method if there is an infinite number of solutions. Note that the number of unknowns equals the number of columns in A, which is given by `size_A(2)`, the second element in `size_A`. Note also that the rank of **A** cannot exceed the number of columns in **A.**

Test Your Understanding

T4.3–2 Type in the script file `lineq.m` given in Table 4.3–2 and run it for the following cases. Hand check the answers.

    a. A = [1, -1;1, 1], b = [3;5]
    b. A = [1, -1;2, -2], b = [3;6]
    c. A = [1, -1;2, -2], b = [3;5]

4.4 Loops

A *loop* is a structure for repeating a calculation a number of times. Each repetition of the loop is a *pass.* MATLAB uses two types of explicit loops: the for loop, when the number of passes is known ahead of time, and the `while` loop, when the looping process must terminate when a specified condition is satisfied, and thus the number of passes is not known in advance.

`for` Loops

A simple example of a `for` loop is

```
for k = 5:10:35
    x = k^2
end
```

The *loop variable* `k` is initially assigned the value 5, and `x` is calculated from `x = k^2`. Each successive pass through the loop increments `k` by 10 and calculates `x` until `k` exceeds 35. Thus `k` takes on the values 5, 15, 25, and 35, and `x` takes on the values 25, 225, 625, and 1225. The program then continues to execute any statements following the `end` statement.

The typical structure of a `for` loop is

```
for loop variable = m:s:n
    statements
end
```

The expression `m:s:n` assigns an initial value of `m` to the loop variable, which is incremented by the value `s`—called the *step value* or *incremental value*. The *statements* are executed once during each pass, using the current value of the loop variable. The looping continues until the loop variable exceeds the *terminating value* `n`. For example, in the expression `for k = 5:10:36`, the final value of `k` is 35. Note that we need not place a semicolon after the `for m:s:n` statement to suppress printing `k`.

Note that a `for` statement needs an accompanying `end` statement. The `end` statement marks the end of the *statements* that are to be executed. A space is required between the `for` and the *loop variable,* which may be a scalar, a vector, or a matrix, although the scalar case is by far the most common.

The `for` loop may be written on a single line; for example:

```
for x = 0:2:10, y = sqrt(x), end
```

However, this form is less readable than the previous form. The usual practice is to indent the *statements* to clarify which statements belong to the `for` and its corresponding `end` and thereby improve readability.

Series Calculation with a `for` Loop

EXAMPLE 4.4–1

Write a script file to compute the sum of the first 15 terms in the series $5k^2 - 2k$, $k = 1, 2, 3, \ldots, 15$.

■ Solution
Because we know how many times we must evaluate the expression $5k^2 - 2k$, we can use a `for` loop. The script file is the following:

```
total = 0;
for k = 1:15
    total = 5*k^2 - 2*k + total;
end
disp ('The sum for 15 terms is:')
disp (total)
```

The answer is 5960.

EXAMPLE 4.4–2   Plotting with a `for` Loop

Write a script file to plot the function:

$$y = \begin{cases} 15\sqrt{4x} + 10 & x \geq 9 \\ 10x + 10 & 0 \leq x < 9 \\ 10 & x < 0 \end{cases}$$

for $-5 \leq x \leq 30$.

■ **Solution**

We choose a spacing $dx = 35/300$ to obtain 301 points, which is sufficient to obtain a smooth plot. The script file is the following:

```
dx = 35/300;
x = [-5:dx:30];
for k = 1:length(x)
   if x(k) >= 9
     y(k) = 15*sqrt(4*x(k)) + 10;
   elseif x(k) >= 0
     y(k) = 10*x(k) + 10;
   else
     y(k) = 10;
   end
end
plot (x,y), xlabel('x'), ylabel('y')
```

Note that we must use the index `k` to refer to `x` within the loop, as `x(k)`.

NESTED LOOPS

We may nest loops and conditional statements, as shown by the following example. (Note that each `for` and `if` statement needs an accompanying `end` statement.)

Suppose we want to create a special square matrix that has ones in the first row and first column, and whose remaining elements are the sum of two elements, the element above and the element to the left, if the sum is less than 20. Otherwise, the element is the maximum of those two element values. The following function creates this matrix. The row index is r; the column index is c. Note how indenting improves the readability.

```
function A = specmat(n)
A = ones(n);
for r = 1:n
    for c = 1:n
        if (r>1)&(c>1)
            s = A(r-1,c) + A(r,c-1);
            if s<20
                A(r,c) = s;
            else
                A(r,c) = max(A(r-1,c),A(r,c-1));
            end
        end
    end
end
```

Typing specmat(5) produces the following matrix

$$\begin{bmatrix} 1 & 1 & 1 & 1 & 1 \\ 1 & 2 & 3 & 4 & 5 \\ 1 & 3 & 6 & 10 & 15 \\ 1 & 4 & 10 & 10 & 15 \\ 1 & 5 & 15 & 15 & 15 \end{bmatrix}$$

Test Your Understanding

T4.4–1 Write a program to produce the following matrix:

$$A = \begin{bmatrix} 4 & 8 & 12 \\ 10 & 14 & 18 \\ 16 & 20 & 24 \\ 22 & 26 & 30 \end{bmatrix}$$

Note the following rules when using for loops with the loop variable expression k = m:s:n:

- The step value s may be negative. For example, k = 10:-2:4 produces k = 10, 8, 6, 4.
- If s is omitted, the step value defaults to one.
- If s is positive, the loop will not be executed if m is greater than n.

- If s is negative, the loop will not be executed if m is less than n.
- If m equals n, the loop will be executed only once.
- If the step value s is not an integer, round-off errors can cause the loop to execute a different number of passes than intended.

When the loop is completed, k retains its last value. You should not alter the value of the loop variable k within the *statements*. Doing so can cause unpredictable results.

A common practice in traditional programming languages like BASIC and FORTRAN is to use the symbols i and j as loop variables. However, this convention is not good practice in MATLAB, which uses these symbols for the imaginary unit $\sqrt{-1}$. For example, what do you think is the result of the following program? Try it and see!

```
x = 1; y = 1;
for i = 1:5
    x = x + 6i
    y = y + 5/i
end
```

The break and continue Statements

It is permissible to use an if statement to "jump" out of the loop before the loop variable reaches its terminating value. The break command, which terminates the loop but does not stop the entire program, can be used for this purpose. For example:

```
for k = 1:10
    x = 50 - k^2;
    if x < 0
        break
    end
    y = sqrt(x)
end
% The program execution jumps to here
% if the break command is executed.
```

However, it is usually possible to write the code to avoid using the break command. This can often be done with a while loop as explained in the next section.

The break statement stops the execution of the loop. There can be applications where we want to not execute the case producing an error but continue executing the loop for the remaining passes. We can use the continue statement to do this. The continue statement passes control to the next iteration of the for or while loop in which it appears, skipping any remaining statements in the body of the loop. In nested loops, continue passes control to the next iteration of the for or while loop enclosing it.

For example, the following code uses a `continue` statement to avoid computing the logarithm of a negative number.

```
x = [10,1000,-10,100];
y = NaN*x;
for k = 1:length(x)
    if x(k) < 0
        continue
    end
    y(k) = log10(x(k));
end
y
```

The result is `y = 1, 3, NaN, 2`.

Using an Array as a Loop Index

It is permissible to use a matrix expression to specify the number of passes. In this case the loop variable is a vector that is set equal to the successive columns of the matrix expression during each pass. For example,

```
A = [1,2,3;4,5,6];
for v = A
    disp(v)
end
```

is equivalent to

```
A = [1,2,3;4,5,6];
n = 3;
for k = 1:n
    v = A(:,k)
end
```

The common expression `k = m:s:n` is a special case of a matrix expression in which the columns of the expression are scalars, not vectors.

For example, suppose we want to compute the distance from the origin to a set of three points specified by their xy coordinates (3,7), (6,6), and (2,8). We can arrange the coordinates in the array `coord` as follows.

$$\begin{bmatrix} 3 & 6 & 2 \\ 7 & 6 & 8 \end{bmatrix}$$

Then `coord = [3,6,2;7,6,8]`. The following program computes the distance and determines which point is farthest from the origin. The first time through the loop the index `coord` is `[3, 7]'`. The second time the index is `[6, 6]'`, and during the final pass it is `[2, 8]'`.

```
k = 0;
for coord = [3,6,2;7,6,8]
```

```
   k = k + 1;
   distance(k) = sqrt(coord'*coord)
end
[max_distance,farthest] = max(distance)
```

The previous program illustrates the use of an array index but the problem can be solved more concisely with the following program, which uses the `diag` function to extract the diagonal elements of an array.

```
coord = [3,6,2;7,6,8];
distance = sqrt(diag(coord'*coord))
[max_distance,farthest] = max(distance)
```

EXAMPLE 4.4–3   Data Sorting

A vector x has been obtained from measurements. Suppose we want to consider any data value in the range $-0.1 < x < 0.1$ as being erroneous. We want to remove all such elements and replace them with zeros at the end of the array. Develop two ways of doing this. An example is given in the following table.

	Before	**After**
x(1)	1.92	1.92
x(2)	0.05	−2.43
x(3)	−2.43	0.85
x(4)	−0.02	0
x(5)	0.09	0
x(6)	0.85	0
x(7)	−0.06	0

■ Solution

The following script file uses a `for` loop with conditional statements. Note how the empty array `[]` is used.

```
x = [1.92,0.05,-2.43,-0.02,0.09,0.85,-0.06];
y = [];z = [];
for k = 1:length(x)
   if abs(x(k)) >= 0.1
      y = [y,x(k)];
   else
      z = [z,x(k)];
   end
end
xnew = [y,zeros(size(z))]
```

The next script file uses the `find` function.

```
x = [1.92,0.05,-2.43,-0.02,0.09,0.85,-0.06];
y = x(find(abs(x) >= 0.1));
z = zeros(size(find(abs(x)<0.1)));
xnew = [y,z]
```

Use of Logical Arrays as Masks

Consider the array **A**.

$$\mathbf{A} = \begin{bmatrix} 0 & -1 & 4 \\ 9 & -14 & 25 \\ -34 & 49 & 64 \end{bmatrix}$$

The following program computes the array **B** by computing the square roots of all the elements of **A** whose value is no less than 0, and adding 50 to each element that is negative.

```
A = [0, -1, 4; 9, -14, 25; -34, 49, 64];
for m = 1:size(A,1)
    for n = 1:size(A,2)
        if A(m,n) >= 0
            B(m,n) = sqrt(A(m,n));
        else
            B(m,n) = A(m,n) + 50;
        end
    end
end
B
```

The result is

$$\mathbf{B} = \begin{bmatrix} 0 & 49 & 2 \\ 3 & 36 & 5 \\ 16 & 7 & 8 \end{bmatrix}$$

When a logical array is used to address another array, it extracts from that array the elements in the locations where the logical array has 1s. We can often avoid the use of loops and branching and thus create simpler and faster programs by using a logical array as a *mask* that selects elements of another array. Any elements not selected will remain unchanged.

MASK

The following session creates the logical array C from the numeric array A given previously.

```
>>A = [0, -1, 4; 9, -14, 25; -34, 49, 64];
>>C = (A >= 0);
```

The result is

$$C = \begin{bmatrix} 1 & 0 & 1 \\ 1 & 0 & 1 \\ 0 & 1 & 1 \end{bmatrix}$$

We can use this technique to compute the square root of only those elements of A given in the previous program that are no less than 0 and add 50 to those elements that are negative. The program is

```
A = [0, -1, 4; 9, -14, 25; -34, 49, 64];
C = (A >= 0);
A(C)  = sqrt(A(C))
A(~C) = A(~C) + 50
```

The result after the third line is executed is

$$A = \begin{bmatrix} 0 & -1 & 2 \\ 3 & -14 & 25 \\ -34 & 49 & 64 \end{bmatrix}$$

The result after the last line is executed is

$$A = \begin{bmatrix} 0 & 49 & 2 \\ 3 & 36 & 5 \\ 16 & 7 & 8 \end{bmatrix}$$

`while` Loops

WHILE LOOP

The `while` loop is used when the looping process terminates because a specified condition is satisfied, and thus the number of passes is not known in advance. A simple example of a `while` loop is

```
x = 5;
while x < 25
    disp(x)
    x = 2*x - 1;
end
```

The results displayed by the `disp` statement are 5, 9, and 17. The *loop variable* x is initially assigned the value 5, and it has this value until the statement x = 2*x - 1 is encountered the first time. The value then changes to 9. Before each pass through the loop, x is checked to see whether its value is less than 25. If so, the pass is made. If not, the loop is skipped and the program continues to execute any statements following the `end` statement. A principal application of `while` loops is when we want the loop to continue as long as a certain statement is true. Such a task is often more difficult to do with a `for` loop.

The typical structure of a `while` loop follows.

```
while logical expression
    statements
end
```

MATLAB first tests the truth of the *logical expression*. A loop variable must be included in the *logical expression*. For example, x is the loop variable in the statement `while x < 25`. If the *logical expression* is true, the *statements* are executed. For the `while` loop to function properly, the following two conditions must occur:

1. The loop variable must have a value before the `while` statement is executed.
2. The loop variable must be changed somehow by the *statements.*

The *statements* are executed once during each pass, using the current value of the loop variable. The looping continues until the *logical expression* is false. Always make sure that the loop variable has a value assigned to it before the start of the loop.

Each `while` statement must be matched by an accompanying `end`. As with `for` loops, the *statements* should be indented to improve readability. You may nest `while` loops, and you may nest them with `for` loops and `if` statements.

It is possible to create an *infinite loop,* which is a loop that never ends. For example:

```
x = 8;
while x ~= 0
    x = x - 3;
end
```

Within the loop the variable x takes on the values $5, 2, -1, -4, \ldots$, and the condition `x ~= 0` is always satisfied, so the loop never stops.

Series Calculation with a `while` Loop

EXAMPLE 4.4–4

Write a script file to determine how many terms are required for the sum of the series $5k^2 - 2k, k = 1, 2, 3, \ldots$ to exceed 10,000. What is the sum for this many terms?

■ Solution
Because we do not know how many times we must evaluate the expression $5k^2 - 2k$, we use a `while` loop. The script file is the following:

```
total = 0;
k = 0;
while total < 1e+4
    k = k + 1;
    total = 5*k^2 - 2*k + total;
end
disp('The number of terms is:')
```

```
disp(k)
disp('The sum is:')
disp(total)
```

The sum is 10,203 after 18 terms.

EXAMPLE 4.4–5   Growth of a Bank Account

Determine how long it will take to accumulate at least $10,000 in a bank account if you deposit $500 initially and $500 at the end of each year, if the account pays 5 percent annual interest.

■ Solution

Because we do not know how many years it will take, a `while` loop should be used. The script file is the following.

```
amount = 500;
k=0;
while amount < 10000
    k = k+1;
    amount = amount*1.05 + 500;
end
amount
k
```

The final results are `amount` = `1.0789e+004`, or $10,789, and `k` = `14`, or 14 years.

The Editor/Debugger is capable of automatically indenting to improve the readability of a file. For example, `if`, `else`, `elseif`, `for`, and `while` structures do not require indenting, but doing so enables the reader to identify the structure more easily. The Editor/Debugger automatically indents the lines after `if`, `else`, `elseif`, `for`, and `while` statements when you press the **Enter** key. It continues to indent until the corresponding `end` statement is reached. It also uses *syntax highlighting* to identify key statements by displaying them in different colors. Table 4.4–1 summarizes these statements.

Table 4.4–1 Some MATLAB programming statements

Command	Description
else	Delineates an alternate block of commands.
elseif	Conditionally executes an alternate block of commands.
end	Terminates `for`, `while`, and `if` statements.
find(x)	Computes an array containing the indices of the nonzero elements of the array x.
for	Repeats commands a specified number of times.
if	Executes commands conditionally.
while	Repeats commands an indefinite number of times.

Test Your Understanding

T4.4–2 Write a script file using conditional statements to evaluate the following function, assuming that the scalar variable x has a value. The function is $y = \sqrt{x^2 + 1}$ for $x < 0$, $y = 3x + 1$ for $0 \le x < 10$, and $y = 9$ sin $(5x - 50) + 31$ for $x \ge 10$. Use your file to evaluate y for $x = -5, x = 5$, and $x = 15$, and check the results by hand.

T4.4–3 Use a for loop to determine the sum of the first 20 terms in the series $3k^2$, $k = 1, 2, 3, \ldots 20$. (Answer: 8610.)

T4.4–4 Use a while loop to determine how many terms in the series $3k^2$, $k = 1, 2, 3, \ldots$ are required for the sum of the terms to exceed 2000. What is the sum for this number of terms? (Answer: 13 terms, with a sum of 2457.)

T4.4–5 Rewrite the following code using a while loop to avoid using the break command.

```
for k = 1:10
    x = 50 - k^2;
    if x < 0
        break
    end
    y = sqrt(x)
end
```

T4.4–6 Find to two decimal places the largest value of x before the error in the series approximation $e^x \approx 1 + x + x^2/2 + x^3/6$ exceeds 1 percent. (Answer: $x = 0.83$.)

4.5 The switch Structure

The switch *structure* provides an alternative to using the if, elseif, and else commands. Anything programmed using switch can also be programmed using if structures. However, for some applications the switch structure is more readable than code using the if structure. The syntax is

SWITCH STRUCTURE

switch *input expression* (scalar or string)
    case *value1*
        statement group 1
    case *value2*
        statement group 2

.
.
.

```
   otherwise
      statement group n
end
```

The *input expression* is compared to each `case` value. If they are the same, then the statements following that `case` statement are executed and processing continues with any statements after the `end` statement. If the *input expression* is a string, then it is equal to the `case` *value* if `strcmp` returns a value of 1 (true). Only the *first* matching `case` is executed. If no match occurs, the statements following the `otherwise` statement are executed. However, the `otherwise` statement is optional. If it is absent, execution continues with the statements following the `end` statement if no match exists. Each `case` *value* statement must be on a single line.

For example, suppose the variable `angle` has an integer value that represents an angle measured in degrees from North. The following `switch` block displays the point on the compass that corresponds to that angle.

```
switch angle
   case 45
      disp('Northeast')
   case 135
      disp('Southeast')
   case 225
      disp('Southwest')
   case 315
      disp('Northwest')
   otherwise
      disp('Direction Unknown')
end
```

The use of a string variable for the *input expression* can result in very readable programs. For example, in the following code the numeric vector x has values, and the user enters the value of the string variable `response`; its intended values are `min`, `max`, or `sum`. The code then either finds the minimum or maximum value of x or sums the elements of x, as directed by the user.

```
t = [0:100]; x = exp(-t).*sin(t);
response = input('Type min, max, or sum.','s')
response = lower(response);
switch response
   case 'min'
      minimum = min(x)
   case 'max'
      maximum = max(x)
```

```
      case 'sum'
         total = sum(x)
      otherwise
         disp('You have not entered a proper choice.')
end
```

The switch statement can handle multiple conditions in a single case statement by enclosing the case *value* in a cell array (Cell arrays are denoted by curly braces and are treated in Section 2.7.). For example, the following switch block displays the corresponding point on the compass, given the integer angle measured from North.

```
switch angle
   case {0,360}
      disp('North')
   case {-180,180}
      disp('South')
   case {-270,90}
      disp('East')
   case {-90,270}
      disp('West')
   otherwise
      disp('Direction Unknown')
end
```

Test Your Understanding

T4.5–1 Write a program using the switch structure to input one angle, whose value may be 45, −45, 135, or −135°, and display the quadrant (1, 2, 3, or 4) containing the angle.

Using the switch Structure for Calendar Calculations

EXAMPLE 4.5-1

Use the switch structure to compute the total elapsed days in a year, given the number (1–12) of the month, the day, and an indication of whether or not the year is a leap year.

■ Solution

Note that February has an extra day if the year is a leap year. The following function computes the total elapsed number of days in a year, given the month, the day of the month, and the value of extra_day, which is 1 for a leap year, and 0 otherwise.

```
function total_days = total(month,day,extra_day)
total_days = day;
for k = 1:month - 1
   switch k
```

```
   case {1,3,5,7,8,10,12}
      total_days = total_days + 31;
   case {4,6,9,11}
      total_days = total_days + 30;
   case 2
      total_days = total_days + 28 + extra_day;
   end
end
```

The function can be used as shown in the following program.

```
month = input('Enter month (1 - 12): ');
day = input('Enter day (1 - 31): ');
extra_day = input('Enter 1 for leap year; 0 otherwise: ');
total_days = total(month,day,extra_day)
```

4.6 Debugging MATLAB Programs

Use of the MATLAB Editor/Debugger as an M-file *editor* was discussed in Section 1.4 of Chapter 1. Figure 1.4–1 (in Chapter 1) shows the Editor/Debugger screen. Figure 4.6–1 shows the Debugger containing two programs to be analyzed. Here we discuss its use as a *debugger*. Before you use the Debugger, try to debug your program using the common sense guidelines presented under **Debugging Script Files** in Section 1.4. MATLAB programs are often short because of the power of its commands, and you may not need to use the Debugger unless you are writing

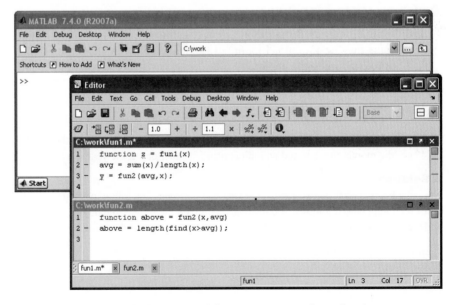

Figure 4.6–1 The Editor/Debugger containing two programs to be analyzed.

large programs. However, the cell mode discussed in this section is useful even for short programs. The Editor/Debugger menu bar contains the following items: **File, Edit, Text, Cell, Tools, Debug, Desktop, Window,** and **Help.** The **File, Edit, Desktop, Window,** and **Help** menus are similar to those in the Desktop. The **Cell** menu will be discussed shortly. The **Tools** menu involves advanced topics that will not be treated in this text. The **Desktop** menu is similar to that in the Command window. It enables you to dock and undock windows, arrange the Editor window, and turn the Editor toolbar on and off.

Below the menu bar is the Editor/Debugger toolbar. It enables you to access several of the items in the menus with one click of the mouse. Hold the mouse cursor over a button on the toolbar to see its function. For example, clicking the button with the binoculars icon is equivalent to selecting **Find and Replace** from the **Edit** menu. One item on the toolbar that is not in the menus is the function button with the script f icon (f). Use this button to go to a particular function in the M-file. The list of functions that you will see includes only those functions whose function statements are in the program. The list does not include functions that are called from the M-file.

The **Text** menu supplements the **Edit** menu for creating M-files. With the **Text** menu you can insert or remove comments, increase or decrease the amount of indenting, turn on smart indenting, and evaluate and display the values of selected variables in the Command window. Click *anywhere* in a previously typed line, and then click **Comment** in the **Text** menu. This makes the entire line a comment. To turn a commented line into an executable line, click *anywhere* in the line, and then click **Uncomment** in the **Text** menu.

Cell Mode

The cell mode can be used to debug programs. It can also be used to generate a report. See the end of Section 5.2 for a discussion of the latter usage. A cell is a group of commands. (Such a cell should not be confused with the cell array data type covered in Section 2.7.) The double percent character (%%) is used to mark the beginning of a new cell, and it is called a *cell divider*. To use the cell mode, first enter your program into the Editor. Then click the **Cell** button and select **Enable Cell Mode.** The cell toolbar then appears, as shown in Figure 4.6–2.

Consider the following simple program that plots either a quadratic or a cubic function.

```
%% Evaluate a quadratic and a cubic.
clear, clc
x = linspace(0, 10, 300);
%% Quadratic
y1 = polyval([1, -8, 6], x); plot(x,y1)
%% Cubic
y2 = polyval([1, -11, 9, 9], x); plot(x,y2)
```

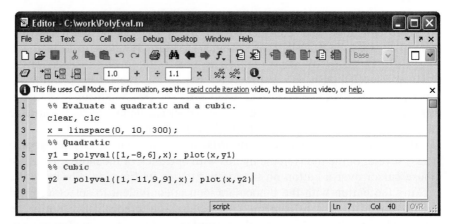

Figure 4.6–2   The cell mode of the Editor/Debugger.

After entering and saving the program, and enabling cell mode, you can click on one of the evaluation icons shown on the left-hand side of the cell toolbar (see Figure 4.6–2). These enable you to evaluate the current single cell (where the cursor is currently), to evaluate the current cell and advance to the next cell, or to evaluate the entire program.

A useful feature of cell mode is that it enables you to evaluate the results of changing a parameter. For example, in Figure 4.6–2, suppose the cursor is next to the number -8. If you click the plus $(+)$ or minus $(-)$ sign in the cell toolbar, the parameter (-8) will be decremented or incremented by the increment shown in the window (1.0 is the default, which you can change). If you have already run the program and the quadratic plot is on the screen, click the minus sign once to change the parameter from -8 to -9 and watch the plot change.

You can also change the parameter by a divisive or multiplicative factor (1.1 is the default). Click the divide or multiply symbol on the cell toolbar.

The Debug Menu

BREAKPOINT

Breakpoints are points in the file where execution stops temporarily so that you can examine the values of the variables up to that point. You set breakpoints with the **Set/Clear Breakpoint** item on the **Debug** menu. Use the **Step, Step In,** and **Step Out** items on the **Debug** menu to step through your file after you have set breakpoints and run the file. Click **Step** to watch the script execute one step at a time. Click **Step In** to step into the first executable line in a function being called. Click **Step Out** in a called function to run the rest of the function and then return to the calling program.

The solid green arrow to the left of the line text indicates the next line to be executed. When this arrow changes to a hollow green arrow, MATLAB control is now in a function being called. Execution returns to the line with the solid green arrow after the function completes its operation. The arrow turns yellow at a line where execution pauses or where a function completes its operation. When the program pauses you can assign new values to a variable, using either the Command window or the Array Editor.

Click on the **Go Until Cursor** item to run the file until it reaches the line where the cursor is; this process sets a temporary breakpoint at the cursor. You can save and execute your program directly from the **Debug** menu if you want, by clicking on **Run** (or **Save and Run** if you have made changes). You need not set any breakpoints beforehand. Click **Exit Debug Mode** to return to normal editing. To save any changes you have made to the program, first exit the debug mode, and then save the file.

Using Breakpoints

Most debugging sessions start by setting a breakpoint. A breakpoint stops M-file execution at a specified line and allows you to view or change values in the function's workspace before resuming execution. To set a breakpoint, position the cursor in the line of text and click on the breakpoint icon in the toolbar or select **Set/Clear Breakpoints** from the **Debug** menu. You can also set a breakpoint by right-clicking on the line of text to bring up the **Context** menu and choose **Set/Clear Breakpoint.** A red circle next to a line indicates that a breakpoint is set at that line. If the line selected for a breakpoint is not an executable line, then the breakpoint is set at the next executable line. The **Debug** menu enables you to clear all the breakpoints (select **Clear Breakpoints in All Files**). The **Debug** menu also lets you halt M-file execution if your code generates a warning, an error, or a NaN or Inf value (select **Stop if Errors/Warnings**).

For more details about the menus and examples of using the Debugger, see Chapter 4 of Palm, (2005a).

4.7 Summary

Now that you have finished this chapter, you should be able to write programs that can perform decision-making procedures; that is, the program's operations depend on results of the program's calculations or on input from the user. Sections 4.1 through 4.3 covered the required features: the relational operators, the logical operators and functions, and the conditional statements.

You should also be able to use MATLAB loop structures to write programs that repeat calculations a specified number of times or until some condition is satisfied. This feature enables you to solve problems of great complexity or requiring numerous calculations. The for loop and while loop structures were covered in Section 4.4. Section 4.5 covered the switch structure.

Section 4.6 gave an overview of how to debug programs using the Editor/ Debugger. Tables summarizing the MATLAB commands introduced in this chapter are located throughout the chapter. Table 4.7–1 will help you locate these tables. It also summarizes those commands not found in the other tables.

Table 4.7–1 Guide to MATLAB commands introduced in Chapter 4

Relational operators	Table 4.1–1
Logical operators	Table 4.2–1
Order of precedence for operator types	Table 4.2–2
Truth table	Table 4.2–3
Logical functions	Table 4.2–4
Some MATLAB programming statements	Table 4.4–1

Miscellaneous commands

Command	Description	Section
break	Terminates the execution of a `for` or a `while` loop.	4.4
case	Used with `switch` to direct program execution.	4.5
continue	Passes control to the next iteration of a `for` or `while` loop.	4.4
double	Converts a logical array to class double.	4.1
else	Delineates an alternate block of statements.	4.3
elseif	Conditionally executes statements.	4.3
end	Terminates `for`, `while`, and `if` statements.	4.3, 4.4
for	Repeats statements a specific number of times.	4.4
if	Executes statements conditionally.	4.3
input('s1', 's')	Display the prompt string `s1` and stores user input as a string.	4.3
logical	Converts numeric values to logical values.	4.1
lower('s')	Converts the string `s` to all lowercase.	4.5
nargin	Determines the number of input arguments of a function.	4.3
nargout	Determines the number of output arguments of a function.	4.3
switch	Directs program execution by comparing the input expression with the associated `case` expressions.	4.5
while	Repeats statements an indefinite number of times.	4.4
xor	Exclusive OR function.	4.2

Key Terms with Page References

Breakpoint,   186
Conditional statement,   163
Flowchart,   169
for loop,   171
Logical operator,   156
Mask,   177

Nested loops,   172
Relational operator,   153
switch structure,   181
Truth table,   159
while loop,   178

Problems

You can find answers to problems marked with an asterisk at the end of the text.

Section 4.1

1.*   Suppose that x = 6. Find the results of the following operations by hand and use MATLAB to check your results.

  a. z = (x<10)
  b. z = (x==10)

c. z = (x>=4)
d. z = (x~=7)

2.* Find the results of the following operations by hand and use MATLAB to check your results.

a. z = 6>3+8
b. z = 6+3>8
c. z = 4>(2+9)
d. z = (4<7)+3
e. z = 4<7+3
f. z = (4<7)*5
g. z = 4<(7*5)
h. z = 2/5>=5

3.* Suppose that x = [10, -2, 6, 5, -3] and y = [9,-3, 2, 5,-1]. Find the results of the following operations by hand and use MATLAB to check your results.

a. z = (x<6)   *b.* z = (x<=y)
c. z = (x==y)   *d.* z = (x~=y)

4. For the arrays x and y given below, use MATLAB to find all the elements in x that are greater than the corresponding elements in y.

x = [-3, 0, 0, 2, 6, 8] y = [-5, -2, 0, 3, 4, 10]

5. The array price given below contains the price in dollars of a certain stock over 10 days. Use MATLAB to determine how many days the price was above $20.

price = [19, 18, 22, 21, 25, 19, 17, 21, 27, 29]

6. The arrays price_A and price_B given below contain the price in dollars of two stocks over 10 days. Use MATLAB to determine how many days the price of stock A was above the price of stock B.

price_A = [19, 18, 22, 21, 25, 19, 17, 21, 27, 29]
price_B = [22, 17, 20, 19, 24, 18, 16, 25, 28, 27]

7. The arrays price_A, price_B, and price_C given below contain the price in dollars of three stocks over 10 days.

a. Use MATLAB to determine how many days the price of stock A was above both the price of stock B and the price of stock C.
b. Use MATLAB to determine how many days the price of stock A was above either the price of stock B or the price of stock C.
c. Use MATLAB to determine how many days the price of stock A was above either the price of stock B or the price of stock C, but not both.

price_A = [19, 18, 22, 21, 25, 19, 17, 21, 27, 29]
price_B = [22, 17, 20, 19, 24, 18, 16, 25, 28, 27]
price_C = [17, 13, 22, 23, 19, 17, 20, 21, 24, 28]

Section 4.2

8.* Suppose that x $= $ [-3, 0, 0, 2, 5, 8] and y $= $ [-5, -2, 0, 3, 4, 10]. Find the results of the following operations by hand and use MATLAB to check your results.

a. z = y<~x   *b.* z = x&y

c. z = x|y   *d.* z = xor(x,y)

9. The height and speed of a projectile (such as a thrown ball) launched with a speed of v_0 at an angle A to the horizontal are given by

$$h(t) = v_0 t \sin A - 0.5gt^2$$
$$v(t) = \sqrt{v_0^2 - 2v_0 gt \sin A + g^2 t^2}$$

where g is the acceleration due to gravity. The projectile will strike the ground when $h(t) = 0$, which gives the time to hit $t_{\text{hit}} = 2(v_0/g) \sin A$.

Suppose that $A = 30°$, $v_0 = 40$ m/s, and $g = 9.81$ m/s². Use the MATLAB relational and logical operators to find the times when

a. The height is no less than 15 m.

b. The height is no less than 15 m and the speed is simultaneously no greater than 36 m/s.

c. The height is less than 5 m or the speed is greater than 35 m/s.

10.* The price, in dollars, of a certain stock over a 10-day period is given in the following array.

```
price = [19, 18, 22, 21, 25, 19, 17, 21, 27, 29]
```

Suppose you owned 1000 shares at the start of the 10-day period, and you bought 100 shares every day the price was below $20 and sold 100 shares every day the price was above $25. Use MATLAB to compute (*a*) the amount you spent in buying shares, (*b*) the amount you received from the sale of shares, (*c*) the total number of shares you own after the 10th day, and (*d*) the net increase in the worth of your portfolio.

11. Let e1 and e2 be logical expressions. DeMorgan's laws for logical expressions state that

NOT(e1 AND e2) implies that (NOT e1) OR (NOT e2)

and

NOT(e1 OR e2) implies that (NOT e1) AND (NOT e2)

Use these laws to find an equivalent expression for each of the following expressions and use MATLAB to verify the equivalence.

a. ~((x < 10)&(x>=6))

b. ~((x == 2)|(x > 5))

12. Are these following expressions equivalent? Use MATLAB to check your answer for specific values of *a, b, c,* and *d.*
    a. 1. `(a==b)&((b==c)|(a==c))`
       2. `(a==b)|((b==c)&(a==c))`
    b. 1. `(a<b)&((a>c)|(a>d))`
       2. `(a<b)&(a>c)|((a<b)&(a>d))`

Section 4.3

13. Write a script file using conditional statements to evaluate the following function, assuming that the scalar variable x has a value. The function is $y = e^{x+1}$ for $x < -1$, $y = 2 + \cos(\pi x)$ for $-1 \le x < 5$, and $y = 10(x - 5) + 1$ for $x \ge 5$. Use your file to evaluate y for $x = -5$, $x = 3$, and $x = 15$, and check the results by hand.

14. Rewrite the following statements to use only one `if` statement.

```
if x < y
   if z < 10
      w = x*y*z
   end
end
```

15. Write a program that accepts a numerical value x from 0 to 100 as input and computes and displays the corresponding letter grade given by the following table.
    A $x \ge 90$
    B $80 \le x \le 89$
    C $70 \le x \le 79$
    D $60 \le x \le 69$
    F $x < 60$
    a. Use nested `if` statements in your program (do not use `elseif`).
    b. Use only `elseif` clauses in your program.

16. Write a program that accepts a year and determines whether or not the year is a leap year. Use the `mod` function. The output should be the variable `extra_day`, which should be 1 if the year is a leap year and 0 otherwise. The rules for determining leap years in the Gregorian calendar are:

    1. All years evenly divisible by 400 are leap years.
    2. Years evenly divisible by 100 but not by 400 are not leap years.
    3. Years divisible by 4 but not by 100 are leap years.
    4. All other years are not leap years.
       For example, the years 1800, 1900, 2100, 2300, and 2500 are not leap years, but 2400 is a leap year.

17. Figure P17 shows a mass-spring model of the type used to design packaging systems and vehicle suspensions, for example. The springs exert a force that is proportional to their compression, and the proportionality constant is the spring constant k. The two side springs provide additional resistance if the weight W is too heavy for the center spring. When the weight W is gently placed, it moves through a distance x before coming to rest. From statics, the weight force must balance the spring forces at this new position. Thus

$$W = k_1 x \qquad\qquad \text{if } x < d$$
$$W = k_1 x + 2k_2(x - d) \qquad\qquad \text{if } x \geq d$$

a. Create a function file that computes the distance x, using the input parameters W, k_1, k_2, and d. Test your function for the following two cases, using the values $k_1 = 10^4$ N/m; $k_2 = 1.5 \times 10^4$ N/m; $d = 0.1$ m.

$$W = 500 \text{ N}$$
$$W = 2000 \text{ N}$$

b. Use your function to plot x versus W for $0 \leq W \leq 3000$ N for the values of k_1, k_2, and d given in part *a*.

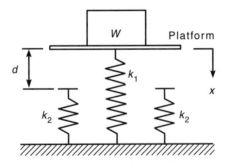

Figure P17

Section 4.4

18. Use a `for` loop to plot the function given in Problem 13 over the interval $-2 \leq x \leq 6$. Properly label the plot. The variable y represents height in kilometers, and the variable x represents time in seconds.

19. Plot the function $y = 10(1 - e^{-x/4})$ over the interval $0 \leq x \leq x_{max}$, using a `while` loop to determine the value of x_{max} such that $y(x_{max}) = 9.8$. Properly label the plot. The variable y represents force in newtons, and the variable x represents time in seconds.

20. Use a `for` loop to determine the sum of the first 10 terms in the series $5k^3$, $k = 1, 2, 3, \ldots 10$.

21. The (x, y) coordinates of a certain object as a function of time t are given by

$$x(t) = 5t - 10 \qquad y(t) = 25t^2 - 120t + 144$$

for $0 \le t \le 4$. Write a program to determine the time at which the object is the closest to the origin at $(0, 0)$. Determine also the minimum distance. Do this two ways:

a. By using a `for` loop.
b. By not using a `for` loop.

22. Consider the array **A**.

$$\mathbf{A} = \begin{bmatrix} 3 & 5 & -4 \\ -8 & -1 & 33 \\ -17 & 6 & -9 \end{bmatrix}$$

Write a program that computes the array **B** by computing the natural logarithm of all the elements of **A** whose value is no less than 1, and adding 20 to all the other elements. Do this two ways:

a. By using a `for` loop with conditional statements.
b. By using a logical array as a mask.

23. We want to analyze the mass-spring system discussed in Problem 17 for the case in which the weight W is dropped onto the platform attached to the center spring. If the weight is dropped from a height h above the platform, we can find the maximum spring compression x by equating the weight's gravitational potential energy $W(h + x)$ with the potential energy stored in the springs. Thus

$$W(h + x) = \tfrac{1}{2}k_1x^2 \qquad \text{if } x < d$$

which can be solved for x as

$$x = \frac{W \pm \sqrt{W^2 + 2k_1Wh}}{k_1} \qquad \text{if } x < d$$

and

$$W(h + x) = \tfrac{1}{2}k_1x^2 + \tfrac{1}{2}(2k_2)(x - d)^2 \qquad \text{if } x \ge d$$

which gives the following quadratic equation to solve for x:

$$(k_1 + 2k_2)x^2 - (4k_2d + 2W)x + 2k_2d^2 - 2Wh = 0 \qquad \text{if } x \ge d$$

a. Create a function file that computes the maximum compression x due to the falling weight. The function's input parameters are k_1, k_2, d, W, and h. Test your function for the following two cases, using the values $k_1 = 10^4$ N/m; $k_2 = 1.5 \times 10^4$ N/m; and $d = 0.1$ m.

$$W = 100 \text{ N}, h = 0.5 \text{ m}$$
$$W = 2000 \text{ N}, h = 0.5 \text{ m}$$

b. Use your function file to generate a plot of x versus h for $0 \le h \le 2\text{m}$. Use $W = 100$ N and the preceding values for k_1, k_2, and d.

24. Electrical resistors are said to be connected "in series" if the same current passes through each and "in parallel" if the same voltage is applied across each. If in series, they are equivalent to a single resistor whose resistance is given by

$$R = R_1 + R_2 + R_3 + \cdots + R_n$$

If in parallel, their equivalent resistance is given by

$$\frac{1}{R} = \frac{1}{R_1} + \frac{1}{R_2} + \frac{1}{R_3} + \cdots + \frac{1}{R_n}$$

Write an M-file that prompts the user for the type of connection (series or parallel) and the number of resistors n and then computes the equivalent resistance.

25. a. An *ideal* diode blocks the flow of current in the direction opposite that of the diode's arrow symbol. It can be used to make a *half-wave rectifier* as shown in Figure P25a. For the ideal diode, the voltage v_L across the load R_L is given by

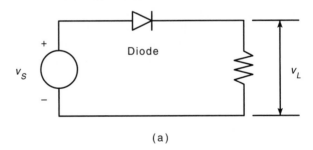

(a)

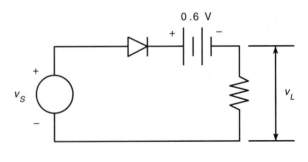

(b)

Figure P25

$$v_L = \begin{cases} v_S & \text{if } v_S > 0 \\ 0 & \text{if } v_S \le 0 \end{cases}$$

Suppose the supply voltage is

$$v_S(t) = 3e^{-t/3} \sin(\pi t) \text{ volts}$$

where time t is in seconds. Write a MATLAB program to plot the voltage v_L versus t for $0 \le t \le 10$.

b. A more accurate model of the diode's behavior is given by the *offset diode* model, which accounts for the offset voltage inherent in semiconductor diodes. The offset model contains an ideal diode and a battery whose voltage equals the offset voltage (which is approximately 0.6 V for silicon diodes) [Rizzoni, 1996]. The half-wave rectifier using this model is shown in Figure P25b. For this circuit,

$$v_L = \begin{cases} v_S - 0.6 & \text{if } v_S > 0.6 \\ 0 & \text{if } v_S \le 0.6 \end{cases}$$

Using the same supply voltage given in part a, plot the voltage v_L versus t for $0 \le t \le 10$; then compare the results with the plot obtained in part a.

26.* A company wants to locate a distribution center that will serve six of its major customers in a 30 × 30 mi area. The locations of the customers

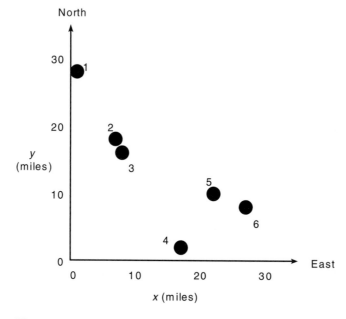

Figure P26

relative to the southwest corner of the area are given in the following table in terms of (x, y) coordinates (the x direction is east; the y direction is north) (see Figure P26). Also given is the volume in tons per week that must be delivered from the distribution center to each customer. The weekly delivery cost c_i for customer i depends on the volume V_i and the distance d_i from the distribution center. For simplicity we will assume that this distance is the straight-line distance. (This assumes that the road network is dense.) The weekly cost is given by $c_i = 0.5d_iV_i$; $i = 1, \ldots, 6$. Find the location of the distribution center (to the nearest mile) that minimizes the total weekly cost to service all six customers.

Customer	x location (miles)	y location (miles)	Volume (tons/week)
1	1	28	3
2	7	18	7
3	8	16	4
4	17	2	5
5	22	10	2
6	27	8	6

27. A company has the choice of producing up to four different products with its machinery, which consists of lathes, grinders, and milling machines. The number of hours on each machine required to produce a product is given in the following table, along with the number of hours available per week on each type of machine. Assume that the company can sell everything it produces. The profit per item for each product appears in the last line of the table.

	Product				
	1	2	3	4	Hours available
Hours required					
Lathe	1	2	0.5	3	40
Grinder	0	2	4	1	30
Milling	3	1	5	2	45
Unit profit ($)	100	150	90	120	

a. Determine how many units of each product the company should make to maximize its total profit and then compute this profit. Remember, the company cannot make fractional units, so your answer must be in integers. (Hint: First estimate the upper limits on the number of products that can be produced without exceeding the available capacity.)

b. How sensitive is your answer? How much does the profit decrease if you make one more or one less item than the optimum?

28. A certain company makes televisions, stereo units, and speakers. Its parts inventory includes chassis, picture tubes, speaker cones, power supplies,

and electronics. The inventory, required components, and profit for each product appear in the following table. Determine how many of each product to make in order to maximize the profit.

	Product			
	Television	Stereo unit	Speaker unit	Inventory
Requirements				
Chassis	1	1	0	450
Picture Tube	1	0	0	250
Speaker Cone	2	2	1	800
Power Supply	1	1	0	450
Electronics	2	2	1	600
Unit profit ($)	80	50	40	

29. Use a `while` loop to determine how many terms in the series 2^k, $k = 1$, $2, 3, \ldots$, are required for the sum of the terms to exceed 2000. What is the sum for this number of terms?

30. One bank pays 5.5 percent annual interest, while a second bank pays 4.5 percent annual interest. Determine how much longer it will take to accumulate at least $50,000 in the second bank account if you deposit $1000 initially, and $1000 at the end of each year.

31.* Use a loop in MATLAB to determine how long it will take to accumulate $1,000,000 in a bank account if you deposit $10,000 initially and $10,000 at the end of each year; the account pays 6 percent annual interest.

32. A weight W is supported by two cables anchored a distance D apart (see Figure P32). The cable length L_{AB} is given, but the length L_{AC} is to be selected. Each cable can support a maximum tension force equal to W. For the weight to remain stationary, the total horizontal force and total vertical force must each be zero. This principle gives the equations

$$-T_{AB} \cos \theta + T_{AC} \cos \phi = 0$$
$$T_{AB} \sin \theta + T_{AC} \sin \phi = W$$

We can solve these equations for the tension forces T_{AB} and T_{AC} if we know the angles θ and ϕ. From the law of cosines

$$\theta = \cos^{-1}\left(\frac{D^2 + L_{AB}^2 - L_{AC}^2}{2DL_{AB}}\right)$$

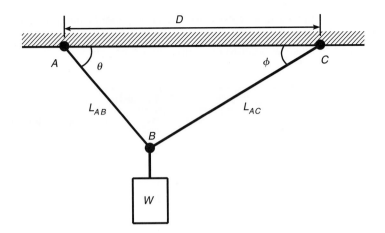

Figure P32

From the law of sines

$$\phi = \sin^{-1}\left(\frac{L_{AB}\sin\theta}{L_{AC}}\right)$$

For the given values $D = 6$ ft, $L_{AB} = 3$ ft, and $W = 2000$ lb, use a loop in MATLAB to find $L_{AC\,min}$, the shortest length L_{AC} we can use without T_{AB} or T_{AC} exceeding 2000 lb. Note that the largest L_{AC} can be is 6.7 ft (which corresponds to $\theta = 90°$). Plot the tension forces T_{AB} and T_{AC} on the same graph versus L_{AC} for $L_{AC\,min} \le L_{AC} \le 6.7$.

33.* In the structure in Figure P33a, six wires support three beams. Wires 1 and 2 can support no more than 1200 N each, wires 3 and 4 can support no more than 400 N each, and wires 5 and 6 no more than 200 N each. Three equal weights W are attached at the points shown. Assuming that the structure is stationary and that the weights of the wires and the beams are very small compared to W, the principles of statics applied to a particular beam state that the sum of vertical forces is zero and that the sum of moments about any point is also zero. Applying these principles to each beam using the free-body diagrams shown in Figure P33b, we obtain the following equations. Let the tension force in wire i be T_i. For beam 1

$$T_1 + T_2 = T_3 + T_4 + W + T_6$$
$$-T_3 - 4T_4 - 5W - 6T_6 + 7T_2 = 0$$

For beam 2

$$T_3 + T_4 = W + T_5$$
$$-W - 2T_5 + 3T_4 = 0$$

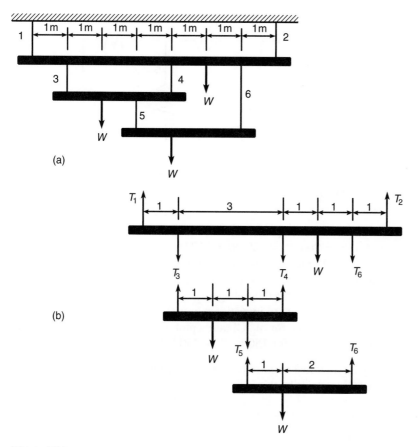

Figure P33

For beam 3

$$T_5 + T_6 = W$$
$$-W + 3T_6 = 0$$

Find the maximum value of the weight W the structure can support. Remember that the wires cannot support compression, so T_i must be nonnegative.

34. The equations describing the circuit shown in Figure P34 are:

$$-v_1 + R_1i_1 + R_4i_4 = 0$$
$$-R_4i_4 + R_2i_2 + R_5i_5 = 0$$
$$-R_5i_5 + R_3i_3 + v_2 = 0$$
$$i_1 = i_2 + i_4$$
$$i_2 = i_3 + i_5$$

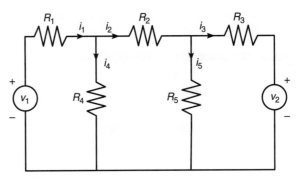

Figure P34

a. The given values of the resistances and the voltage v_1 are $R_1 = 5, R_2 = 100, R_3 = 200, R_4 = 150, R_5 = 250$ kΩ, and $v_1 = 100$ V. (Note that 1 kΩ $= 1000$ Ω.) Suppose that each resistance is rated to carry a current of no more than 1 mA $(= 0.001$ A). Determine the allowable range of positive values for the voltage v_2.

b. Suppose we want to investigate how the resistance R_3 limits the allowable range for v_2. Obtain a plot of the allowable limit on v_2 as a function of R_3 for $150 \leq R_3 \leq 250$ kΩ.

35. Many applications require us to know the temperature distribution in an object. For example, this information is important for controlling the material properties, such as hardness, when cooling an object formed from molten metal. In a heat transfer course, the following description of the temperature distribution in a flat, rectangular metal plate is often derived. The temperature is held constant at T_1 on three sides, and at T_2 on

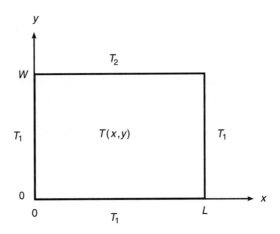

Figure P35

the fourth side (see Figure P35). The temperature $T(x, y)$ as a function of the xy coordinates shown is given by

$$T(x, y) = (T_2 - T_1)w(x, y) + T_1$$

where

$$w(x, y) = \frac{2}{\pi} \sum_{n \text{ odd}}^{\infty} \frac{2}{n} \sin\left(\frac{n\pi x}{L}\right) \frac{\sinh(n\pi y/L)}{\sinh(n\pi W/L)}$$

Use the following data: $T_1 = 70°F$, $T_2 = 200°F$, and $W = L = 2$ ft.

a. The terms in the preceding series become smaller in magnitude as n increases. Write a MATLAB program to verify this fact for $n = 1, \ldots,$ 19 for the center of the plate ($x = y = 1$).

b. Using $x = y = 1$, write a MATLAB program to determine how many terms are required in the series to produce a temperature calculation that is accurate to within 1 percent. (That is, for what value of n will the addition of the next term in the series produce a change in T of less than 1 percent.) Use your physical insight to determine whether this answer gives the correct temperature at the center of the plate.

c. Modify the program from part b to compute the temperatures in the plate; use a spacing of 0.2 for both x and y.

36. Consider the following script file. Fill in the lines of the following table with the values that would be displayed immediately after the `while` statement if you ran the script file. Write in the values the variables have each time the `while` statement is executed. You might need more or fewer lines in the table. Then type in the file, and run it to check your answers.

Pass	k	b	x	y
First				
Second				
Third				
Fourth				
Fifth				

```
k = 1;b = -2;x = -1;y = -2;
while k <= 3
    k, b, x, y
    y = x^2 - 3;
    if y < b
        b = y;
```

```
      end
      x = x + 1;
      k = k + 1;
   end
```

37. Assume that the human player makes the first move against the computer in a game of Tic-Tac-Toe, which has a 3 × 3 grid. Write a MATLAB function that lets the computer respond to that move. The function's input argument should be the cell location of the human player's move. The function's output should be the cell location of the computer's first move. Label the cells as 1, 2, 3 across the top row; 4, 5, 6 across the middle row, and 7, 8, 9 across the bottom row.

38. Suppose you project that you will be able to deposit the following monthly amounts into a savings account for a period of five years. The account initially has no money in it.

Year	1	2	3	4	5
Monthly deposit ($)	300	350	350	350	400

At the end of each year in which the account balance is at least $3000, you withdraw $2000 to buy a certificate of deposit (CD), which pays 6 percent interest compounded annually.

   Write a MATLAB program to compute how much money will accumulate in five years in the account and in any CDs you buy. Run the program for two different savings interest rates: 4 percent and 5 percent.

39.* A certain company manufactures and sells golf carts. At the end of each week, the company transfers the carts produced that week into storage (inventory). All carts that are sold are taken from the inventory. A simple model of this process is

$$I(k + 1) = P(k) + I(k) - S(k)$$

where

$$P(k) = \text{the number of carts produced in week } k$$

$$I(k) = \text{the number of carts in inventory in week } k$$

$$S(k) = \text{the number of carts sold in week } k$$

The projected weekly sales for 10 weeks are

Week	1	2	3	4	5	6	7	8	9	10
Sales	50	55	60	70	70	75	80	80	90	55

Suppose the weekly production is based on the previous week's sales so that $P(k) = S(k - 1)$. Assume that the first week's production is 50 carts; that is, $P(1) = 50$. Write a MATLAB program to compute and plot the number of carts in inventory for each of the 10 weeks or until the inventory drops below zero. Run the program for two cases: *a.* an initial inventory of 50 carts so that $I(1) = 50$, and *b.* an initial inventory of 30 carts so that $I(1) = 30$.

40. Redo Problem 39 with the restriction that the next week's production is set to zero if the inventory exceeds 40 carts.

Section 4.5

41. The following table gives the approximate values of the static coefficient of friction μ for various materials.

Materials	μ
Metal on metal	0.20
Wood on wood	0.35
Metal on wood	0.40
Rubber on concrete	0.70

To start a weight W moving on a horizontal surface, you must push with a force F, where $F = \mu W$. Write a MATLAB program that uses the `switch` structure to compute the force F. The program should accept as input the value of W and the type of materials.

42. The height and speed of a projectile (such as a thrown ball) launched with a speed of v_0 at an angle A to the horizontal are given by

$$h(t) = v_0 t \sin A - 0.5 g t^2$$
$$v(t) = \sqrt{v_0^2 - 2 v_0 g t \sin A + g^2 t^2}$$

where g is the acceleration due to gravity. The projectile will strike the ground when $h(t) = 0$, which gives the time to hit $t_{hit} = 2(v_0/g) \sin A$.

Use the `switch` structure to write a MATLAB program to compute either the maximum height reached by the projectile, the total horizontal distance traveled, or the time to hit. The program should accept as input the user's choice of which quantity to compute and the values of v_0, A, and g. Test the program for the case where $v_0 = 40$ m/s, $A = 30°$, and $g = 9.81$ m/s^2.

43. Use the `switch` structure to write a MATLAB program to compute how much money accumulates in a savings account in one year. The program should accept the following input: the initial amount of money deposited in the account; the frequency of interest compounding (monthly, quarterly, semiannually, or annually); and the interest rate. Run your program for a $1000 initial deposit for each case; use a 5 percent interest rate. Compare the amounts of money that accumulate for each case.

44. We often need to estimate the pressures and volumes of a gas in a container. The *van der Waals* equation is often used for this purpose. It is

$$P = \frac{RT}{\hat{V} - b} - \frac{a}{\hat{V}^2}$$

where the term b is a correction for the volume of the molecules, and the term $a/\hat{V}^2$ is a correction for molecular attractions. The gas constant is R, the *absolute* temperature is T, and the gas specific volume is $\hat{V}$. The value of R is the same for all gases; it is $R = 0.08206$ L-atm/mol-K. The values of a and b depend on the type of gas. Some values are given in the following table. Write a user-defined function using the switch structure that computes the pressure P on the basis of the van der Waals equation. The function's input arguments should be T, $\hat{V}$, and a string variable containing the name of a gas listed in the table. Test your function for chlorine (Cl_2) for $T = 300$ K and $\hat{V} = 20$ L/mol.

Gas	a (L^2-atm/mol^2)	b (L/mol)
Helium, He	0.0341	0.0237
Hydrogen, H_2	0.244	0.0266
Oxygen, O_2	1.36	0.0318
Chlorine, Cl_2	6.49	0.0562
Carbon dioxide, CO_2	3.59	0.0427

45. Using the program developed in Problem 16, write a program that uses the switch structure to compute the number of days in a year up to a given date, given the year, the month, and the day of the month.

5

Advanced Plotting and Model Building

In this chapter you will learn additional features to use to create a variety of two-dimensional plots, which are also called *xy plots,* and three-dimensional plots called *xyz plots,* or *surface* plots. Two-dimensional plots are discussed in Sections 5.1 through 5.3. Section 5.7 discusses three-dimensional plots. These plotting functions are described in the `graph2d` and `graph3d` help categories, so typing `help graph2d` or `help graph3d` will display a list of the relevant plotting functions.

An important application of plotting is *function discovery,* the technique for using data plots to obtain a mathematical function or "mathematical model" that describes the process that generated the data (Section 5.4). Another method for obtaining models is *regression,* covered in Sections 5.5 and 5.6.

5.1 xy Plotting Functions

The "anatomy" and nomenclature of a typical xy plot is shown in Figure 5.1–1, in which the plot of a data set and a curve generated from an equation appear. A plot can be made from measured data or from an equation. When data is plotted,

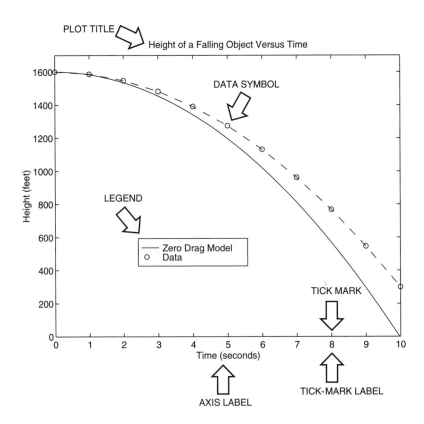

Figure 5.1–1 Nomenclature for a typical xy plot.

each data point is plotted with a *data symbol,* or *point marker,* such as the small circles shown in Figure 5.1–1. An exception to this rule would be when there are so many data points that the symbols would be too densely packed. In that case, the data points should be plotted with a dot. However, when the plot is generated from a function, data symbols must *never* be used! Lines between closely spaced points are always used to plot a function.

The MATLAB basic xy plotting function is plot(x,y) as we saw in Chapter 1. If x and y are vectors, a single curve is plotted with the x values on the abscissa and the y values on the ordinate. The xlabel and ylabel commands put labels on the abscissa and the ordinate, respectively. The syntax is xlabel('text'), where text is the text of the label. Note that you must enclose the label's text in single quotes. The syntax for ylabel is the same. The title command puts a title at the top of the plot. Its syntax is title ('text'), where text is the title's text.

The plot(x,y) function in MATLAB automatically selects a tick-mark spacing for each axis and places appropriate tick labels. This feature is called autoscaling. MATLAB also chooses limits for the x and y axes. The order of the xlabel, ylabel, and title commands does not matter, but we must place them *after* the `plot` command, either on separate lines using ellipses or on the same line separated by commas.

After the `plot` command is executed, the plot will appear in the Figure window. You can obtain a hard copy of the plot in one of several ways:

1. Use the menu system. Select **Print** on the **File** menu in the Figure window. Answer **OK** when you are prompted to continue the printing process.
2. Type `print` at the command line. This command sends the current plot directly to the printer.
3. Save the plot to a file to be printed later or imported into another application such as a word processor. You need to know something about graphics file formats to use this file properly. See the subsection **Exporting Figures** later in this section.

Type `help print` to obtain more information.

MATLAB assigns the output of the `plot` command to figure window number 1. When another `plot` command is executed, MATLAB overwrites the contents of the existing figure window with the new plot. Although you can keep more than one figure window active, we do not use this feature in this text.

When you have finished with the plot, close the figure window by selecting **Close** from the **File** menu in the figure window. If you do not close the window, it will not reappear when a new `plot` command is executed. However, the figure will still be updated.

`grid` and `axis` Commands

The `grid` command displays gridlines at the tick marks corresponding to the tick labels. You can use the `axis` command to override the MATLAB selections for the axis limits. The basic syntax is `axis([xmin xmax ymin ymax])`. This command sets the scaling for the *x*- and *y*-axes to the minimum and maximum values indicated. Note that, unlike an array, this command does not use commas to separate the values.

AXIS LIMITS

The `axis` command has the following variants:

- `axis square`, which selects the axes' limits so that the plot will be square.
- `axis equal`, which selects the scale factors and tick spacing to be the same on each axis. This variation makes `plot(sin(x),cos(x))` look like a circle, instead of an oval.
- `axis auto`, which returns the axis scaling to its default autoscaling mode in which the best axes limits are computed automatically.

Type `help axis` to see the full list of variants.

Plots of Complex Numbers

With only one argument, say, `plot(y)`, the `plot` function will plot the values in the vector `y` versus their indices 1, 2, 3, . . . , and so on. If `y` is complex, plot(y) plots the imaginary parts versus the real parts. Thus plot(y) in this case is equivalent to plot(real(y),imag(y)). This situation is the only time when the plot function handles the imaginary parts; in all other variants of the plot function, it ignores the imaginary parts. For example, the script file

```
z = 0.1 + 0.9i;
n = [0:0.01:10];
plot(z.^n),xlabel('Real'),ylabel('Imaginary')
```

generates a spiral plot.

The Function Plot Command `fplot`

MATLAB has a "smart" command for plotting functions. The `fplot` command automatically analyzes the function to be plotted and decides how many plotting points to use so that the plot will show all the features of the function. Its syntax is `fplot(function, [xmin xmax])`, where `function` is a function handle to the function to be plotted and `[xmin xmax]` specifies the minimum and maximum values of the independent variable. The range of the dependent variable can also be specified. In this case the syntax is `fplot(function, [xmin xmax ymin ymax])`.

For example, the session

```
>>f = @(x) (cos(tan(x)) - tan(sin(x)));
>>fplot(f,[1 2])
```

produces the plot shown in Figure 5.1–2. The `fplot` command automatically chooses enough plotting points to display all the variations in the function. We can achieve the same results using the `plot` command, but we need to know how many values to compute to generate the plot.

Another form is `[x,y] = fplot(function, limits)`, where `limits` may be either `[xmin xmax]` or `[xmin xmax ymin ymax]`. With this form the command returns the abscissa and ordinate values in the column vectors x and y, but no plot is produced. The returned values can then be used for other purposes, such as plotting multiple curves, which is the topic of the next section. Other commands can be used with the `fplot` command to enhance a plot's appearance, for example, the `title`, `xlabel`, and `ylabel` commands and the line type commands to be introduced in the next section.

Plotting Polynomials

We can plot polynomials more easily by using the `polyval` function. For example, to plot the polynomial $3x^5 + 2x^4 - 100x^3 + 2x^2 - 7x + 90$ over the range $-6 \leq x \leq 6$ with a spacing of 0.01, you type

```
>>x = [-6:0.01:6];
>>p = [3,2,-100,2,-7,90];
>>plot(x,polyval(p,x)),xlabel('x'),ylabel('p')
```

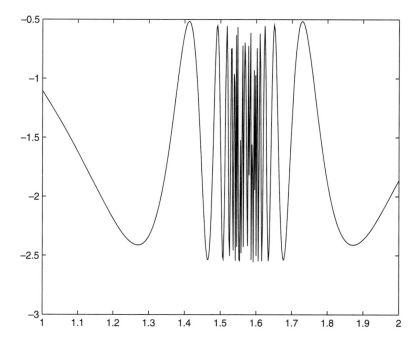

Figure 5.1–2 A plot generated with the `fplot` command.

Table 5.1–1 summarizes the xy plotting commands discussed in this section.

Table 5.1–1 Basic xy plotting commands

Command	Description
`axis([xmin xmax ymin ymax])`	Sets the minimum and maximum limits of the *x*- and *y*-axes.
`fplot(function,[xmin xmax])`	Performs intelligent plotting of functions, where `function` is a function handle that describes the function to be plotted and `[xmin xmax]` specifies the minimum and maximum values of the independent variable. The range of the dependent variable can also be specified. In this case the syntax is `fplot(function, [xmin xmax ymin ymax])`.
`grid`	Displays gridlines at the tick marks corresponding to the tick labels.
`plot(x,y)`	Generates a plot of the array `y` versus the array `x` on rectilinear axes.
`plot(y)`	Plots the values of `y` versus their indices if `y` is a vector. Plots the imaginary parts of `y` versus the real parts if `y` is a vector having complex values.
`print`	Prints the plot in the Figure window.
`title('text')`	Puts text in a title at the top of a plot.
`xlabel('text')`	Adds a text label to the *x*-axis (the abscissa).
`ylabel('text')`	Adds a text label to the *y*-axis (the ordinate).

Test Your Understanding

T5.1–1 Plot the equation $y = 0.4\sqrt{1.8x}$ for $0 \le x \le 35$ and $0 \le y \le 3.5$.

T5.1–2 Use the `fplot` command to investigate the function $\tan(\cos x) - \sin(\tan x)$ for $0 \le x \le 2\pi$. How many values of x are needed to obtain the same plot using the `plot` command? (Answer: 292 values.)

T5.1–3 Plot the imaginary part versus the real part of the function $(0.2 + 0.8i)^n$ for $0 \le n \le 20$. Choose enough points to obtain a smooth curve. Label each `axis` and put a title on the plot. Use the `axis` command to change the tick-label spacing.

Saving Figures

When you create a plot, the Figure window appears. This window has eight menus, which are discussed in detail in Section 5.4. The **File** menu is used for saving and printing the figure. You can save your figure in a format that can be opened during another MATLAB session or in a format that can be used by other applications.

To save a figure that can be opened in subsequent MATLAB sessions, save it in a figure file with the .fig file name extension. To do this, select **Save** from the Figure window **File** menu or click the **Save** button (the disk icon) on the toolbar. If this is the first time you are saving the file, the **Save As** dialog box appears. Make sure that the type is MATLAB Figure (*.fig). Specify the name you want assigned to the figure file. Click OK. You can also use the `saveas` command.

To open a figure file, select **Open** from the **File** menu or click the **Open** button (the opened folder icon) on the toolbar. Select the figure file you want to open and click OK. The figure file appears in a new figure window.

Exporting Figures

If you want to save the figure in a format that can be used by another application, such as the standard graphics file formats TIFF or EPS, perform these steps.

1. Select **Export Setup** from the **File** menu. This dialog provides options you can specify for the output file, such as the figure size, fonts, line size and style, and output format.
2. Select **Export** from the **Export Setup** dialog. A standard **Save As** dialog appears.
3. Select the format from the list of formats in the **Save As** type menu. This selects the format of the exported file and adds the standard file name extension given to files of that type.
4. Enter the name you want to give the file, less the extension.
5. Click **Save.**

You can also export the figure from the command line, by using the `print` command. See MATLAB help for more information about exporting figures in different formats.

You can also copy a figure to the clipboard and then paste it into another application:

1. Select **Copy Options** from the **Edit** menu of the Figure window. The **Copying Options** page of the **Preferences** dialog box appears.
2. Complete the fields on the **Copying Options** page and click **OK.**
3. Select **Copy Figure** from the **Edit** menu.

The figure is copied to the Windows clipboard and can be pasted into another application.

MATLAB also enables you to save figures in formats compatible with PowerPoint and MSWord. See the MATLAB help for more information.

The graphics functions covered in this section and in Section 5.3 can be placed in script files that can be reused to create similar plots. This feature gives them an advantage over the interactive plotting tools that are discussed in Section 5.3.

5.2 Additional Commands and Plot Types

MATLAB can create figures that contain an array of plots, called *subplots.* These are useful when you want to compare the same data plotted with different axis types, for example. The MATLAB `subplot` command creates such figures. We frequently need to plot more than one curve or data set on a single plot. Such a plot is called an *overlay plot.* This section describes these plots and several other types of plots.

SUBPLOT

Subplots

You can use the `subplot` command to obtain several smaller "subplots" in the same figure. The syntax is `subplot(m,n,p)`. This command divides the Figure window into an array of rectangular panes with *m* rows and *n* columns. The variable p tells MATLAB to place the output of the `plot` command following the `subplot` command into the *p*th pane. For example, `subplot(3,2,5)` creates an array of six panes, three panes deep and two panes across, and directs the next plot to appear in the fifth pane (in the bottom-left corner). The following script file created Figure 5.2–1, which shows the plots of the functions $y = e^{-1.2x} \sin(10x + 5)$ for $0 \le x \le 5$ and $y = |x^3 - 100|$ for $-6 \le x \le 6$.

```
x = [0:0.01:5];
y = exp(-1.2*x).*sin(10*x+5);
subplot(1,2,1)
plot(x,y),xlabel('x'),ylabel('y'),axis([0 5 -1 1])
x = [-6:0.01:6];
y = abs(x.^3-100);
subplot(1,2,2)
plot(x,y),xlabel('x'),ylabel('y'),axis([-6 6 0 350])
```

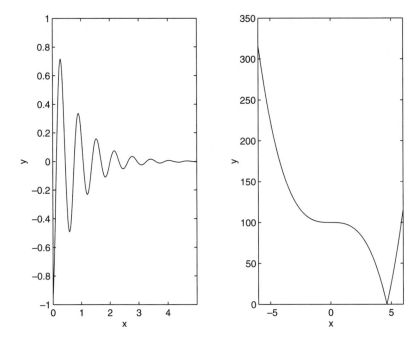

Figure 5.2–1 Application of the `subplot` command.

Test Your Understanding

T5.2–1 Pick a suitable spacing for t and v , and use the `subplot` command to plot the function $z = e^{-0.5t} \cos(20t - 6)$ for $0 \le t \le 8$ and the function $u = 6 \log_{10}(v^2 + 20)$ for $-8 \le v \le 8$. Label each axis.

Overlay Plots

OVERLAY PLOT

You can use the following variants of the MATLAB basic plotting functions `plot(x,y)` and `plot(y)` to create *overlay plots*:

■ `plot(A)` plots the columns of A versus their indices and generates *n* curves where A is a matrix with *m* rows and *n* columns.

■ `plot(x,A)` plots the matrix A versus the vector x, where x is either a row vector or column vector and A is a matrix with *m* rows and *n* columns. If the length of x is *m*, then each *column* of A is plotted versus the vector x. There will be as many curves as there are columns of A. If x has length *n*, then each *row* of A is plotted versus the vector x. There will be as many curves as there are rows of A.

■ `plot(A,x)` plots the vector x versus the matrix A. If the length of x is *m*, then x is plotted versus the *columns* of A. There will be as many curves as

Table 5.2–1 Specifiers for data markers, line types, and colors

Data markers[†]		Line types		Colors	
Dot (·)	·	Solid line	-	Black	k
Asterisk (*)	*	Dashed line	- -	Blue	b
Cross (×)	×	Dash-dotted line	-.	Cyan	c
Circle (o)	o	Dotted line	:	Green	g
Plus sign (+)	+			Magenta	m
Square (□)	s			Red	r
Diamond (◇)	d			White	w
Five-pointed star (★)	p			Yellow	y

[†]Other data markers are available. Search for "markers" in MATLAB help.

there are columns of A. If the length of x is *n*, then x is plotted versus the *rows* of A. There will be as many curves as there are rows of A.

- plot(A,B) plots the columns of the matrix B versus the columns of the matrix A.

Data Markers and Line Types

To plot the vector y versus the vector x and mark each point with a data marker, enclose the symbol for the marker in single quotes in the plot function. Table 5.2–1 shows the symbols for some of the available data markers. For example, to use a small circle, which is represented by the lowercase letter o, type plot(x,y, 'o'). This notation results in a plot like the one on the left in Figure 5.2–2. To connect each data marker with a straight line, we must plot the data twice, by typing plot(x,y,x,y,'o'). See the plot on the right in Figure 5.2–2.

Suppose we have two curves or data sets stored in the vectors x, y, u, and v. To plot y versus x and v versus u on the same plot, type plot(x,y,u,v). Both sets will be plotted with a solid line, which is the default line style. To distinguish the sets, we can plot them with different line types. To plot y versus x with a solid line and u versus v with a dashed line, type plot(x,y,u,v,'− −'), where the symbols '− −' represent a dashed line. Table 5.2–1 gives the symbols for other line types. To plot y versus x with asterisks (*) connected with a dotted line, you must plot the data twice by typing plot(x,y,'*',x,y,':').

You can obtain symbols and lines of different colors by using the color symbols shown in Table 5.2–1. The color symbol can be combined with the data-marker symbol and the line-type symbol. For example, to plot y versus x with green asterisks (*) connected with a red dashed line, you must plot the data twice by typing plot(x,y,'g*',x,y,'r− −'). (Do not use colors if you are going to print the plot on a black-and-white printer.)

Labeling Curves and Data

When more than one curve or data set is plotted on a graph, we must distinguish between them. If we use different data symbols or different line types, then we must

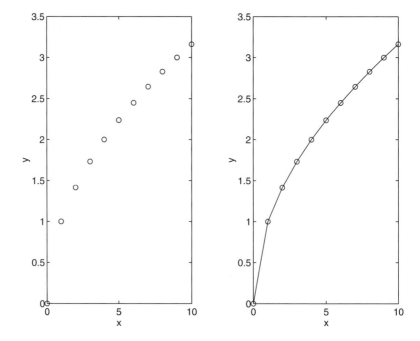

Figure 5.2–2 Use of data markers.

either provide a legend or place a label next to each curve. To create a legend, use the `legend` command. The basic form of this command is `legend` (`'string1'`,`'string2'`), where `string1` and `string2` are text strings of your choice. The `legend` command automatically obtains from the plot the line type used for each data set and displays a sample of this line type in the legend box next to the string you selected. The following script file produced the plot in Figure 5.2–3.

```
x = [0:0.01:2];
y = sinh(x);
z = tanh(x);
plot(x,y,x,z,'--'),xlabel('x'),...
ylabel('Hyperbolic Sine and Tangent'),...
   legend('sinh(x)','tanh(x)')
```

The `legend` command must be placed somewhere after the `plot` command. When the plot appears in the Figure window, use the mouse to position the legend box. (Hold down the left button on the mouse to move the box.)

Another way to distinguish curves is to place a label next to each. The label can be generated with either the `gtext` command, which lets you place the label using the mouse, or with the `text` command, which requires you to specify the coordinates of the label. The syntax of the `gtext` command is `gtext('string')`, where `string` is a text string that specifies the label of

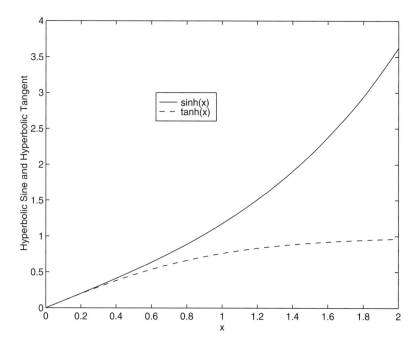

Figure 5.2–3 Application of the `legend` command.

your choice. When this command is executed, MATLAB waits for a mouse button or a key to be pressed while the mouse pointer is within the Figure window; the label is placed at that position of the mouse pointer. You may use more than one `gtext` command for a given plot. The `text` command, `text(x,y,'string')`, adds a text string to the plot at the location specified by the coordinates `x,y`. These coordinates are in the same units as the plot's data. Of course, finding the proper coordinates to use with the `text` command usually requires some trial and error.

The `hold` Command

The `hold` command creates a plot that needs two or more `plot` commands. Suppose we wanted to plot $y_2 = 4 + e^{-x} \cos 6x$ versus $y_1 = 3 + e^{-x} \sin 6x$, $-1 \leq x \leq 1$ on the same plot with $z = (0.1 + 0.9i)^n$, where $0 \leq n \leq 10$. This plot requires two plot commands. The script file to create this plot using the `hold` command follows.

```
x = [-1:0.01:1];
y1 = 3+exp(-x).*sin(6*x);
y2 = 4+exp(-x).*cos(6*x);
plot((0.1+0.9i).^[0:0.01:10]),hold,plot(y1,y2),...
gtext('y2 versus y1'),gtext('Imag(z) versus Real(z)')
```

Figure 5.2–4 shows the result. When more than one `plot` command is used, do not place any of the `gtext` commands before any `plot` command. Because the

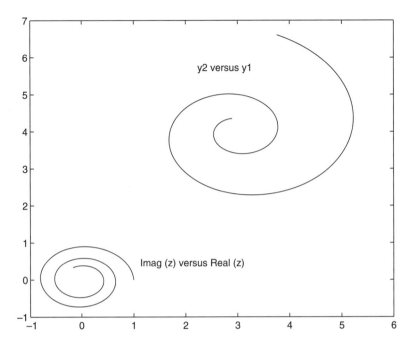

y2 versus y1

Imag (z) versus Real (z)

Figure 5.2–4 Application of the hold command.

scaling changes as each plot command is executed, the label placed by the gtext command might end up in the wrong position. Table 5.2–2 summarizes the plot enhancement introduced in this section.

Table 5.2–2 Plot enhancement commands

Command	Description
gtext ('text')	Places the string text in the Figure window at a point specified by the mouse.
hold	Freezes the current plot for subsequent graphics commands.
legend ('leg1','leg2',...)	Creates a legend using the strings leg1, leg2, and so on and specifies its placement with the mouse.
plot (x,y,u,v)	Plots, on rectilinear axes, four arrays: y versus x and v versus u.
plot (x,y,'type')	Plots the array y versus the array x on rectilinear axes, using the line type, data marker, and colors specified in the string type. See Table 5.2–1.
plot (A)	Plots the columns of the $m \times n$ array A versus their indices and generates n curves.
plot (P,Q)	Plots array Q versus array P. See the text for a description of the possible variants involving vectors and/or matrices: plot (x,A), plot (A,x), and plot (A,B).
subplot (m,n,p)	Splits the Figure window into an array of subwindows with m rows and n columns and directs the subsequent plotting commands to the pth subwindow.
text (x,y,'text')	Places the string text in the Figure window at a point specified by coordinates x, y.

Test Your Understanding

T5.2–2 Plot the following two data sets on the same plot. For each set, $x = 0, 1, 2, 3, 4, 5$. Use a different data marker for each set. Connect the markers for the first set with solid lines. Connect the markers for the second set with dashed lines. Use a legend, and label the plot axes appropriately. The first set is $y = 11, 13, 8, 7, 5, 9$. The second set is $y = 2, 4, 5, 3, 2, 4$.

T5.2–3 Plot $y = \cosh x$ and $y = 0.5e^x$ on the same plot for $0 \le x \le 2$. Use different line types and a legend to distinguish the curves. Label the plot axes appropriately.

T5.2–4 Plot $y = \sinh x$ and $y = 0.5e^x$ on the same plot for $0 \le x \le 2$. Use a solid line type for each, the `gtext` command to label the sinh x curve, and the `text` command to label the $0.5e^x$ curve. Label the plot axes appropriately.

T5.2–5 Use the `hold` command and the `plot` command twice to plot $y = \sin x$ and $y = x - x^3/3$ on the same plot for $0 \le x \le 1$. Use a solid line type for each and use the `gtext` command to label each curve. Label the plot axes appropriately.

Annotating Plots

You can create text, titles, and labels that contain mathematical symbols, Greek letters, and other effects such as italics. The features are based on the T_EX typesetting language. For more information, including a list of the available characters, search the online help for "text properties."

You can create a title having the mathematical function $Ae^{-t/\tau} \sin(\omega t)$ by typing

```
>>title('{\it Ae}^{-{\it t/\tau}}\sin({\it \omega t})')
```

The backslash character \ precedes all T_EX character sequences. Thus the strings `\tau` and `\omega` represent the Greek letters τ and ω. Superscripts are created by typing ^; subscripts are created by typing _. To set multiple characters as superscripts or subscripts, enclose them in braces. For example, type `x_{13}` to produce x_{13}. In mathematical text variables are usually set in italic, and functions, like sin, are set in roman type. To set a character, say, x, in italic using the T_EX commands, you type `{\it x}`.

Logarithmic Plots

Logarithmic scales—abbreviated log scales—are widely used (1) to represent a data set that covers a wide range of values and (2) to identify certain trends in data. Certain types of functional relationships appear as straight lines when plotted using a log scale. This method makes it easier to identify the function. A *log-log* plot has log scales on both axes. A *semilog* plot has a log scale on only one axis.

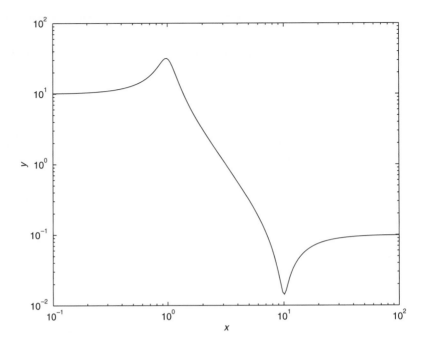

Figure 5.2–5 Example of a log-log plot. Note the wide range of values of both x and y.

Figure 5.2–5 shows a log-log plot of the function

$$y = \sqrt{\frac{100(1 - 0.01x^2)^2 + 0.02x^2}{(1 - x^2)^2 + 0.1x^2}} \qquad 0.1 \le x \le 100$$

Because of the wide range in values on both the abscissa and ordinate, rectilinear scales would not reveal the important features. The following program produced Figure 5.2–5.

```
x = logspace(-1, 2, 500); u = x.^2;
num = 100*(1-0.01*u).^2 + 0.02*u;
den = (1-u).^2 + 0.1*u;
y = sqrt(num./den);
loglog(x,y), xlabel('x'), ylabel('y')
```

It is important to remember the following points when using log scales:

1. You cannot plot negative numbers on a log scale, because the logarithm of a negative number is not defined as a real number.
2. You cannot plot the number 0 on a log scale, because $\log_{10} 0 = \ln 0 = -\infty$. You must choose an appropriately small number as the lower limit on the plot.
3. The tick-mark labels on a log scale are the actual values being plotted; they are not the logarithms of the numbers. For example, the range of x values in the plot in Figure 5.2–5 is from $10^{-1} = 0.1$ to $10^2 = 100$.

Table 5.2–3 Specialized plot commands

Command	Description
bar(x,y)	Creates a bar chart of y versus x.
loglog(x,y)	Produces a log-log plot of y versus x.
plotyy(x1,y1,x2,y2)	Produces a plot with two y-axes, y1 on the left and y2 on the right.
polar(theta,r,'type')	Produces a polar plot from the polar coordinates theta and r, using the line type, data marker, and colors specified in the string type.
semilogx(x,y)	Produces a semilog plot of y versus x with logarithmic abscissa scale.
semilogy(x,y)	Produces a semilog plot of y versus x with logarithmic ordinate scale.
stairs(x,y)	Produces a stairs plot of y versus x.
stem(x,y)	Produces a stem plot of y versus x.

MATLAB has three commands for generating plots having log scales. The appropriate command depends on which axis must have a log scale. Follow these rules:

1. Use the loglog(x,y) command to have both scales logarithmic.

2. Use the semilogx(x,y) command to have the *x* scale logarithmic and the *y* scale rectilinear.

3. Use the semilogy(x,y) command to have the *y* scale logarithmic and the *x* scale rectilinear.

Table 5.2–3 summarizes these functions. For other 2D plot types, type help specgraph.

We can plot multiple curves with these commands just as with the plot command. In addition, we can use the other commands, such as grid, xlabel, and axis, in the same manner.

Stem, Stairs, and Bar Plots

MATLAB has several other plot types that are related to xy plots. These include the stem, stairs, and bar plots. Their syntax is very simple; namely, stem(x,y), stairs(x,y), and bar(x,y). See Table 5.2–3.

Separate *y*-Axes

The plotyy function generates a graph with two *y*-axes. The syntax plotyy (x1,y1,x2,y2) plots y1 versus x1 with *y*-axis labeling on the left, and plots y2 versus x2 with *y*-axis labeling on the right. The syntax plotyy(x1,y1,x2, y2,'type1','type2') generates a 'type1' plot of y1 versus x1 with *y*-axis labeling on the left, and generates a 'type2' plot of y2 versus x2 with *y*-axis labeling on the right. For example, plotyy(x1,y1,x2,y2,'plot','stem') uses `plot(x1,y1)` to generate a plot for the left axis, and `stem(x2,y2)` to generate a plot for the right axis. To see other variations of the `plotyy` function, type `help plotyy`.

Polar Plots

Polar plots are two-dimensional plots made using polar coordinates. If the polar coordinates are (θ, r), where θ is the angular coordinate and r is the radial coordinate of a point, then the command `polar(theta,r)` will produce the polar plot. A grid is automatically overlaid on a polar plot. This grid consists of concentric circles and radial lines every 30°. The `title` and `gtext` commands can be used to place a title and text. The variant command `polar(theta,r,'type')` can be used to specify the line type or data marker, just as with the `plot` command.

EXAMPLE 5.2–1   Plotting Orbits

The equation

$$r = \frac{p}{1 - \epsilon \cos \theta}$$

describes the polar coordinates of an orbit measured from one of the orbit's two focal points. For objects in orbit around the sun, the sun is at one of the focal points. Thus r is the distance of the object from the sun. The parameters p and ϵ determine the size of the orbit and its eccentricity, respectively. Obtain the polar plot that represents an orbit having $\epsilon = 0.5$ and $p = 2$ AU (AU stands for "astronomical unit"; 1 AU is the mean distance from the sun to Earth). How far away does the orbiting object get from the sun? How close does it approach Earth's orbit?

■ **Solution**

Figure 5.2–6 shows the polar plot of the orbit. The plot was generated by the following session.

```
>>theta = [0:pi/90:2*pi];
>>r = 2./(1-0.5*cos(theta));
>>polar(theta,r),title('Orbital Eccentricity = 0.5')
```

The sun is at the origin, and the plot's concentric circular grid enables us to determine that the closest and farthest distances the object is from the sun are approximately 1.3 and 4 AU. Earth's orbit, which is nearly circular, is represented by the innermost circle. Thus the closest the object gets to Earth's orbit is approximately 0.3 AU. The radial gridlines allow us to determine that when $\theta = 90°$ and 270°, the object is 2 AU from the sun.

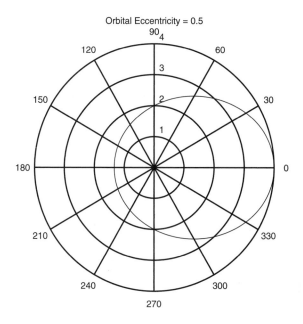

Figure 5.2–6 A polar plot showing an orbit having an eccentricity of 0.5.

Test Your Understanding

T5.2–6 Obtain the plots shown in Figure 5.2–7. The power function is $y = 2x^{-0.5}$, and the exponential function is $y = 10^{1-x}$.

T5.2–7 Plot the function $y = 8x^3$ for $-1 \le x \le 1$ with a tick spacing of 0.25 on the x-axis and 2 on the y-axis.

T5.2–8 The *spiral of Archimedes* is described by the polar coordinates (θ, r), where $r = a\theta$. Obtain a polar plot of this spiral for $0 \le \theta \le 4\pi$, with the parameter $a = 2$.

Publishing Reports Containing Graphics

MATLAB 7 provides the `publish` function for creating reports, which may have embedded graphics. Reports generated by the `publish` function may be exported to a variety of common formats including HTML (Hyper Text Markup Language), which is used for Web-based reports, MS Word, PowerPoint, and LATEX. To publish a report, do the following.

1. Open the Editor, type in the m-file that forms the basis of the report, and save it. Use the double percent character (%%) to indicate a section heading in the report. This character marks the beginning of a new cell, which is a group of commands. (Such a cell should not be confused with the cell array data type covered in Section 2.7.) Enter any blank lines you wish to

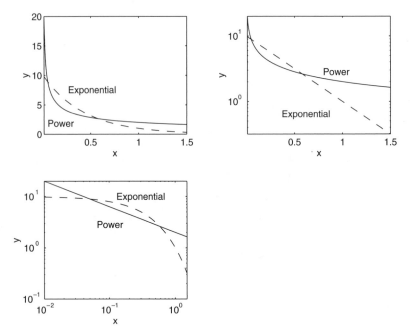

Figure 5.2–7 The power function $y = 2x^{-0.5}$ and the exponential function $y = 10^{1-x}$.

appear in the report. Consider, as a very simple example, the following sample file `polyplot.m`.

```
%% Example of Report Publishing:
% Plotting the cubic y = x^3 - 6x^2 + 10x+4.

%% Create the independent variable.
x = linspace(0, 4, 300); % Use 300 points between 0 and 4.
%% Define the cubic from its coefficients.
P = [1, -6, 10, 4]; % p contains the coefficients.

%% Plot the cubic
plot (x,polyval(p,x)), xlabel('x'),ylabel('y')
```

2. Run the file to check it for errors. (To do this for a larger file, you may use the cell mode of the Debugger to execute its each cell one at a time; see Section 4.6)

3. Use the `publish` and `open` functions to create the report in the desired format. Using our sample file, we can obtain a report in HTML format by typing

```
>>publish ('polyplot','html')
>>open html/polyplot.html
```

You should see a report like the one shown in Figure 5.2–8.

Example of Report Publishing:

Plotting the cubic y = x^3 − 6 x^2 + 10x+ 4 .

Contents

- Create the independent variable.
- Define the cubic.
- Plot the cubic.

Create the independent variable.

```
x = linspace(0, 4, 300); % Use 300 points between 0 and 4.
```

Define the cubic.

```
p = [1, -6, 10, 4];  % p contains the coefficients.
```

Plot the cubic.

```
plot(x,polyval(p,x)),xlabel('x'),ylabel('y')
```

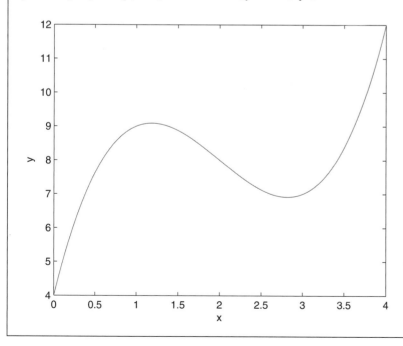

Figure 5.2–8 A sample report published from MATLAB.

Instead of using the `publish` and `open` functions, you may select **Publish to HTML** from the **File** menu in the Editor window. To publish to another format, select instead **Publish to** and then choose the desired format from the menu.

Once it is published in HTML, you may click on a section heading in the Contents to go to that section. This is useful for larger reports.

If you want the equation to look professionally typeset, you may edit the resulting report in the appropriate editor (say, MS Word, or LaTeX). For example, to set the cubic polynomial in the resulting LaTeX file, use the commands presented earlier in this section to replace the equation in the second line of the report with

```
y = {\it x}^3 - 6{\it x}^2 + 10{\it x} + 4
```

5.3 Interactive Plotting in MATLAB

The interactive plotting environment in MATLAB is a set of tools for:

- Creating different types of graphs,
- Selecting variables to plot directly from the Workspace Browser,
- Creating and editing subplots,
- Adding annotations such as lines, arrows, text, rectangles, and ellipses, and
- Editing properties of graphics objects, such as their color, line weight, and font.

The Plot Tools interface includes the following three panels associated with a given figure.

- **The Figure Palette:** Use this to create and arrange subplots, to view and plot workspace variables, and to add annotations.
- **The Plot Browser:** Use this to select and control the visibility of the axes or graphics objects plotted in the figure, and to add data for plotting.
- **The Property Editor:** Use this to set basic properties of the selected object and to obtain access to all properties through the Property Inspector.

The Figure Window

When you create a plot, the Figure window appears with the Figure toolbar visible (see Figure 5.3–1). This window has eight menus.

The File Menu   The **File** menu is used for saving and printing the figure. This menu was discussed in Section 5.1 under **Saving Figures** and **Exporting Figures.**

The Edit Menu   You can use the **Edit** menu to cut, copy, and paste items, such as legend or title text, that appear in the figure. Click on **Figure Properties** to open the Property Editor—Figure dialog box to change certain properties of the figure.

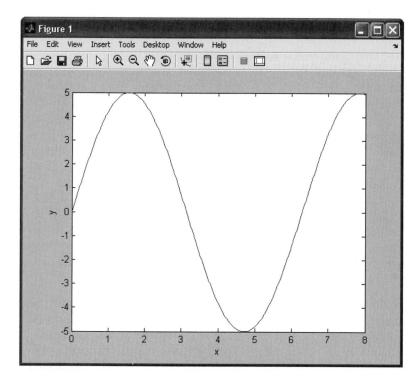

Figure 5.3–1 The Figure window with the Figure toolbar displayed.

Three items on the **Edit** menu are very useful for editing the figure. Clicking the **Axes Properties** item brings up the Property Editor—Axes dialog box. Double-clicking on any axis also brings up this box. You can change the scale type (linear, log, etc.), the labels, and the tick marks by selecting the tab for the desired axis or the font to be edited.

The **Current Object Properties** item enables you to change the properties of an object in the figure. To do this, first click on the object, such as a plotted line, then click on **Current Object Properties** in the **Edit** menu. You will see the Property Editor—Lineseries dialog box that lets you change properties such as line weight and color, data-marker type, and plot type.

Clicking on any text, such as that placed with the `title`, `xlabel`, `ylabel`, `legend`, or `gtext` commands, then selecting **Current Object Properties** in the **Edit** menu brings up the Property Editor—Text dialog box, which enables you to edit the text.

The View Menu The items on the **View** menu are the three toolbars (**Figure Toolbar, Plot Edit Toolbar,** and **Camera Toolbar**), the **Figure Palette,** the **Plot Browser,** and the **Property Editor.** These will be discussed later in this section.

The Insert Menu The **Insert** menu enables you to insert labels, legends, titles, text, and drawing objects, rather than using the relevant commands from the

Command window. To insert a label on the *y*-axis, for example, click on the **Y Label** item on the menu; a box will appear on the *y*-axis. Type the label in this box, and then click outside the box to finish.

The **Insert** menu also enables you to insert arrows, lines, text, rectangles, and ellipses in the figure. To insert an arrow, for example, click on the **Arrow** item; the mouse cursor changes to a crosshair style. Then click the mouse button, and move the cursor to create the arrow. The arrowhead will appear at the point where you release the mouse button. Be sure to add arrows, lines, and other annotations only after you are finished moving or resizing your axes, because these objects are not anchored to the axes. (They can be anchored to the plot by *pinning;* see the MATLAB help.)

To delete or move a line or arrow, click on it, then press the **Delete** key to delete it, or press the mouse button and move it to the desired location. The **Axes** item lets you use the mouse to place a new set of axes within the existing plot. Click on the new axes, and a box will surround them. Any further plot commands issued from the Command window will direct the output to these axes.

The **Light** item applies to 3D plots.

The Tools Menu   The **Tools** menu includes items for adjusting the view (by zooming and panning) and the alignment of objects on the plot. The **Edit Plot** item starts the plot editing mode, which can also be started by clicking on the northwest-facing arrow on the Figure toolbar. The **Tools** menu also gives access to the **Data Cursor,** which is discussed later in this section. The last two items, **Basic Fitting** and **Data Statistics,** will be discussed in Sections 5.6 and 6.1, respectively.

Other Menus   The **Desktop** menu enables you to dock the Figure window within the desktop. The **Window** menu lets you switch between the Command window and any other Figure windows. The **Help** menu accesses the general MATLAB Help System, as well as help features specific to plotting.

There are three toolbars available in the Figure window: the Figure toolbar, the Plot Edit toolbar, and the Camera toolbar. The **View** menu lets you select which ones you want to appear. We will discuss the Figure toolbar and the Plot Edit toolbar in this section. The Camera toolbar is useful for 3D plots, which are discussed at the end of this chapter.

The Figure Toolbar

To activate the Figure toolbar, select it from the **View** menu (see Figure 5.3–1). The four left-most buttons are for opening, saving, and printing the figure. Clicking on the northwest-facing arrow button toggles the plot edit mode on and off.

The **Zoom-in** and **Zoom-out** buttons let you obtain a close-up or faraway view of the figure. The **Pan** and **Rotate 3D** buttons are used for 3D plots.

The **Data Cursor** button enables you to read data directly from a graph by displaying the values of points you select on plotted lines, surfaces, images, and so on.

The **Insert Colorbar** button inserts a color map strip in the graph and is useful for 3D surface plots. The **Insert Legend** button enables you to insert a legend

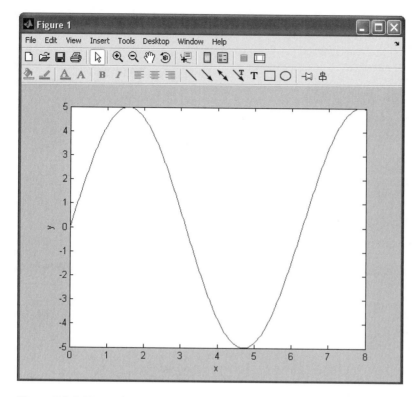

Figure 5.3–2 The Figure window with the Figure and Plot Edit toolbars displayed.

in the plot. The last two buttons hide or show the plot tools and dock the figure if it is undocked.

The Plot Edit Toolbar

Once a plot is in the window, you can enable plot editing by clicking on the northwest-facing arrow on the Figure toolbar. Then double-click on an axis, a plotted line, or a label to activate the appropriate property editor. Select **Plot Edit toolbar** from the **View** menu (see Figure 5.3–2). To add text that is not a label, title, or legend, click the button labeled **T,** move the cursor to the desired location for the text, click the mouse button, and type the text. When finished, click outside the text box and note that the nine left-most buttons become highlighted and available. These enable you to modify the color, font, and other attributes of the text.

To insert arrows, lines, rectangles, and ellipses, click on the appropriate button and follow the instructions given previously for the **Insert** menu.

The Plot Tools

Once a figure has been created you can display any or all of the three Plot Tools (Figure Palette, Plot Browser, and Property Editor) by selecting them from the

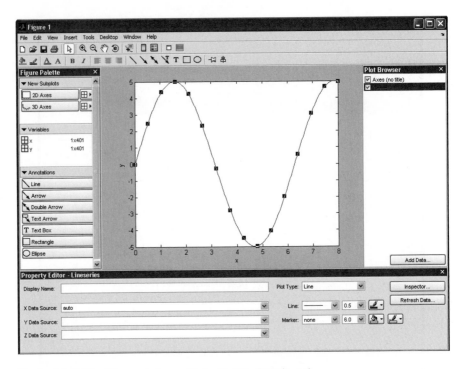

Figure 5.3–3 The Figure window with the Plot Tools activated.

View menu. You can also start the environment by first creating a plot and then clicking on the **Show Plot Tools** icon in the Figure toolbar (see Figure 5.3–3), or by creating a figure with the plotting tools attached by using the `plottools` command. Remove the tools by clicking on the **Hide Tools** icon.

Figure 5.3–3 shows the result of clicking on the plotted line after clicking the **Show Plot Tools** icon. The plotting interface then displays the Property Editor—Lineseries.

The Figure Palette

The Figure Palette contains three panels, which are selected and expanded by clicking the appropriate button. Click on the grid icon in the **New Subplots** panel to display the selector grid that enables you to specify the layout of the subplots. In the Variables panel you can select a graphics function to plot the variable by selecting the variable and right-clicking to display the context menu. This menu contains a list of possible plot types based on the type of variable you select. You can also drag the variable into an axes set and MATLAB will select an appropriate plot type.

Selecting **More Plots** from the context menu activates the Plot Catalog tool, which provides access to most of the plotting functions. After selecting a plot category, and a plot type from that category, you will see its description in the right-most display. Type the name of one or more variables in the Plotted Variables field, separated by commas, and they will be passed to the selected plotting function as arguments. You can also type a MATLAB expression that uses any workspace variables shown in the Figure Palette.

Click on the **Annotations** panel to display a menu of objects such as lines, arrows, etc. Click on the desired object and use the mouse to position and size it.

The Plot Browser

The Plot Browser provides a legend of all the graphs in the figure. For example, if you plot an array with multiple rows and columns, the Browser lists each axis and the objects (lines, surfaces, etc.) used to create the graph. To set the properties of an individual line, double-click on the line. Its properties are displayed in the Property Editor—Lineseries box, which opens on the bottom of the figure.

If you select a line in the graph, the corresponding entry in the Plot Browser is highlighted, indicating which column in the variable produced the line. The check box next to each item in the Browser controls the object's visibility. For example, if you want to plot only certain columns of data, you can uncheck the columns not wanted. The graph updates as you uncheck each box and rescales the axes as required.

The Property Editor

The Property Editor enables you to access a subset of the selected object's properties. When no object is selected, the Property Editor displays the figure's properties. There are several ways to display the Property Editor.

1. Double-click an object when plot edit mode is enabled.
2. Select an object and right-click to display its context menu, then select **Properties.**
3. Select **Property Editor** from the **View** menu.
4. Use the `propertyeditor` command.

The Property Editor enables you to change the most commonly used object properties. If you want to access all object properties, use the Property Inspector. To display the Property Inspector, click the **Inspector** button on any Property Editor panel. Use of this feature requires detailed knowledge of object properties and handle graphics, and thus will not be covered here.

Recreating Graphs from M-Files

Once your graph is finished, you can generate MATLAB code to reproduce the graph by selecting **Generate M-File** from the **File** menu. MATLAB creates a function that recreates the graph and opens the generated M-File in the editor. This feature is particularly useful for capturing property settings and other modifications made in the plot editor. You can also use the `makemcode` function.

Adding Data to Axes

The Plot Browser provides the mechanism by which you add data to axes. The procedure is as follows:

1. Select a 2D or 3D axis from the New Subplots subpanel.
2. After creating the axis, select it in the Plot Browser panel to enable the **Add Data** button at the bottom of the panel.
3. Click the **Add Data** button to display the Add Data to Axes dialog box. The Add Data to Axes dialog enables you to select a plot type and specify the workspace variables to pass to the plotting function. You can also specify a MATLAB expression, which is evaluated to produce the data to plot.

5.4 Function Discovery

Function discovery is the process of finding, or "discovering," a function that can describe a particular set of data. The following three function types can often describe physical phenomena.

1. The *linear* function: $y(x) = mx + b$. Note that $y(0) = b$.
2. The *power* function: $y(x) = bx^m$. Note that $y(0) = 0$ if $m \geq 0$, and $y(0) = \infty$ if $m < 0$.
3. The *exponential* function: $y(x) = b(10)^{mx}$ or its equivalent form $y = be^{mx}$, where e is the base of the natural logarithm ($\ln e = 1$). Note that $y(0) = b$ for both forms.

Each function gives a straight line when plotted using a specific set of axes:

1. The linear function $y = mx + b$ gives a straight line when plotted on rectilinear axes. Its slope is m and its intercept is b.
2. The power function $y = bx^m$ gives a straight line when plotted on log-log axes.
3. The exponential function $y = b(10)^{mx}$ and its equivalent form $y = be^{mx}$ give a straight line when plotted on a semilog plot whose y-axis is logarithmic.

The last two properties are illustrated in Figure 5.2–7, which shows the power function $y = 2x^{-0.5}$ and the exponential function $y = 10^{1-x}$. We look for a straight line on the plot because it is relatively easy to recognize, and therefore we can easily tell whether the function will fit the data well.

Use the following procedure to find a function that describes a given set of data. We assume that one of the function types (linear, exponential, or power) can describe the data.

1. Examine the data near the origin. The exponential function can never pass through the origin (unless of course $b = 0$, which is a trivial case). (See Figure 5.4–1 for examples with $b = 1$.) The linear function can pass through the origin only if $b = 0$. The power function can pass through the origin but only if $m > 0$. (See Figure 5.4–2 for examples with $b = 1$.)

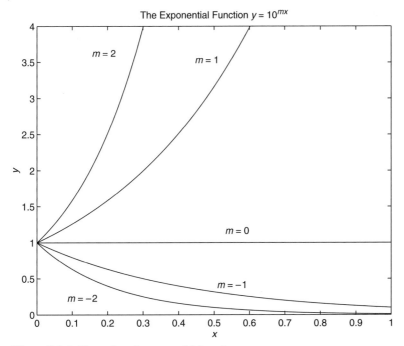

Figure 5.4–1 Examples of exponential functions.

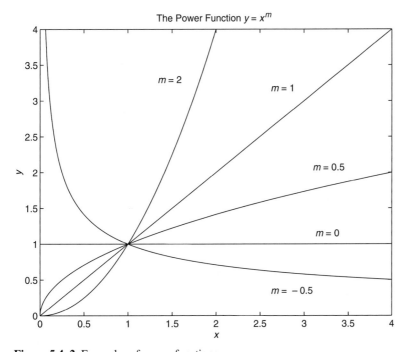

Figure 5.4–2 Examples of power functions.

2. Plot the data using rectilinear scales. If it forms a straight line, then it can be represented by the linear function and you are finished. Otherwise, if you have data at $x = 0$, then
    a. If $y(0) = 0$, try the power function.
    b. If $y(0) \neq 0$, try the exponential function.

    If data is not given for $x = 0$, proceed to step 3.

3. If you suspect a power function, plot the data using log-log scales. Only a power function will form a straight line on a log-log plot. If you suspect an exponential function, plot the data using the semilog scales. Only an exponential function will form a straight line on a semilog plot.

4. In function discovery applications, we use the log-log and semilog plots *only* to identify the function type, but not to find the coefficients b and m. The reason is that it is difficult to interpolate on log scales.

We can find the values of b and m with the MATLAB `polyfit` function. This function finds the coefficients of a polynomial of specified degree n that best fits the data, in the so-called least-squares sense. The syntax appears in Table 5.4–1. The least-squares method is presented in Sections 2.5 and 5.5.

Because we are assuming that our data will form a straight line on either a rectilinear, semilog, or log-log plot, we are interested only in a polynomial that corresponds to a straight line; that is, a first-degree polynomial, which we will denote as $w = p_1 z + p_2$. Thus, referring to Table 5.4–1, we see that the vector p will be $[p_1, p_2]$ if n is 1. This polynomial has a different interpretation in each of the three cases:

■ **The linear function:** $y = mx + b$. In this case the variables w and z in the polynomial $w = p_1 z + p_2$ are the original data variables x and y, and we can find the linear function that fits the data by typing `p = polyfit(x,y,1)`. The first element p_1 of the vector p will be m, and the second element p_2 will be b.

■ **The power function:** $y = bx^m$. In this case $\log_{10} y = m \log_{10} x + \log_{10} b$, which has the form $w = p_1 z + p_2$, where the polynomial variables w and z are related to the original data variables x and y by $w = \log_{10} y$ and $z = \log_{10} x$. Thus we can find the power function that fits the data by typing `p = polyfit(log10(x), log10(y),1)`. The first element p_1 of the vector p will be m, and the second element p_2 will be $\log_{10} b$. We can find b from $b = 10^{p_2}$.

Table 5.4–1 The `polyfit` function

Command	Description
p = polyfit(x,y,n)	Fits a polynomial of degree n to data described by the vectors x and y, where x is the independent variable. Returns a row vector p of length $n + 1$ that contains the polynomial coefficients in order of descending powers.

■ **The exponential function:** $y = b(10)^{mx}$. In this case $\log_{10}y = mx + \log_{10}b$, which has the form $w = p_1z + p_2$, where the polynomial variables w and z are related to the original data variables x and y by $w = \log_{10}y$ and $z = x$. Thus we can find the exponential function that fits the data by typing p = polyfit(x, log10(y),1). The first element p_1 of the vector p will be m, and the second element p_2 will be $\log_{10}b$. We can find b from $b = 10^{p_2}$.

Temperature Dynamics

EXAMPLE 5.4–1

The temperature of coffee cooling in a porcelain mug at room temperature (68°F) was measured at various times. The data follows.

Time t (sec)	Temperature T (°F)
0	145
620	130
2266	103
3482	90

Develop a model of the coffee's temperature as a function of time and use the model to estimate how long it took the temperature to reach 120°F.

■ Solution

Because $T(0)$ is finite but nonzero, the power function cannot describe this data, so we do not bother to plot the data on log-log axes. Common sense tells us that the coffee will cool and its temperature will eventually equal the room temperature. So we subtract the room temperature from the data and plot the relative temperature, $T - 68$, versus time. If the relative temperature is a linear function of time, the model is $T - 68 = mt + b$. If the relative temperature is an exponential function of time, the model is $T - 68 = b(10)^{mt}$. Figure 5.4–3 shows the plots used to solve the problem. The following MATLAB script file generates the top two plots. The time data is entered in the array time, and the temperature data is entered in temp.

```
% Enter the data.
time = [0,620,2266,3482];
temp = [145,130,103,90];
% Subtract the room temperature.
temp = temp - 68;
% Plot the data on rectilinear scales.
subplot(2,2,1)
plot(time,temp,time,temp,'o'),xlabel('Time (sec)'),...
   ylabel('Relative Temperature (deg F)')
%
% Plot the data on semilog scales.
subplot(2,2,2)
semilogy(time,temp,time,temp,'o'),xlabel('Time (sec)'),...
   ylabel('Relative Temperature (deg F)')
```

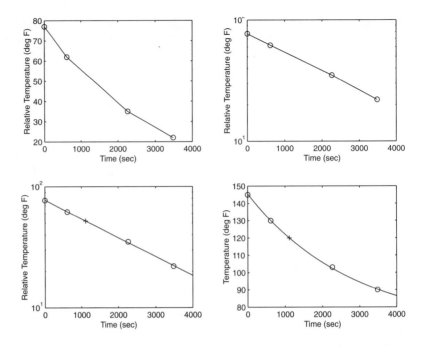

Figure 5.4–3 Temperature of a cooling cup of coffee, plotted on various coordinates.

The data forms a straight line on the semilog plot only (the top right plot). Thus it can be described with the exponential function $T = 68 + b(10)^{mt}$. Using the `polyfit` command, the following lines can be added to the script file.

```
% Fit a straight line to the transformed data.
p = polyfit(time,log10(temp),1);
m = p(1)
b = 10^p(2)
```

The computed values are $m = -1.5557 \times 10^{-4}$ and $b = 77.4469$. Thus our derived model is $T = 68 + b(10)^{mt}$. To estimate how long it will take for the coffee to cool to 120°F, we must solve the equation $120 = 68 + b(10)^{mt}$ for t. The solution is $t = [(\log_{10}(120-68)-\log_{10}b)]/m$. The MATLAB command for this calculation is shown in the following script file, which is a continuation of the previous script and produces the bottom two subplots shown in Figure 5.4–3.

```
% Compute the time to reach 120 degrees.
t_120 = (log10(120-68)-log10(b))/m
% Show the derived curve and estimated point on semilog
scales.
t = [0:10:4000];
T = 68+b*10.^(m*t);
subplot(2,2,3)
semilogy(t,T-68,time,temp,'o',t_120,120-68,'+'),
```

```
xlabel('Time (sec)'),...
   ylabel('Relative Temperature (deg F)')
%
% Show the derived curve and estimated point on rectilinear scales.
subplot(2,2,4)
plot(t,T,time,temp+68,'o',t_120,120,'+'),xlabel('Time
(sec)'),...
   ylabel('Temperature (deg F)')
```

The computed value of `t_120` is 1112. Thus the time to reach 120° F is 1112 sec. The plot of the model, along with the data and the estimated point (1112, 120) marked with a + sign, is shown in the bottom two subplots in Figure 5.4–3. Because the graph of our model lies near the data points, we can treat its prediction of 1112 sec with some confidence.

Hydraulic Resistance

EXAMPLE 5.4–2

A 15-cup coffee pot (see Figure 5.4–4) was placed under a water faucet and filled to the 15-cup line. With the outlet valve open, the faucet's flow rate was adjusted until the water level remained constant at 15 cups, and the time for one cup to flow out of the pot was measured. This experiment was repeated with the pot filled to the various levels shown in the following table:

Liquid volume V (cups)	Time to fill one cup t (sec)
15	6
12	7
9	8
6	9

Figure 5.4–4 An experiment to verify Torricelli's principle.

(a) Use the preceding data to obtain a relation between the flow rate and the number of cups in the pot. (b) The manufacturer wants to make a 36-cup pot using the same outlet valve but is concerned that a cup will fill too quickly, causing spills. Extrapolate the relation developed in part (a) and predict how long it will take to fill one cup when the pot contains 36 cups.

■ Solution

(a) Torricelli's principle in hydraulics states that $f = rV^{1/2}$, where f is the flow rate through the outlet valve in cups per second, V is the volume of liquid in the pot in cups, and r is a constant whose value is to be found. We see that this relation is a power function where the exponent is 0.5. Thus if we plot $\log_{10}(f)$ versus $\log_{10}(V)$, we should obtain a straight line. The values for f are obtained from the reciprocals of the given data for t. That is, $f = 1/t$ cups per second.

The MATLAB script file follows. The resulting plots appear in Figure 5.4–5. The volume data is entered in the array cups, and the time data is entered in meas_times.

```
% Data for the problem.
cups = [6,9,12,15];
meas_times = [9,8,7,6];
meas_flow = 1./meas_times;
%
% Fit a straight line to the transformed data.
```

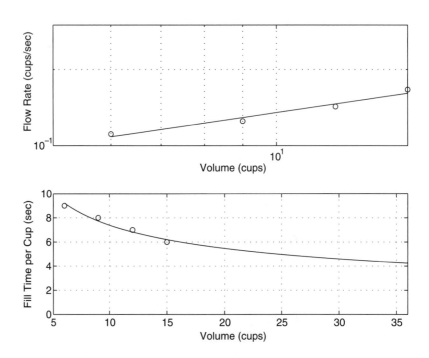

Figure 5.4–5 Flow rate and fill time for a coffee pot.

```
p = polyfit(log10(cups),log10(meas_flow),1);
coeffs = [p(1),10^p(2)];
m = coeffs(1)
b = coeffs(2)
%
% Plot the data and the fitted line on a loglog plot to see
% how well the line fits the data.
x = [6:0.01:40];
y = b*x.^m;
subplot(2,1,1)
loglog(x,y,cups,meas_flow,'o'),grid,xlabel('Volume (cups)'),...
    ylabel('Flow Rate (cups/sec)'),axis([5 15 0.1 0.3])
```

The computed values are $m = 0.433$ and $b = 0.0499$, and our derived relation is $f = 0.0499V^{0.433}$. Because the exponent is 0.433, not 0.5, our model does not agree exactly with Torricelli's principle, but it is close. Note that the first plot in Figure 5.4–5 shows that the data points do not lie exactly on the fitted straight line. In this application it is difficult to measure the time to fill one cup with an accuracy greater than an integer second, so this inaccuracy could have caused our result to disagree with that predicted by Torricelli.

(b) Note that the fill time is $1/f$, the reciprocal of the flow rate. The remainder of the MATLAB script uses the derived flow rate relation $f = 0.0499V^{0.433}$ to plot the extrapolated fill-time curve $1/f$ versus t.

```
% Plot the fill time curve extrapolated to 36 cups.
subplot(2,1,2)
plot(x,1./y,cups,meas_times,'o'),grid,xlabel('Volume(cups)'),...
    ylabel('Fill Time per Cup (sec)'),axis([5 36 0 10])
%
% Compute the fill time for V = 36 cups.
fill_time = 1/(b*36^m)
```

The predicted fill time for one cup is 4.2 sec. The manufacturer must now decide if this time is sufficient for the user to avoid overfilling. (In fact, the manufacturer did construct a 36-cup pot, and the fill time is approximately 4 sec, which agrees with our prediction.)

5.5 Regression

In the previous section we used the MATLAB function `polyfit` to perform regression analysis with functions that are linear or could be converted to linear form by a logarithmic or other transformation. The `polyfit` function is based on the least squares method, which is also called *regression*. We now show how to use this function to develop polynomial and other types of functions.

REGRESSION

The Least Squares Method

Suppose we have the three data points given in the following table, and we need to determine the coefficients of the straight line $y = mx + b$ that best fit the following data in the least squares sense.

x	y
0	2
5	6
10	11

RESIDUALS

According to the least squares criterion, the line that gives the best fit is the one that minimizes J, the sum of the squares of the vertical differences between the line and the data points. These differences are called the *residuals*. Here there are three data points, and J is given by

$$J = \sum_{i=1}^{3}(mx_i + b - y_i)^2 = (0m + b - 2)^2 + (5m + b - 6)^2 + (10m + b - 11)^2$$

The values of m and b that minimize J are found by setting the partial derivatives $\partial J/\partial m$ and $\partial J/\partial b$ equal to zero.

$$\frac{\partial J}{\partial m} = 250m + 30b - 280 = 0$$

$$\frac{\partial J}{\partial b} = 30m + 6b - 38 = 0$$

These conditions give two equations that must be solved for the two unknowns m and b. The solution is $m = 0.9$ and $b = 11/6$. The best straight line in the least squares sense is $y = 0.9x + 11/6$. If we evaluate this equation at the data values $x = 0, 5,$ and 10, we obtain the values $y = 1.833, 6.333, 10.8333$. These values are different from the given data values $y = 2, 6,$ and 11 because the line is not a perfect fit to the data. The value of J is $J = (1.833-2)^2 + (6.333-6)^2 + (10.8333-11)^2 = 0.16656689$. No other straight line will give a lower value of J for this data.

In general, for the polynomial $a_1x^n + a_2x^{n-1} + \cdots + a_nx + a_{n+1}$, the sum of the squares of the residuals for m data points is

$$J = \sum_{i=1}^{m}(a_1x^n + a_2x^{n-1} + \cdots + a_nx + a_{n+1} - y_i)^2$$

The values of the $n + 1$ coefficients a_i that minimize J can be found by solving a set of $n + 1$ linear equations. The `polyfit` function provides this solution. Its syntax is `p = polyfit(x,y,n)`. Table 5.5–1 summarizes the `polyfit` and `polyval` functions.

Consider the data set where $x = 1, 2, 3, \ldots, 9$ and $y = 5, 6, 10, 20, 28, 33, 34, 36, 42$. The following script file computes the coefficients of the first- through fourth-degree polynomials for this data and evaluates J for each polynomial.

Table 5.5–1 Functions for polynomial regression

Command	Description
`p = polyfit(x,y,n)`	Fits a polynomial of degree `n` to data described by the vectors `x` and `y`, where `x` is the independent variable. Returns a row vector `p` of length `n+1` that contains the polynomial coefficients in order of descending powers.
`[p,s,mu] = polyfit(x,y,n)`	Fits a polynomial of degree `n` to data described by the vectors `x` and `y`, where `x` is the independent variable. Returns a row vector `p` of length `n+1` that contains the polynomial coefficients in order of descending powers and a structure `s` for use with `polyval` to obtain error estimates for predictions. The optional output variable `mu` is a two-element vector containing the mean and standard deviation of `x`.
`[y,delta] = polyval(p,x,s,mu)`	Uses the optional output structure `s` generated by `[p,s,mu] = polyfit(x,y,n)` to generate error estimates. If the errors in the data used with `polyfit` are independent and normally distributed with constant variance, at least 50 percent of the data will lie within the band $y \pm$ `delta`.

```
x = [1:9];
y = [5,6,10,20,28,33,34,36,42];
for k = 1:4
    coeff = polyfit(x,y,k)
    J(k) = sum((polyval(coeff,x)-y).^2)
end
```

The J values are, to two significant figures, 72, 57, 42, and 4.7. Thus the value of J decreases as the polynomial degree is increased, as we would expect.

It is tempting to use a high-degree polynomial to obtain the best possible fit. However, there are two dangers in using high-degree polynomials. High-degree polynomials often exhibit large excursions between the data points and thus should be avoided if possible. The second danger with using high-degree polynomials is that they can produce large errors if their coefficients are not represented with a large number of significant figures. In some cases it might not be possible to fit the data with a low-degree polynomial. In such cases we might be able to use several cubic polynomials. This method, called cubic splines, is covered in Chapter 6.

Test Your Understanding

T5.5–1 Obtain and plot the first- through fourth-degree polynomials for the following data: $x = 0, 1, \ldots, 5$ and $y = 0, 1, 60, 40, 41,$ and 47. Find the coefficients and the J values.

(Answer: The polynomials are $9.5714x + 7.5714$; $-3.6964x^2 + 28.0536x - 4.7500$; $0.3241x^3 - 6.1270x^2 + 32.4934x - 5.7222$; and $2.5208x^4 - 24.8843x^3 + 71.2986x^2 - 39.5304x - 1.4008$. The corresponding J values are 1534, 1024, 1017, and 495, respectively.)

Fitting Other Functions

Given the data (y, z), the logarithmic function $y = m \ln z + b$ can be converted to a first-degree polynomial by transforming the z values into x values by the transformation $x = \ln z$. The resulting function is $y = mx + b$.

Given the data (y, z), the function $y = b(10)^{m/z}$ can be converted to an exponential function by transforming the z values by the transformation $x = 1/z$.

Given the data (v, x), the function $v = 1/(mx + b)$ can be converted to a first-degree polynomial by transforming the v data values with the transformation $y = 1/v$. The resulting function is $y = mx + b$.

The Quality of a Curve Fit

The least squares criterion used to fit a function $f(x)$ is the sum of the squares of the residuals J. It is defined as

$$J = \sum_{i=1}^{m}[f(x_i) - y_i]^2 \qquad (5.5\text{–}1)$$

We can use the J value to compare the quality of the curve fit for two or more functions used to describe the same data. The function that gives the smallest J value gives the best fit.

We denote the sum of the squares of the deviation of the y values from their mean $\bar{y}$ by S, which can be computed from

$$S = \sum_{i=1}^{m}(y_i - \bar{y})^2 \qquad (5.5\text{–}2)$$

COEFFICIENT OF DETERMINATION

This formula can be used to compute another measure of the quality of the curve fit, the *coefficient of determination,* also known as the *r-squared value.* It is defined as

$$r^2 = 1 - \frac{J}{S} \qquad (5.5\text{–}3)$$

For a perfect fit, $J = 0$ and thus $r^2 = 1$. Thus the closer r^2 is to 1, the better the fit. The largest r^2 can be is 1. The value of S indicates how much the data is spread around the mean, and the value of J indicates how much of the data spread is unaccounted for by the model. Thus the ratio J/S indicates the fractional variation unaccounted for by the model. It is possible for J to be larger than S, and thus it is possible for r^2 to be negative. Such cases, however, are indicative of a very poor model that should not be used. As a rule of thumb, a good fit accounts for at least 99 percent of the data variation. This value corresponds to $r^2 \geq 0.99$.

For example, the following table gives the values of J, S, and r^2 for the first-through fourth-degree polynomials used to fit the data $x = 1, 2, 3, \ldots, 9$ and $y = 5, 6, 10, 20, 28, 33, 34, 36, 42$.

Degree n	J	S	r^2
1	72	1562	0.9542
2	57	1562	0.9637
3	42	1562	0.9732
4	4.7	1562	0.9970

Because the fourth-degree polynomial has the largest r^2 value, it represents the data better than the representation from first- through third-degree polynomials, according to the r^2 criterion.

To calculate the values of S and r^2, add the following lines to the end of the script file shown on page 239.

```
mu = mean(y);
for k=1:4
    S(k) = sum((y-mu).^2);
    r2(k) = 1 - J(k)/S(k);
end
S
r2
```

Scaling the Data

The effect of computational errors in computing the coefficients can be lessened by properly scaling the x values. When the function `polyfit(x,y,n)` is executed, it will issue a warning message if the polynomial degree n is greater than or equal to the number of data points (because there will not be enough equations for MATLAB to solve for the coefficients), or if the vector x has repeated, or nearly repeated, points, or if the vector x needs centering and/or scaling. The alternate syntax `[p, s, mu] = polyfit(x,y,n)` finds the coefficients p of a polynomial of degree n in terms of the variable

$$\hat{x} = (x - \mu_x)/\sigma_x$$

The output variable mu is a two-element vector, $[\mu_x, \sigma_x]$, where μ_x is the mean of x, and σ_x is the standard deviation of x (the standard deviation is discussed in Chapter 6).

You can scale the data yourself before using `polyfit`. Some common scaling methods are

$$\hat{x} = x - x_{min} \quad \text{or} \quad \hat{x} = x - \mu_x$$

if the range of x is small, or

$$\hat{x} = \frac{x}{x_{max}} \quad \text{or} \quad \hat{x} = \frac{x}{x_{mean}}$$

if the range of x is large.

| EXAMPLE 5.5–1 | Estimation of Traffic Flow |

The following data gives the number of vehicles (in millions) crossing a bridge each year for 10 years. Fit a cubic polynomial to the data and use the fit to estimate the flow in the year 2008.

Year	1998	1999	2000	2001	2002	2003	2004	2005	2006	2007
Vehicle flow (millions)	2.1	3.4	4.5	5.3	6.2	6.6	6.8	7	7.4	7.8

■ Solution

If we attempt to fit a cubic to this data, as in the following session, we get a warning message.

```
>>Year = [1998:2007];
>>Veh_Flow = [2.1,3.4,4.5,5.3,6.2,6.6,6.8,7,7.4,7.8];
>>p = polyfit(Year,Veh_Flow,3)
Warning: Polynomial is badly conditioned.
```

The problem is caused by the large values of the independent variable `Year`. Because their range is small, we can simply subtract 1998 from each value. Continue the session as follows.

```
>>x = Year-1998; y = Veh_Flow;
>>p = polyfit(x,y,3)
p =
    0.0087    -0.1851   1.5991  2.0362
>>J = sum((polyval(p,x)-y).^2);
>>S = sum((y-mean(y)).^2);
>>r2 = 1 - J/S
r2 =
    0.9972
```

Thus the polynomial fit is good because the coefficient of determination is 0.9972. The corresponding polynomial is

$$f = 0.0087(t - 1998)^3 - 0.1851(t - 1998)^2 + 1.5991(t - 1998) + 2.0362$$

where f is the traffic flow in millions of vehicles, and t is the time in years measured from 0. We can use this equation to estimate the flow at the year 2008 by substituting $t = 2008$, or by typing in MATLAB `polyval(p,10)`. Rounded to one decimal place, the answer is 8.2 million vehicles.

Using Residuals

We now show how to use the residuals as a guide to choosing an appropriate function to describe the data. In general, if you see a pattern in the plot of the residuals, it indicates that another function can be found to describe the data better.

Modeling Bacteria Growth

EXAMPLE 5.5–2

The following table gives data on the growth of a certain bacteria population with time. Fit an equation to this data.

Time (min)	Bacteria (ppm)	Time (min)	Bacteria (ppm)
0	6	10	350
1	13	11	440
2	23	12	557
3	33	13	685
4	54	14	815
5	83	15	990
6	118	16	1170
7	156	17	1350
8	210	18	1575
9	282	19	1830

■ Solution

We try three polynomial fits (linear, quadratic, and cubic), and an exponential fit. The script file is given below. Note that we can write the exponential form as $y = b(10)^{mt} = 10^{mt+a}$, where $b = 10^a$.

```
% Time data
x = [0:19];
% Population data
y = [6,13,23,33,54,83,118,156,210,282,...
   350,440,557,685,815,990,1170,1350,1575,1830];
% Linear fit
p1 = polyfit(x,y,1);
% Quadratic fit
p2 = polyfit(x,y,2);
% Cubic fit
p3 = polyfit(x,y,3);
% Exponential fit
p4 = polyfit(x,log10(y),1);
% Residuals
res1 = polyval(p1,x)-y;
res2 = polyval(p2,x)-y;
res3 = polyval(p3,x)-y;
res4 = 10.^polyval(p4,x)-y;
```

You can then plot the residuals as shown in Figure 5.5–1. Note that there is a definite pattern in the residuals of the linear fit. This indicates that the linear function cannot match the curvature of the data. The residuals of the quadratic fit are much smaller, but there is still a pattern, with a random component. This indicates that the quadratic function also cannot match the curvature of the data. The residuals of the cubic fit are

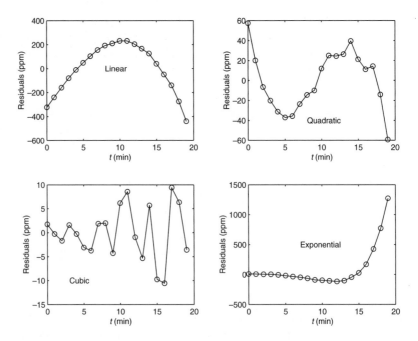

Figure 5.5–1 Residual plots for the four models.

even smaller, with no strong pattern and a large random component. This indicates that a polynomial degree higher than three will not be able to match the data curvature any better than the cubic. The residuals for the exponential are the largest of all, and indicate a poor fit. Note also how the residuals systematically increase with t, indicating that the exponential cannot describe the data's behavior after a certain time.

Thus the cubic is the best fit of the four models considered. Its coefficient of determination is $r^2 = 0.9999$. The model is

$$y = 0.1916t^3 + 1.2082t^2 + 3.607t + 7.7307$$

where y is the bacteria population in ppm and t is time in minutes.

MULTIPLE LINEAR REGRESSION

Multiple Linear Regression

Suppose that y is a linear function of two or more variables, $x_1, x_2, \ldots,$ for example, $y = a_0 + a_1 x_1 + a_2 x_2$. To find the coefficient values $a_0, a_1,$ and a_2 to fit a set of data (y, x_1, x_2) in the least squares sense, we can make use of the fact that the left-division method for solving linear equations uses the least squares method when the equation set is overdetermined (see Section 2.5). To use this method, let n be the number of data points and write the linear equation in matrix form as $\mathbf{Xa} = \mathbf{y}$ where

$$\mathbf{a} = \begin{bmatrix} a_0 \\ a_1 \\ a_2 \end{bmatrix} \qquad \mathbf{X} = \begin{bmatrix} 1 & x_{11} & x_{21} \\ 1 & x_{12} & x_{22} \\ 1 & x_{13} & x_{23} \\ & \cdots & \\ 1 & x_{1n} & x_{2n} \end{bmatrix} \qquad \mathbf{y} = \begin{bmatrix} y_1 \\ y_2 \\ y_3 \\ \cdots \\ y_n \end{bmatrix}$$

where x_{1i}, x_{2i}, and y_i are the data, $i = 1, \ldots, n$. The solution for the coefficients is given by `a = X\y`.

Breaking Strength and Alloy Composition

EXAMPLE 5.5–3

We want to predict the strength of metal parts as a function of their alloy composition. The tension force y required to break a steel bar is a function of the percentage x_1 and x_2 of each of two alloying elements present in the metal. The following table gives some pertinent data. Obtain a linear model $y = a_0 + a_1 x_1 + a_2 x_2$ to describe the relationship.

Breaking strength (kN) y	% of element 1 x_1	% of element 2 x_2
7.1	0	5
19.2	1	7
31	2	8
45	3	11

■ Solution
The script file is as follows:

```
x1 = [0:3]';x2 = [5,7,8,11]';
y = [7.1,19.2,31,45]';
X = [ones(size(x1)), x1, x2];
a = X\y
yp = X*a;
Max_Percent_Error = 100*max(abs((yp-y)./y))
```

The vector `yp` is the vector of breaking-strength values predicted by the model. The scalar `Max_Percent_Error` is the maximum percent error in the four predictions. The results are `a = [0.8000, 10.2429, 1.2143]'` and `Max_Percent_Error = 3.2193`. Thus the model is $y = 0.8 + 10.2429x_1 + 1.2143x_2$. The maximum percent error of the model's predictions, as compared to the given data, is 3.2193 percent.

Linear-in-the-Parameters Regression

LINEAR-IN-PARAMETERS

Sometimes we want to fit an expression that is neither a polynomial nor a function that can be converted to linear form by a logarithmic or other transformation. In some cases we can still do a least squares fit if the function is a linear expression in terms of its parameters. The following example illustrates the method.

EXAMPLE 5.5–4

Response of a Biomedical Instrument

Engineers developing instrumentation often need to obtain a *response* curve that describes how fast the instrument can make measurements. The theory of instrumentation shows that often the response can be described by one of the following equations, where v is the voltage output, and t is time. In both models, the voltage reaches a steady-state constant value as $t \rightarrow \infty$, and T is the time required for the voltage to equal 95 percent of the steady-state value.

$$v(t) = a_1 + a_2 e^{-3t/T} \qquad \text{(first-order model)}$$
$$v(t) = a_1 + a_2 e^{-3t/T} + a_3 t e^{-3t/T} \qquad \text{(second-order model)}$$

The following data gives the output voltage of a certain device as a function of time. Obtain a function that describes this data.

t (s)	0	0.3	0.8	1.1	1.6	2.3	3
v (V)	0	0.6	1.28	1.5	1.7	1.75	1.8

■ Solution

Plotting the data we estimate that it takes approximately 3 seconds for the voltage to become constant. Thus we estimate that $T = 3$. The first-order model written for each of the n data points results in n equations, which can be expressed as follows:

$$\begin{bmatrix} 1 & e^{-t_1} \\ 1 & e^{-t_2} \\ \cdots & \cdots \\ 1 & e^{-t_n} \end{bmatrix} \begin{bmatrix} a_1 \\ a_2 \end{bmatrix} = \begin{bmatrix} y_1 \\ y_2 \\ \cdots \\ y_n \end{bmatrix}$$

or, in matrix form,

$$\mathbf{Xa} = \mathbf{y}'$$

which can be solved for the coefficient vector $\mathbf{a}$ using left division. The following MATLAB script solves the problem.

```
t = [0,0.3,0.8,1.1,1.6,2.3,3];
y = [0,0.6,1.28,1.5,1.7,1.75,1.8];
X = [ones(size(t));exp(-t)]';
a = X\y'
```

The answer is $a_1 = 2.0258$ and $a_2 = -1.9307$.

A similar procedure can be followed for the second-order model.

$$\begin{bmatrix} 1 & e^{-t_1} & t_1 e^{-t_1} \\ 1 & e^{-t_2} & t_2 e^{-t_2} \\ \cdots & \cdots & \\ 1 & e^{-t_n} & t_n e^{-t_n} \end{bmatrix} \begin{bmatrix} a_1 \\ a_2 \\ a_3 \end{bmatrix} = \begin{bmatrix} y_1 \\ y_2 \\ \cdots \\ y_n \end{bmatrix}$$

Continue the previous script as follows.

```
X = [ones(size(t));exp(-t);t.*exp(-t)]';
a = X\y'
```

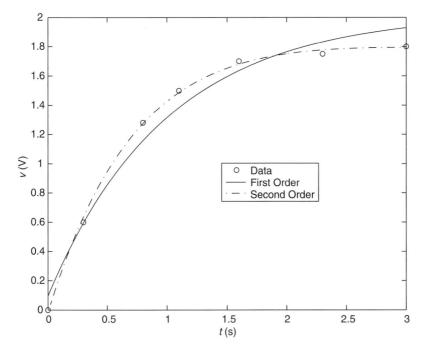

Figure 5.5–2 Comparison of first- and second-order model fits.

The answer is $a_1 = 1.7496$, $a_2 = -1.7682$, and $a_3 = 0.8885$. The two models are plotted with the data in Figure 5.5–2. Clearly the second-order model gives the better fit.

5.6 The Basic Fitting Interface

MATLAB supports curve fitting through the Basic Fitting interface. Using this interface, you can quickly perform basic curve fitting tasks within the same easy-to-use environment. The interface is designed so that you can:

- Fit data using a cubic spline or a polynomial up to degree 10.
- Plot multiple fits simultaneously for a given data set.
- Plot the residuals.
- Examine the numerical results of a fit.
- Interpolate or extrapolate a fit.
- Annotate the plot with the numerical fit results and the norm of residuals.
- Save the fit and evaluated results to the MATLAB workspace.

Depending on your specific curve fitting application, you can use the Basic Fitting interface, the command line functions, or both. Note: you can use the Basic Fitting interface only with two-dimensional data. However, if you plot

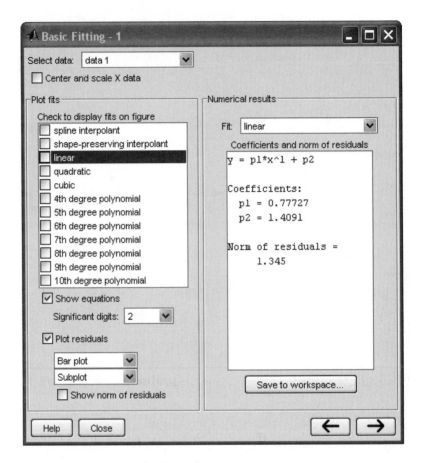

Figure 5.6–1 The Basic Fitting interface.

multiple data sets as a subplot, and at least one data set is two-dimensional, then the interface is enabled.

Two panes of the Basic Fitting interface are shown in Figure 5.6–1. To reproduce this state:

1. Plot some data.
2. Select **Basic Fitting** from the **Tools** menu of the Figure window.
3. When the first pane of the Basic Fitting interface appears, click the right arrow button once.

The third pane is used for interpolating or extrapolating a fit. It appears when you click the right arrow button a second time.

At the top of the first pane is the **Select data** window which contains the names of all the data sets you display in the Figure window associated with the Basic Fitting interface. Use this menu to select the data set to be fit. You can perform multiple fits for the current data set. Use the Plot Editor to change the name of a data set. The remaining items on the first pane are used as follows.

- **Center and scale X data.** If checked, the data is centered at zero mean and scaled to unit standard deviation. You may need to center and scale your data to improve the accuracy of the subsequent numerical computations. As described in the previous section, a warning is returned to the Command window if a fit produces results that may be inaccurate.

- **Plot fits.** This panel allows you to visually explore one or more fits to the current data set.

- **Check to display fits on figure.** Select the fits you want to display for the current data set. You can choose as many fits for a given data set as you want. However, if your data set has *n* points, then you should use polynomials with, at most, *n* coefficients. If you fit using polynomials with more than *n* coefficients, the interface will automatically set a sufficient number of coefficients to 0 during the calculation so that a solution can be obtained.

- **Show equations.** If checked, the fit equation is displayed on the plot.

- **Significant digits.** Select the significant digits associated with the fit coefficient display.

- **Plot residuals.** If checked, the residuals are displayed. You can display the residuals as a bar plot, a scatter plot, a line plot using either the same figure window as the data or using a separate figure window. If you plot multiple data sets as a subplot, then residuals can be plotted only in a separate figure window. See Figure 5.6–2.

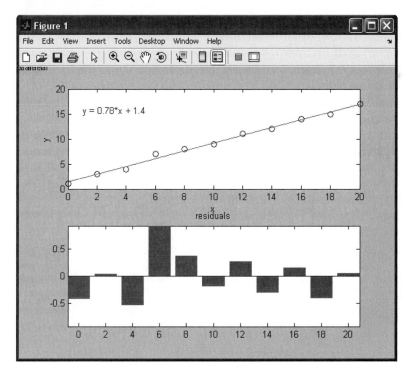

Figure 5.6–2 A figure produced by the Basic Fitting interface.

- **Show norm of residuals.** If checked, the norm of residuals is displayed. The norm of residuals is a measure of the goodness of fit, where a smaller value indicates a better fit. The norm is the square root of the sum of the squares of the residuals.

The second pane of the Basic Fitting Interface is labeled *Numerical Results.* This pane enables you to explore the numerical results of a single fit to the current data set without plotting the fit. It contains three items.

- **Fit.** Use this menu to select an equation to fit to the current data set. The fit results are displayed in the box below the menu. Note that selecting an equation in this menu does not affect the state of the **Plot fits** selection. Therefore, if you want to display the fit in the data plot, you may need to check the relevant check box in **Plot fits**.

- **Coefficients and norm of residuals.** Displays the numerical results for the equation selected in **Fit.** Note that when you first open the **Numerical Results** panel, the results of the last fit you selected in **Plot fits** are displayed.

- **Save to workspace.** Launches a dialog box that allows you to save the fit results to workspace variables.

The third pane of the Basic Fitting interface contains three items.

- **Find** $Y = f(X)$. Use this to interpolate or extrapolate the current fit. Enter a scalar or a vector of values corresponding to the independent variable (X). The current fit is evaluated after you click on the **Evaluate** button, and the results are displayed in the associated window. The current fit is displayed in the **Fit** window.

- **Save to workspace.** Launches a dialog box that allows you to save the evaluated results to workspace variables.

- **Plot evaluated results.** If checked, the evaluated results are displayed on the data plot.

5.7 Three-Dimensional Plots

MATLAB provides many functions for creating three-dimensional plots. Here we will summarize the basic functions to create three types of plots: line plots, surface plots, and contour plots. Information about the related functions is available in MATLAB help (category `graph3d`).

Three-Dimensional Line Plots

Lines in three-dimensional space can be plotted with the `plot3` function. Its syntax is `plot3(x,y,z)`. For example, the following equations generate a three-dimensional curve as the parameter *t* is varied over some range:

$$x = e^{-0.05t} \sin t$$
$$y = e^{-0.05t} \cos t$$
$$z = t$$

Test Your Understanding

T5.7–1 Create a surface plot and a contour plot of the function $z = (x - 2)^2 + 2xy + y^2$.

5.8 Summary

This chapter explained how to use the powerful MATLAB commands to create effective and pleasing two-dimensional and three-dimensional plots. You learned an important application of plotting—function discovery—which is the technique for using data plots to obtain a mathematical function that describes the data. Regression can be used to develop a model for cases where there is considerable scatter in the data.

The following guidelines will help you create plots that effectively convey the desired information:

■ Label each axis with the name of the quantity being plotted *and its units!*
■ Use regularly spaced tick marks at convenient intervals along each axis.
■ If you are plotting more than one curve or data set, label each on its plot or use a legend to distinguish them.
■ If you are preparing multiple plots of a similar type or if the axes' labels cannot convey enough information, use a title.
■ If you are plotting measured data, plot each data point in a given set with the same symbol, such as a circle, square, or cross.
■ If you are plotting points generated by evaluating a function (as opposed to measured data), do *not* use a symbol to plot the points. Instead, connect the points with solid lines.

Key Terms with Page References

Axis limits, 207	Overlay plot, 212
Coefficient of determination, 240	Polar plot, 220
Contour plot, 253	Regression, 237
Data symbol, 206	Residuals, 238
Linear-in-parameters, 245	Subplot, 211
Multiple linear regression, 244	Surface mesh plot, 251

Problems

You can find the answers to problems marked with an asterisk at the end of the text.

Sections 5.1, 5.2, and 5.3

1.* *Breakeven analysis* determines the production volume at which the total production cost is equal to the total revenue. At the breakeven point, there is neither profit nor loss. In general, production costs consist of fixed costs and variable costs. Fixed costs include salaries of those not directly involved with production, factory maintenance costs, insurance costs, and so on. Variable costs depend on production volume and include material costs, labor costs, and energy costs. In the following analysis, assume that we produce only what we can sell; thus the production quantity equals the sales. Let the production quantity be Q, in gallons per year.

Consider the following costs for a certain chemical product:
- Fixed cost: $3 million per year.
- Variable cost: 2.5 cents per gallon of product.
- The selling price is 5.5 cents per gallon.

Use this data to plot the total cost and the revenue versus Q, and graphically determine the breakeven point. Fully label the plot and mark the breakeven point. For what range of Q is production profitable? For what value of Q is the profit a maximum?

2. Consider the following costs for a certain chemical product:

Fixed cost: $2.045 million/year.

Variable costs:
- Material cost: 62 cents per gallon of product.
- Energy cost: 24 cents per gallon of product.
- Labor cost: 16 cents per gallon of product.

Assume that we produce only what we sell. Let P be the selling price in dollars per gallon. Suppose that the selling price and the sales quantity Q are interrelated as follows: $Q = 6 \times 10^6 - 1.1 \times 10^6 P$. Accordingly, if we raise the price, the product becomes less competitive and sales drop.

Use this information to plot the fixed and total variable costs versus Q, and graphically determine the breakeven point(s). Fully label the plot and mark the breakeven points. For what range of Q is the production profitable? For what value of Q is the profit a maximum?

3.* a) Estimate the roots of the equation

$$x^3 - 3x^2 + 5x \sin\left(\frac{\pi x}{4} - \frac{5\pi}{4}\right) + 3 = 0$$

by plotting the equation. b) Use the estimates found in Part (a) to find the roots more accurately with the `fzero` function.

4. To compute the forces in structures, sometimes we must solve equations similar to the following. Use the `fplot` function to find all the positive roots of this equation:

$$x \tan x = 7$$

5.* Cables are used to suspend bridge decks and other structures. If a heavy uniform cable hangs suspended from its two endpoints, it takes the shape of a *catenary* curve whose equation is

$$y = a \cosh\left(\frac{x}{a}\right)$$

where a is the height of the lowest point on the chain above some horizontal reference line, x is the horizontal coordinate measured to the right from the lowest point, and y is the vertical coordinate measured up from the reference line.

Let $a = 10$ m. Plot the catenary curve for $-20 \leq x \leq 30$ m. How high is each endpoint?

6. Using estimates of rainfall, evaporation, and water consumption, the town engineer developed the following model of the water volume in the reservoir as a function of time.

$$V(t) = 10^9 + 10^8(1 - e^{-t/100}) - 10^7 t$$

where V is the water volume in liters, and t is time in days. Plot $V(t)$ versus t. Use the plot to estimate how many days it will take before the water volume in the reservoir is 50 percent of its initial volume of 10^9 L.

7. It is known that the following Leibniz series converges to the value $\pi/4$ as $n \to \infty$.

$$S(n) = \sum_{k=0}^{n}(-1)^k \frac{1}{2k + 1}$$

Plot the difference between $\pi/4$ and the sum $S(n)$ versus n for $0 \leq n \leq 200$.

8. A certain fishing vessel is initially located in a horizontal plane at $x = 0$ and $y = 10$ mi. It moves on a path for 10 hr such that $x = t$ and $y = 0.5t^2 + 10$, where t is in hours. An international fishing boundary is described by the line $y = 2x + 6$.

a. Plot and label the path of the vessel and the boundary.

b. The perpendicular distance of the point (x_1, y_1) from the line $Ax + By + C = 0$ is given by

$$d = \frac{Ax_1 + By_1 + C}{\pm\sqrt{A^2 + B^2}}$$

where the sign is chosen to make $d \geq 0$. Use this result to plot the distance of the fishing vessel from the fishing boundary as a function of time for $0 \leq t \leq 10$ hr.

9. Plot columns 2 and 3 of the following matrix **A** versus column 1. The data in column 1 is time (seconds). The data in columns 2 and 3 is force (newtons).

$$\mathbf{A} = \begin{bmatrix} 0 & -8 & 6 \\ 5 & -4 & 3 \\ 10 & -1 & 1 \\ 15 & 1 & 0 \\ 20 & 2 & -1 \end{bmatrix}$$

10.* Many applications use the following "small angle" approximation for the sine to obtain a simpler model that is easy to understand and analyze. This approximation states that $\sin x \approx x$, where x must be in radians. Investigate the accuracy of this approximation by creating three plots. For the first, plot $\sin x$ and x versus x for $0 \le x \le 1$. For the second, plot the approximation error $\sin x - x$ versus x for $0 \le x \le 1$. For the third, plot the relative error $[\sin(x) - x]/\sin(x)$ versus x for $0 \le x \le 1$. How small must x be for the approximation to be accurate within 5 percent?

11. You can use trigonometric identities to simplify the equations that appear in many applications. Confirm the identity $\tan(2x) = 2 \tan x/(1 - \tan^2 x)$ by plotting both the left and the right sides versus x over the range $0 \le x \le 2\pi$.

12. The complex number identity $e^{ix} = \cos x + i \sin x$ is often used to convert the solutions of equations into a form that is relatively easy to visualize. Confirm this identity by plotting the imaginary part versus the real part for both the left and right sides over the range $0 \le x \le 2\pi$.

13. Use a plot over the range $0 \le x \le 5$ to confirm that $\sin(ix) = i \sinh x$.

14.* The function $y(t) = 1 - e^{-bt}$, where t is time and $b > 0$, describes many processes, such as the height of liquid in a tank as it is being filled and the temperature of an object being heated. Investigate the effect of the parameter b on $y(t)$. To do this, plot y versus t for several values of b on the same plot. How long will it take for $y(t)$ to reach 98 percent of its steady-state value?

15. The following functions describe the oscillations in electrical circuits and the vibrations of machines and structures. Plot these functions on the same plot. Because they are similar, decide how best to plot and label them to avoid confusion.

$$x(t) = 10e^{-0.5t} \sin(3t + 2)$$

$$y(t) = 7e^{-0.4t} \cos(5t - 3)$$

16. In certain kinds of structural vibrations, a periodic force acting on the structure will cause the vibration amplitude to repeatedly increase and

decrease with time. This phenomenon, called *beating,* also occurs in musical sounds. A particular structure's displacement is described by

$$y(t) = \frac{1}{f_1^2 - f_2^2}[\cos(f_2 t) - \cos(f_1 t)]$$

where y is the displacement in inches and t is the time in seconds. Plot y versus t over the range $0 \leq t \leq 20$ for $f_1 = 8$ rad/sec and $f_2 = 1$ rad/sec. Be sure to choose enough points to obtain an accurate plot.

17.* The height $h(t)$ and horizontal distance $x(t)$ traveled by a ball thrown at an angle A with a speed v are given by

$$h(t) = vt \sin A - \frac{1}{2}gt^2$$

$$x(t) = vt \cos A$$

At Earth's surface the acceleration due to gravity is $g = 9.81$ m/s^2.

a. Suppose the ball is thrown with a velocity $v = 10$ m/s at an angle of 35°. Use MATLAB to compute how high the ball will go, how far it will go, and how long it will take to hit the ground.

b. Use the values of v and A given in part a to plot the ball's *trajectory;* that is, plot h versus x for positive values of h.

c. Plot the trajectories for $v = 10$ m/s corresponding to five values of the angle A: 20°, 30°, 45°, 60°, and 70°.

d. Plot the trajectories for $A = 45°$ corresponding to five values of the initial velocity v: 10, 12, 14, 16, and 18 m/s.

18. The perfect gas law relates the pressure p, absolute temperature T, mass m, and volume V of a gas. It states that

$$pV = mRT$$

The constant R is the *gas constant.* The value of R for air is 286.7 N · m/kg · K. Suppose air is contained in a chamber at room temperature (20°C = 293 K). Create a plot having three curves of the gas pressure in N/m^2 versus the container volume V in m^3 for $20 \leq V \leq 100$. The three curves correspond to the following masses of air in the container: $m = 1$ kg; $m = 3$ kg; and $m = 7$ kg.

19. Oscillations in mechanical structures and electric circuits can often be described by the function

$$y(t) = e^{-t/\tau} \sin(\omega t + \phi)$$

where t is time and ω is the oscillation frequency in radians per unit time. The oscillations have a period of $2\pi/\omega$, and their amplitudes decay in time at a rate determined by τ, which is called the *time constant.* The smaller τ is, the faster the oscillations die out.

   a. Use these facts to develop a criterion for choosing the spacing of the *t* values and the upper limit on *t* to obtain an accurate plot of *y(t)*. (Hint: Consider two cases: $4\tau > 2\pi/\omega$ and $4\tau < 2\pi/\omega$.)

   b. Apply your criterion, and plot *y(t)* for $\tau = 10$, $\omega = \pi$, and $\phi = 2$.

   c. Apply your criterion, and plot *y(t)* for $\tau = 0.1$, $\omega = 8\pi$, and $\phi = 2$.

20. When a constant voltage was applied to a certain motor initially at rest, its rotational speed *s(t)* versus time was measured. The data appears in the following table:

Time (sec)	1	2	3	4	5	6	7	8	10
Speed (rpm)	1210	1866	2301	2564	2724	2881	2879	2915	3010

Determine whether the following function can describe the data. If so, find the values of the constants *b* and *c*.

$$s(t) = b(1 - e^{ct})$$

21. The following table shows the average temperature for each year in a certain city. Plot the data as a stem plot, a bar plot, and a stairs plot.

Year	1990	1991	1992	1993	1994
Temperature (°C)	18	19	21	17	20

22. $10,000 invested at 5 percent interest compounded annually will grow according to the formula

$$y(k) = 10^4(1.05)^k$$

where *k* is the number of years ($k = 0, 1, 2, \ldots$). Plot the amount of money in the account for a 10-year period. Do this problem with four types of plots: the xy plot, the stem plot, the stairs plot, and the bar plot.

23. The volume *V* and surface area *A* of a sphere of radius *r* are given by

$$V = \frac{4}{3}\pi r^3 \qquad A = 4\pi r^2$$

   a. Plot *V* and *A* versus *r* in two subplots, for $0.1 \le r \le 100$ m. Choose axes that will result in straight-line graphs for both *V* and *A*.

   b. Plot *V* and *r* versus *A* in two subplots, for $1 \le A \le 10^4$ m². Choose axes that will result in straight-line graphs for both *V* and *r*.

24. The current amount *A* of a principal *P* invested in a savings account paying an annual interest rate *r* is given by

$$A = P\left(1 + \frac{r}{n}\right)^{nt}$$

where n is the number of times per year the interest is compounded. For continuous compounding, $A = Pe^{rt}$. Suppose $10,000 is initially invested at 3.5 percent ($r = 0.035$).

a. Plot A versus t for $0 \leq t \leq 20$ years for four cases: continuous compounding, annual compounding ($n = 1$), quarterly compounding $n = 4$), and monthly compounding ($n = 12$). Show all four cases on the same subplot and label each curve. On a second subplot, plot the difference between the amount obtained from continuous compounding and the other three cases.

b. Redo part a but plot A versus t on log-log and semilog plots. Which plot gives a straight line?

Section 5.4

25. The distance a spring stretches from its "free length" is a function of how much tension force is applied to it. The following table gives the spring length y that the given applied force f produced in a particular spring. The spring's free length is 4.7 in. Find a functional relation between f and x, the extension from the free length ($x = y - 4.7$).

Force f (lb)	Spring length y (in.)
0	4.7
0.47	7.2
1.15	10.6
1.64	12.9

26.* In each of the following problems, determine the best function $y(x)$ (linear, exponential, or power function) to describe the data. Plot the function on the same plot with the data. Label and format the plots appropriately.

a.

x	25	30	35	40	45
y	5	260	480	745	1100

b.

x	2.5	3	3.5	4	4.5	5	5.5	6	7	8	9	10
y	1500	1220	1050	915	810	745	690	620	520	480	410	390

c.

x	550	600	650	700	750
y	41.2	18.62	8.62	3.92	1.86

27.   The population data for a certain country is

Year	2002	2003	2004	2005	2006	2007
Population (millions)	10	10.8	11.7	12.7	13.8	14.9

Obtain a function that describes this data. Plot the function and the data on the same plot. Estimate when the population will be double its 2002 size.

28.*   The *half-life* of a radioactive substance is the time it takes to decay by half. The half-life of carbon 14, which is used for dating previously living things, is 5500 years. When an organism dies, it stops accumulating carbon 14. The carbon 14 present at the time of death decays with time. Let $C(t)/C(0)$ be the fraction of carbon 14 remaining at time t. In radioactive carbon dating, scientists usually assume that the remaining fraction decays exponentially according to the following formula:

$$\frac{C(t)}{C(0)} = e^{-bt}$$

a.   Use the half-life of carbon 14 to find the value of the parameter b, and plot the function.

b.   If 90 percent of the original carbon 14 remains, estimate how long ago the organism died.

c.   Suppose our estimate of b is off by ± 1 percent. How does this error affect the age estimate in b?

29.   *Quenching* is the process of immersing a hot metal object in a bath for a specified time to obtain certain properties such as hardness. A copper sphere 25 mm in diameter, initially at 300°C, is immersed in a bath at 0°C. The following table gives measurements of the sphere's temperature versus time. Find a functional description of this data. Plot the function and the data on the same plot.

Time (s)	0	1	2	3	4	5	6
Temperature (°C)	300	150	75	35	12	5	2

30.   The useful life of a machine bearing depends on its operating temperature, as the following data shows. Obtain a functional description of this data. Plot the function and the data on the same plot. Estimate a bearing's life if it operates at 150°F.

Temperature (°F)	100	120	140	160	180	200	220
Bearing life (hours × 10^3)	28	21	15	11	8	6	4

31. A certain electric circuit has a resistor and a capacitor. The capacitor is initially charged to 100 V. When the power supply is detached, the capacitor voltage decays with time, as the following data table shows. Find a functional description of the capacitor voltage v as a function of time t. Plot the function and the data on the same plot.

Time (s)	0	0.5	1	1.5	2	2.5	3	3.5	4
Voltage (V)	100	62	38	21	13	7	4	2	3

Sections 5.5 and 5.6

32.* The distance a spring stretches from its "free length" is a function of how much tension force is applied to it. The following table gives the spring length y that was produced in a particular spring by the given applied force f. The spring's free length is 4.7 in. Find a functional relation between f and x, the extension from the free length ($x = y - 4.7$).

Force f (lb)	Spring length y (in.)
0	4.7
0.47	7.2
1.15	10.6
1.64	12.9

33. The following data gives the drying time T of a certain paint as a function of the amount of a certain additive A.

  a. Find the first-, second-, third-, and fourth-degree polynomials that fit the data and plot each polynomial with the data. Determine the quality of the curve fit for each by computing J, S, and r^2.

  b. Use the polynomial giving the best fit to estimate the amount of additive that minimizes the drying time.

A (oz)	0	1	2	3	4	5	6	7	8	9
T (min)	130	115	110	90	89	89	95	100	110	125

34.* The following data gives the stopping distance d as a function of initial speed v, for a certain car model. Find a quadratic polynomial that fits the data. Determine the quality of the curve fit by computing J, S, and r^2.

v (mi/hr)	20	30	40	50	60	70
d (ft)	45	80	130	185	250	330

35.* The number of twists y required to break a certain rod is a function of the percentage x_1 and x_2 of each of two alloying elements present in the rod. The following table gives some pertinent data. Use linear multiple regression to obtain a model $y = a_0 + a_1x_1 + a_2x_2$ of the relationship between the number of twists and the alloy percentages. In addition, find the maximum percent error in the predictions.

Number of twists y	Percentage of element 1 x_1	Percentage of element 2 x_2
40	1	1
51	2	1
65	3	1
72	4	1
38	1	2
46	2	2
53	3	2
67	4	2
31	1	3
39	2	3
48	3	3
56	4	3

36. The following represents pressure samples, in pounds per square inch (psi), taken in a fuel line once every second for 10 sec.

Time (sec)	Pressure (psi)	Time (sec)	Pressure (psi)
1	26.1	6	30.6
2	27.0	7	31.1
3	28.2	8	31.3
4	29.0	9	31.0
5	29.8	10	30.5

a. Fit a first-degree polynomial, a second-degree polynomial, and a third-degree polynomial to this data. Plot the curve fits along with the data points.

b. Use the results from part a to predict the pressure at $t = 11$ sec. Explain which curve fit gives the most reliable prediction. Consider the coefficients of determination and the residuals for each fit in making your decision.

37. A liquid boils when its vapor pressure equals the external pressure acting on the surface of the liquid. This is the reason why water boils at a lower temperature at higher altitudes. This information is important for people who must design processes utilizing boiling liquids. Data on the vapor pressure P of water as a function of temperature T is given in the following table. From theory we know that $\ln P$ is proportional to $1/T$. Obtain a curve fit for $P(T)$ from this data. Use the fit to estimate the vapor pressure at 285 K and at 300 K.

T (K)	P (torr)
273	4.579
278	6.543
283	9.209
288	12.788
293	17.535
298	23.756

38. The solubility of salt in water is a function of the water temperature. Let S represent the solubility of NaCl (sodium chloride) as grams of salt in 100 g of water. Let T be temperature in °C. Use the following data to obtain a curve fit for S as a function of T. Use the fit to estimate S when $T = 25°C$.

T (°C)	S (g NaCl/100 g H_2O)
10	35
20	35.6
30	36.25
40	36.9
50	37.5
60	38.1
70	38.8
80	39.4
90	40

39. The solubility of oxygen in water is a function of the water temperature. Let S represent the solubility of O_2 as millimoles of O_2 per liter of water. Let T be temperature in °C. Use the following data to obtain a curve fit for S as a function of T. Use the fit to estimate S when $T = 8°C$ and $T = 50°C$.

T (°C)	S (millimoles O_2 /L H_2O)
5	1.95
10	1.7
15	1.55
20	1.40
25	1.30
30	1.15
35	1.05
40	1.00
45	0.95

40. The following function is linear in the parameters a_1 and a_2.

$$y(x) = a_1 + a_2 \ln x$$

Use least squares regression with the following data to estimate the values of a_1 and a_2. Use the curve fit to estimate the values of y at $x = 1.5$ and at $x = 11$.

x	1	2	3	4	5	6	7	8	9	10
y	10	14	16	18	19	20	21	22	23	23

41. Chemists and engineers must be able to predict the changes in chemical concentration in a reaction. A model used for many single reactant processes is:

$$\text{Rate of change of concentration} = -kC^n$$

where C is the chemical concentration and k is the rate constant. The order of the reaction is the value of the exponent n. Solution methods for differential equations (which are discussed in Chapter 7) can show that the solution for a first-order reaction ($n = 1$) is

$$C(t) = C(0)e^{-kt}$$

The following data describes the reaction

$$(CH_3)_3CBr + H_2O \rightarrow (CH_3)_3COH + HBr$$

Use this data to obtain a least squares fit to estimate the value of k.

Time t (h)	C(mol of $(CH3)_3$ CBr/L)
0	0.1039
3.15	0.0896
6.20	0.0776
10.0	0.0639
18.3	0.0353
30.8	0.0207
43.8	0.0101

42. Chemists and engineers must be able to predict the changes in chemical concentration in a reaction. A model used for many single reactant processes is:

$$\text{Rate of change of concentration} = -kC^n$$

where C is the chemical concentration and k is the rate constant. The order of the reaction is the value of the exponent n. Solution methods for differential equations (which are discussed in Chapter 7) can show that the solution for a first-order reaction ($n = 1$) is

$$C(t) = C(0)e^{-kt}$$

and the solution for a second-order reaction ($n = 2$) is

$$\frac{1}{C(t)} = \frac{1}{C(0)} + kt$$

The following data (from Brown, 1994) describes the gas-phase decomposition of nitrogen dioxide at 300°C.

$$2NO_2 \rightarrow 2NO + O_2$$

Time t (s)	C(mol NO_2 /L)
0	0.0100
50	0.0079
100	0.0065
200	0.0048
300	0.0038

Determine whether this is a first-order or second-order reaction, and estimate the value of the rate constant k.

43. Chemists and engineers must be able to predict the changes in chemical concentration in a reaction. A model used for many single reactant processes is:

$$\text{Rate of change of concentration} = -kC^n$$

where C is the chemical concentration and k is the rate constant. The order of the reaction is the value of the exponent n. Solution methods for differential equations (which are discussed in Chapter 7) can show that the solution for a first-order reaction ($n = 1$) is

$$C(t) = C(0)e^{-kt}$$

The solution for a second-order reaction ($n = 2$) is

$$\frac{1}{C(t)} = \frac{1}{C(0)} + kt$$

and the solution for a third-order reaction ($n = 3$) is

$$\frac{1}{2C^2(t)} = \frac{1}{2C^2(0)} + kt$$

Time t (min)	C (mol of reactant/L)
5	0.3575
10	0.3010
15	0.2505
20	0.2095
25	0.1800
30	0.1500
35	0.1245
40	0.1070
45	0.0865

The preceding data describes a certain reaction. By examining the residuals, determine whether this is a first-order, second-order, or third-order reaction, and estimate the value of the rate constant k.

Section 5.7

44. The popular amusement ride known as the corkscrew has a helical shape. The parametric equations for a circular helix are

$$x = a \cos t$$

$$y = a \sin t$$

$$z = bt$$

where a is the radius of the helical path and b is a constant that determines the "tightness" of the path. In addition, if $b > 0$, the helix has the shape of a right-handed screw; if $b < 0$, the helix is left-handed.

Obtain the three-dimensional plot of the helix for the following three cases and compare their appearance with one another. Use $0 \le t \le 10\pi$ and $a = 1$.

a. $b = 0.1$

b. $b = 0.2$

c. $b = -0.1$

45. A robot rotates about its base at two revolutions per minute while lowering its arm and extending its hand. It lowers its arm at the rate of 120° per minute and extends its hand at the rate of 5 m/min. The arm is 0.5 m long. The xyz coordinates of the hand are given by

$$x = (0.5 + 5t) \sin \left(\frac{2\pi}{3} t \right) \cos (4\pi t)$$

$$y = (0.5 + 5t) \sin \left(\frac{2\pi}{3} t \right) \sin (4\pi t)$$

$$z = (0.5 + 5t) \cos \left(\frac{2\pi}{3} t \right)$$

where t is time in minutes.

Obtain the three-dimensional plot of the path of the hand for $0 \le t \le 0.2$ min.

46. Obtain the surface and contour plots for the function $z = x^2 - 2xy + 4y^2$, showing the minimum at $x = y = 0$.

47. Obtain the surface and contour plots for the function $z = -x^2 + 2xy + 3y^2$. This surface has the shape of a saddle. At its saddlepoint at $x = y = 0$, the surface has zero slope, but this point does not correspond to either a minimum or a maximum. What type of contour lines correspond to a saddlepoint?

48. Obtain the surface and contour plots for the function $z = (x-y^2)(x-3y^2)$. This surface has a singular point at $x = y = 0$, where the surface has zero slope, but this point does not correspond to either a minimum or a maximum. What type of contour lines correspond to a singular point?

49. A square metal plate is heated to 80°C at the corner corresponding to $x = y = 1$. The temperature distribution in the plate is described by

$$T = 80e^{-(x-1)^2}e^{-3(y-1)^2}$$

Obtain the surface and contour plots for the temperature. Label each axis. What is the temperature at the corner corresponding to $x = y = 0$?

50. The following function describes oscillations in some mechanical structures and electric circuits:

$$z(t) = e^{-t/\tau} \sin(\omega t + \phi)$$

In this function t is time, and ω is the oscillation frequency in radians per unit time. The oscillations have a period of $2\pi/\omega$, and their amplitudes decay in time at a rate determined by τ, which is called the *time constant*. The smaller τ is, the faster the oscillations die out.

Suppose that $\phi = 0$, $\omega = 2$, and τ can have values in the range $0.5 \le \tau \le 10$ sec. Then the preceding equation becomes

$$z(t) = e^{-t/\tau} \sin(2t)$$

Obtain a surface plot and a contour plot of this function to help visualize the effect of τ for $0 \le t \le 15$ sec. Let the x variable be time t and the y variable be τ.

51. The following equation describes the temperature distribution in a flat rectangular metal plate. The temperature on three sides is held constant at T_1, and at T_2 on the fourth side (see Figure P51). The temperature $T(x, y)$ as a function of the xy coordinates shown is given by

$$T(x,y) = (T_2 - T_1)w(x,y) + T_1$$

where

$$w(x,y) = \frac{2}{\pi} \sum_{n \text{ odd}}^{\infty} \frac{2}{n} \sin\left(\frac{n\pi x}{L}\right) \frac{\sinh(n\pi y/L)}{\sinh(n\pi W/L)}$$

The given data for this problem are: $T_1 = 70$°F, $T_2 = 200$°F, and $W = L = 2$ ft.

Using a spacing of 0.2 for both x and y, generate a surface mesh plot and a contour plot of the temperature distribution.

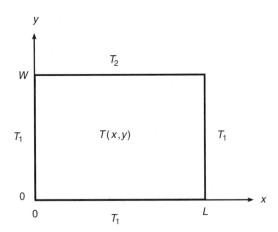

Figure P51

52. The electric potential field V at a point, due to two charged particles, is given by

$$V = \frac{1}{4\pi\epsilon_0}\left(\frac{q_1}{r_1} + \frac{q_2}{r_2}\right)$$

where q_1 and q_2 are the charges of the particles in coulombs (C), r_1 and r_2 are the distances of the charges from the point (in meters), and ϵ_0 is the permittivity of free space, whose value is

$$\epsilon_0 = 8.854 \times 10^{-12}\,\mathrm{C}^2/\mathrm{N}\cdot\mathrm{m}^2$$

Suppose the charges are $q_1 = 2 \times 10^{-10}$ C and $q_2 = 4 \times 10^{-10}$ C. Their respective locations in the xy plane are (0.3, 0) and (−0.3, 0) m. Plot the electric potential field on a 3D surface plot with V plotted on the z-axis over the ranges $-0.25 \le x \le 0.25$ and $-0.25 \le y \le 0.25$. Create the plot two ways: *a.* by using the `surf` function and *b.* by using the `meshc` function.

53. Refer to Problem 23 of Chapter 4. Use the function file created for that problem to generate a surface mesh plot and a contour plot of x versus h and W for $0 \le W \le 500$ N and for $0 \le h \le 2$ m. Use the values: $k_1 = 10^4$ N/m; $k_2 = 1.5 \times 10^4$ N/m; and $d = 0.1$ m.

54. Refer to Problem 26 of Chapter 4. To see how sensitive the cost is to location of the distribution center, obtain a surface plot and a contour plot of the total cost as a function of the x and y coordinates of the distribution center location. How much would the cost increase if we located the center 1 mi in any direction from the optimal location?

55. Refer to Example 3.2–1. Use a surface plot and a contour plot of the perimeter length L as a function of d and θ over the ranges $1 \le d \le 30$ ft and $0.1 \le \theta \le 1.5$ rad. Are there valleys other than the one corresponding to $d = 7.5984$ and $\theta = 1.0472$? Are there any saddle points?

Statistics, Probability, and Interpolation

This chapter begins with an introduction to basic statistics in Section 6.1. You will see how to obtain and interpret *histograms,* which are specialized plots for displaying statistical results. The *normal distribution,* commonly called the *bell-shaped curve,* forms the basis of much of probability theory and many statistical methods. It is covered in Section 6.2. In Section 6.3 you will see how to include random processes in your simulation programs. In Section 6.4 you will see how to use interpolation with data tables to estimate values that are not in the table.

When you have finished this chapter, you should be able to use MATLAB to do the following:

- Solve basic problems in statistics and probability.
- Create simulations incorporating random processes.
- Apply interpolation techniques.

6.1 Statistics and Histograms

With MATLAB you can compute the *mean* (the average), the *mode* (the most frequently occurring value), and the *median* (the middle value) of a set of data. MATLAB provides the `mean(x)`, `mode(x)` and `median(x)` functions to compute the mean, mode, and median of the data values stored in x, if x is a vector. However, if x is a matrix, a row vector is returned containing the mean (or mode or median) value of each column of x. These functions do not require the elements in x to be sorted in ascending or descending order.

The way the data is spread around the mean can be described by a *histogram* plot. A *histogram* is a plot of the frequency of occurrence of data values versus the values themselves. It is a bar plot of the number of data values that occur within each range, with the bar centered in the middle of the range.

To plot a histogram, you must group the data into subranges, called *bins*. The choice of the bin width and bin center can drastically change the shape of the histogram. If the number of data values is relatively small, the bin width cannot be small because some of the bins will contain no data and the resulting histogram might not usefully illustrate the distribution of the data.

To obtain a histogram, first sort the data if it has not yet been sorted (you can use the `sort` function here). Then choose the bin ranges and bin centers and count the number of values in each bin. Use the `bar` function to plot the number of values in each bin versus the bin centers as a bar chart. The function `bar(x,y)` creates a bar chart of y versus x.

MATLAB also provides the `hist` command to generate a histogram. This command has several forms. Its basic form is `hist(y)`, where y is a vector containing the data. This form aggregates the data into 10 bins evenly spaced between the minimum and maximum values in y. The second form is `hist(y,n)`, where n is a user-specified scalar indicating the number of bins. The third form is `hist(y,x)`, where x is a user-specified vector that determines the location of the bin centers; the bin widths are the distances between the centers.

EXAMPLE 6.1–1	Breaking Strength of Thread

To ensure proper quality control, a thread manufacturer selects samples and tests them for breaking strength. Suppose that 20 thread samples are pulled until they break, and the breaking force is measured in newtons rounded off to integer values. The breaking force values recorded were 92, 94, 93, 96, 93, 94, 95, 96, 91, 93, 95, 95, 95, 92, 93, 94, 91, 94, 92, and 93. Plot the histogram of the data.

■ Solution

Store the data in the vector y, which is shown in the following script file. Because there are six outcomes (91, 92, 93, 94, 95, 96 N), we choose six bins. However, if you use `hist(y,6)`, the bins will not be centered at 91, 92, 93, 94, 95, and 96. So use the form

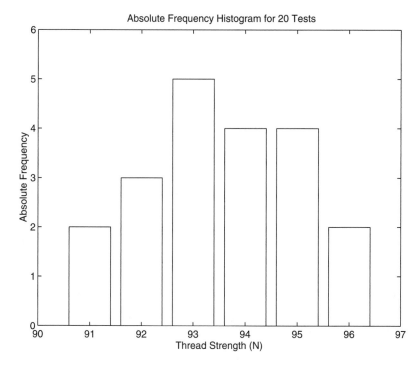

Figure 6.1–1 Histograms for 20 tests of thread strength.

hist(y,x), where x = [91:96]. The following script file generates the histogram shown in Figure 6.1–1.

```
% Thread breaking strength data for 20 tests.
y = [92,94,93,96,93,94,95,96,91,93,...
    95,95,95,92,93,94,91,94,92,93];
% The six possible outcomes are 91,92,93,94,95,96.
x = [91:96];
hist(y,x),axis([90 97 0 6]),ylabel('Absolute Frequency'),...
    xlabel('Thread Strength (N)'),...
    title('Absolute Frequency Histogram for 20 Tests')
```

The *absolute frequency* is the number of times a particular outcome occurs. For example, in 20 tests this data shows that a 95 occurred four times. The absolute frequency is 4, and its *relative frequency* is 4/20, or 20 percent of the time.

When there is a large amount of data, you can avoid typing in every data value by first aggregating the data. The following example shows how this is done using the ones function. The following data was generated by testing 100 thread

ABSOLUTE FREQUENCY

RELATIVE FREQUENCY

samples. The number of times 91, 92, 93, 94, 95, or 96 N was measured is 13, 15, 22, 19, 17, and 14, respectively.

```
% Thread strength data for 100 tests.
y = [91*ones(1,13),92*ones(1,15),93*ones(1,22),...
    94*ones(1,19),95*ones(1,17),96*ones(1,14)];
x = [91:96];
hist(y,x),ylabel('Absolute Frequency'),...
    xlabel('Thread Strength (N)'),...
    title('Absolute Frequency Histogram for 100 Tests')
```

The result appears in Figure 6.1–2.

The `hist` function is somewhat limited in its ability to produce useful histograms. Unless all the outcome values are the same as the bin centers (as is the case with the thread examples), the graph produced by the `hist` function will not be satisfactory. This case occurs when you want to obtain a *relative* frequency histogram. In such cases you can use the `bar` function to generate the histogram. The following script file generates the relative frequency histogram for the 100 thread tests. Note that if you use the `bar` function, you must aggregate the data first.

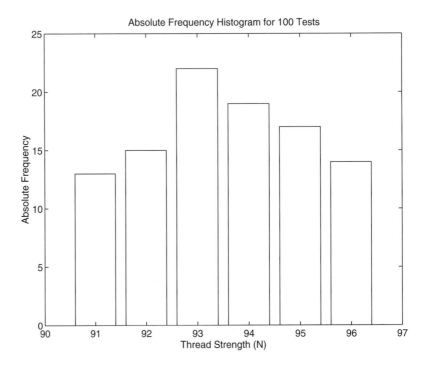

Figure 6.1–2 Absolute frequency histogram for 100 thread tests.

```
% Relative frequency histogram using the bar function.
tests = 100;
y = [13,15,22,19,17,14]/tests;
x = [91:96];
bar(x,y),ylabel('Relative Frequency'),...
    xlabel('Thread Strength (N)'),...
    title('Relative Frequency Histogram for 100 Tests')
```

The result appears in Figure 6.1–3.

The fourth, fifth, and sixth forms of the `hist` function do not generate a plot, but are used to compute the frequency counts and bin locations. The `bar` function can then be used to plot the histogram. The syntax of the fourth form is `[z,x] = hist(y)`, where z is the returned vector containing the frequency count and x is the returned vector containing the bin locations. The fifth and sixth forms are `[z,x] = hist(y,n)` and `[z,x] = hist(y,x)`. In the latter case the returned vector x is the same as the user-supplied vector. The following script file shows how the sixth form can be used to generate a relative frequency histogram for the thread example with 100 tests.

```
tests = 100;
y = [91*ones(1,13),92*ones(1,15),93*ones(1,22),...
```

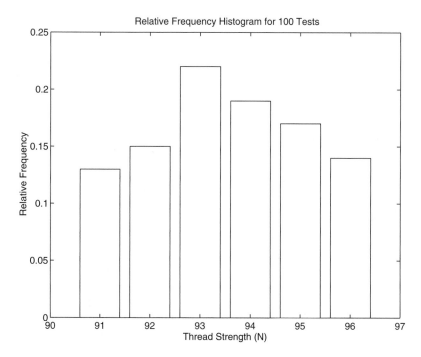

Figure 6.1–3 Relative frequency histogram for 100 thread tests.

Table 6.1–1 Histogram functions

Command	Description
bar(x,y)	Creates a bar chart of y versus x.
hist(y)	Aggregates the data in the vector y into 10 bins evenly spaced between the minimum and maximum values in y.
hist(y,n)	Aggregates the data in the vector y into n bins evenly spaced between the minimum and maximum values in y.
hist(y,x)	Aggregates the data in the vector y into bins whose center locations are specified by the vector x. The bin widths are the distances between the centers.
[z,x] = hist(y)	Same as hist(y) but returns two vectors z and x that contain the frequency count and the bin locations.
[z,x] = hist(y,n)	Same as hist(y,n) but returns two vectors z and x that contain the frequency count and the bin locations.
[z,x] = hist(y,x)	Same as hist(y,x) but returns two vectors z and x that contain the frequency count and the bin locations. The returned vector x is the same as the user-supplied vector x.

```
    94*ones(1,19),95*ones(1,17),96*ones(1,14)];
x = [91:96];
[z,x] = hist(y,x);bar(x,z/tests),...
ylabel('Relative Frequency'),xlabel('Thread Strength (N)'),...
    title('Relative Frequency Histogram for 100 Tests')
```

The plot generated by this M-file will be identical to that shown in Figure 6.1–3. These commands are summarized in Table 6.1–1.

Test Your Understanding

T6.1–1   In 50 tests of thread, the number of times 91, 92, 93, 94, 95, or 96 N was measured was 7, 8, 10, 6, 12, and 7, respectively. Obtain the absolute and relative frequency histograms.

The Data Statistics Tool

With the Data Statistics tool you can calculate statistics for data and add plots of the statistics to a graph of the data. The tool is accessed from the Figure window after you plot the data. Click on the **Tools** menu, then select **Data Statistics.** The menu appears as shown in Figure 6.1–4. To plot the mean of the dependent variable (y), click the box in the row labeled mean under the column labeled Y, as in the figure. You can plot other statistics as well; these are shown in the figure. You can save the statistics to the workspace as a structure by clicking on the **Save to Workspace** button. This opens a dialog

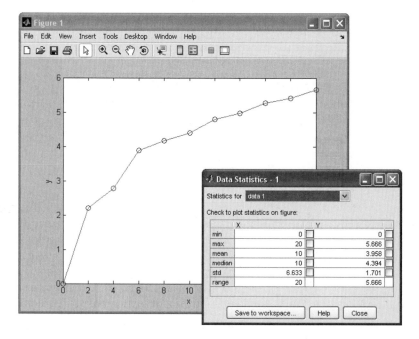

Figure 6.1–4 The Data Statistics tool.

box that prompts you for a name for the structure containing the x data, and a name for the y data structure.

6.2 The Normal Distribution

Rolling a die is an example of a process whose possible outcomes are a limited set of numbers; namely, the integers from 1 to 6. For such processes the probability is a function of a discrete-valued variable, that is, a variable having a limited number of values. For example, Table 6.2–1 gives the measured heights of 100 men 20 years of age. The heights were recorded to the nearest 1/2 in., so the height variable is discrete valued.

Scaled Frequency Histogram

You can plot the data as a histogram using either the absolute or relative frequencies. However, another useful histogram uses data scaled so that the to-
tal area under the histogram's rectangles is 1. This *scaled frequency histogram* is the absolute frequency histogram divided by the total area of that histogram. The area of each rectangle on the absolute frequency histogram equals the bin width times the absolute frequency for that bin. Because all the rectangles have the same width, the total area is the bin width times the sum of the absolute frequencies. The following M-file produces the scaled histogram shown in Figure 6.2–1.

SCALED FREQUENCY HISTOGRAM

Table 6.2–1 Height data for men 20 years of age

Height (in.)	Frequency	Height (in.)	Frequency
64	1	70	9
64.5	0	70.5	8
65	0	71	7
65.5	0	71.5	5
66	2	72	4
66.5	4	72.5	4
67	5	73	3
67.5	4	73.5	1
68	8	74	1
68.5	11	74.5	0
69	12	75	1
69.5	10		

```
% Absolute frequency data.
y_abs=[1,0,0,0,2,4,5,4,8,11,12,10,9,8,7,5,4,4,3,1,1,0,1];
binwidth = 0.5;
% Compute scaled frequency data.
area = binwidth*sum(y_abs);
y_scaled = y_abs/area;
```

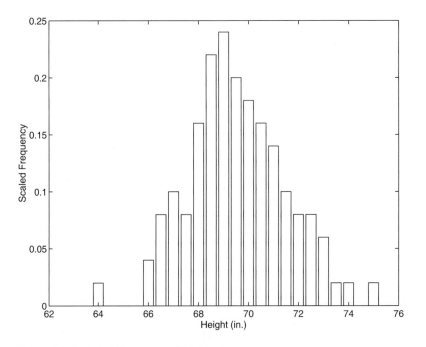

Figure 6.2–1 Scaled histogram of height data.

```
% Define the bins.
bins = [64:binwidth:75];
% Plot the scaled histogram.
bar(bins,y_scaled),...
   ylabel('Scaled Frequency'),xlabel('Height (in.)')
```

Because the total area under the scaled histogram is 1, the fractional area corresponding to a range of heights gives the probability that a randomly selected 20-year-old man will have a height in that range. For example, the heights of the scaled histogram rectangles corresponding to heights of 67 through 69 in. are 0.1, 0.08, 0.16, 0.22, and 0.24. Because the bin width is 0.5, the total area corresponding to these rectangles is $(0.1+0.08+0.16+0.22+0.24)(0.5) = 0.4$. Thus 40 percent of the heights lie between 67 and 69 in.

You can use the `cumsum` function to calculate areas under the scaled frequency histogram, and therefore calculate probabilities. If `x` is a vector, `cumsum(x)` returns a vector the same length as `x`, whose elements are the sum of the previous elements. For example, if `x = [2, 5, 3, 8]`, `cumsum(x) = [2, 7, 10, 18]`. If `A` is a matrix, `cumsum(A)` computes the cumulative sum of each row. The result is a matrix the same size as `A`.

After running the previous script, the last element of `cumsum(y_scaled) * binwidth` is 1, which is the area under the scaled frequency histogram. To compute the probability of a height lying between 67 and 69 in. (that is, above the 6th value up to the 11th value, type)

```
>>prob = cumsum(y_scaled)*binwidth;
>>prob67_69 = prob(11)-prob(6)
```

The result is `prob67_69 = 0.4000`, which agrees with our previous calculation of 40 percent.

Continuous Approximation to the Scaled Histogram

For processes having an infinite number of possible outcomes, the probability is a function of a *continuous* variable and is plotted as a curve rather than as rectangles. It is based on the same concept as the scaled histogram; that is, the total area under the curve is 1, and the fractional area gives the probability of occurrence of a specific range of outcomes. A probability function that describes many processes is the *normal* or *Gaussian* function, which is shown in Figure 6.2–2.

This function is also known as the "bell-shaped curve." Outcomes that can be described by this function are said to be "normally distributed." The normal probability function is a two-parameter function; one parameter, μ, is the mean of the outcomes, and the other parameter, σ, is the *standard deviation*. The mean μ locates the peak of the curve and is the most likely value to occur. The width, or spread, of the curve is described by the parameter σ. Sometimes the term *variance* is used to describe the spread of the curve. The variance is the square of the standard deviation σ.

NORMAL OR GAUSSIAN FUNCTION

NORMALLY DISTRIBUTED

STANDARD DEVIATION

VARIANCE

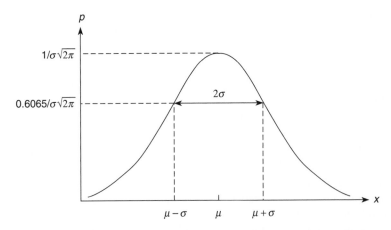

Figure 6.2–2 The basic shape of the normal distribution curve.

The normal probability function is described by the following equation:

$$p(x) = \frac{1}{\sigma\sqrt{2\pi}} e^{-(x-\mu)^2/2\sigma^2} \tag{6.2–1}$$

It can be shown that approximately 68 percent of the area lies between the limits of $\mu - \sigma \le x \le \mu + \sigma$. Consequently, if a variable is normally distributed, there is a 68 percent chance that a randomly selected sample will lie within one standard deviation of the mean. In addition, approximately 96 percent of the area lies between the limits of $\mu - 2\sigma \le x \le \mu + 2\sigma$, and 99.7 percent, or practically 100 percent, of the area lies between the limits of $\mu - 3\sigma \le x \le \mu + 3\sigma$.

The functions mean(x), var(x), and std(x) compute the mean, variance, and standard deviation of the elements in the vector x.

EXAMPLE 6.2–1   Mean and Standard Deviation of Heights

Statistical analysis of data on human proportions is required in many engineering applications. For example, designers of submarine crewquarters need to know how small they can make bunk lengths without eliminating a large percentage of prospective crew members. Use MATLAB to estimate the mean and standard deviation for the height data given in Table 6.2–1.

■ **Solution**

The script file follows. The data given in Table 6.2–1 is the absolute frequency data and is stored in the vector y_abs. A bin width of 1/2 in. is used because the heights were measured to the nearest 1/2 in. The vector bins contains the heights in 1/2 in. increments.

To compute the mean and standard deviation, reconstruct the original (raw) height data from the absolute frequency data. Note that this data has some zero entries. For example, none of the 100 men had a height of 65 in. Thus to reconstruct the raw data, start with a empty vector y_raw and fill it with the height data obtained from the absolute frequencies. The `for` loop checks to see whether the absolute frequency for a particular bin is nonzero. If it is nonzero, append the appropriate number of data values to the vector `y_raw`. If the particular bin frequency is 0, `y_raw` is left unchanged.

```
% Absolute frequency data.
y_abs = [1,0,0,0,2,4,5,4,8,11,12,10,9,8,7,5,4,4,3,1,1,0,1];
binwidth = 0.5;
% Define the bins.
bins = [64:binwidth:75];
% Fill the vector y_raw with the raw data.
% Start with an empty vector.
y_raw = [];
for i = 1:length(y_abs)
   if y_abs(i)>0
      new = bins(i)*ones(1,y_abs(i));
   else
      new = [];
   end
y_raw = [y_raw,new];
end
% Compute the mean and standard deviation.
mu = mean(y_raw),sigma = std(y_raw)
```

When you run this program, you will find that the mean is $\mu = 69.6$ in. and the standard deviation is $\sigma = 1.96$ in.

If you need to compute probabilities based on the normal distribution, you can use the `erf` function. Typing `erf(x)` returns the area to the left of the value $t = x$ under the curve of the function $2e^{-t^2}/\sqrt{\pi}$. This area, which is a function of x, is known as the *error function,* and is written as erf(x). The probability that the random variable x is less than or equal to b is written as $P(x \le b)$ if the outcomes are normally distributed. This probability can be computed from the error function as follows:

ERROR FUNCTION

$$P(x \le b) = \frac{1}{2}\left[1 + \mathrm{erf}\left(\frac{b - \mu}{\sigma\sqrt{2}}\right)\right] \qquad (6.2\text{–}2)$$

The probability that the random variable x is no less than a and no greater than b is written as $P(a \le x \le b)$. It can be computed as follows:

$$P(a \le x \le b) = \frac{1}{2}\left[\mathrm{erf}\left(\frac{b - \mu}{\sigma\sqrt{2}}\right) - \mathrm{erf}\left(\frac{a - \mu}{\sigma\sqrt{2}}\right)\right] \qquad (6.2\text{–}3)$$

Estimation of Height Distribution

EXAMPLE 6.2–2

Use the results of Example 6.2–1 to estimate how many 20-year-old men are no taller than 68 in. How many are within 3 in. of the mean?

■ Solution

In Example 6.2–1 the mean and standard deviation were found to be $\mu = 69.3$ in. and $\sigma = 1.96$ in. In Table 6.2–1, note that few data points are available for heights less than 68 in. However, if you assume that the heights are normally distributed, you can use equation (6.2–2) to estimate how many men are shorter than 68 in. Use (6.2–2) with $b = 68$; that is,

$$P(x \leq 68) = \frac{1}{2} \left[1 + \mathrm{erf}\left(\frac{68 - 69.3}{1.96\sqrt{2}} \right) \right]$$

To determine how many men are within 3 in. of the mean, use (6.2–3) with $a = \mu - 3 = 66.3$ and $b = \mu + 3 = 72.3$; that is,

$$P(66.3 \leq x \leq 72.3) = \frac{1}{2} \left[\mathrm{erf}\left(\frac{3}{1.96\sqrt{2}} \right) - \mathrm{erf}\left(\frac{-3}{1.96\sqrt{2}} \right) \right]$$

In MATLAB these expressions are computed in a script file as follows:

```
mu = 69.3;
sigma = 1.96;
% How many are no taller than 68 inches?
b1 = 68;
P1 = (1+erf((b1-mu)/(sigma*sqrt(2))))/2
% How many are within 3 inches of the mean?
a2 = 66.3;
b2 = 72.3;
P2 = (erf((b2-mu)/(sigma*sqrt(2)))-erf((a2-mu)/(sigma* sqrt(2))))/2
```

When you run this program, you obtain the results P1 = 0.2536 and P2 = 0.8741. Thus 25 percent of 20-year-old men are estimated to be 68 inches or less in height, and 87 percent are estimated to be between 66.3 and 72.3 inches tall.

Test Your Understanding

T6.2–1 Suppose that 10 more height measurements are obtained so that the following numbers must be *added* to Table 6.2–1.

Height (in.)	Additional data
64.5	1
65	2
66	1
67.5	2
70	2
73	1
74	1

(a) Plot the scaled frequency histogram. (b) Find the mean and standard deviation. (c) Use the mean and standard deviation to estimate how many

20-year-old men are no taller than 69 in. (d) Estimate how many are between 68 and 72 in. tall.
(Answers: (b) mean = 69.4 in., standard deviation = 2.14 in.; (c) 43 percent; (d) 63 percent.)

Sums and Differences of Random Variables

It can be proved that the mean of the sum (or difference) of two independent normally distributed random variables equals the sum (or difference) of their means, but the variance is always the sum of the two variances. That is, if x and y are normally distributed with means μ_x and μ_y, and variances σ_x^2 and σ_y^2, and if $u = x + y$ and $v = x - y$, then

$$\mu_u = \mu_x + \mu_y \qquad (6.2-4)$$

$$\mu_v = \mu_x - \mu_y \qquad (6.2-5)$$

$$\sigma_u^2 = \sigma_v^2 = \sigma_x^2 + \sigma_y^2 \qquad (6.2-6)$$

These properties are applied in some of the homework problems.

6.3 Random Number Generation

We often do not have a simple probability distribution to describe the distribution of outcomes in many engineering applications. For example, the probability that a circuit consisting of many components will fail is a function of the number and the age of the components, but we often cannot obtain a function to describe the failure probability. In such cases we often resort to simulation to make predictions. The simulation program is executed many times, using a random set of numbers to represent the failure of one or more components, and the results are used to estimate the desired probability.

Uniformly Distributed Numbers

In a sequence of *uniformly distributed* random numbers, all values within a given interval are equally likely to occur. The MATLAB function `rand` generates random numbers uniformly distributed over the interval [0,1]. Type `rand` to obtain a single random number in the interval [0,1]. Typing `rand` again generates a different number because the MATLAB algorithm used for the `rand` function requires a "state" to start. MATLAB obtains this state from the computer's CPU clock. Thus every time the `rand` function is used, a different result will be obtained. For example,

UNIFORMLY DISTRIBUTED

```
rand
ans =
     0.6161
rand
ans =
     0.5184
```

Type `rand(n)` to obtain an $n \times n$ matrix of uniformly distributed random numbers in the interval [0, 1]. Type `rand(m,n)` to obtain an $m \times n$ matrix of random numbers. For example, to create a 1×100 vector `y` having 100 random values in the interval [0, 1], type `y = rand(1,100)`. Using the `rand` function this way is equivalent to typing `rand` 100 times. Even though there is a single call to the `rand` function, the `rand` function's calculation has the effect of using a different state to obtain each of the 100 numbers so that they will be random.

Use `Y = rand(m,n,p,...)` to generate a multidimensional array `Y` having random elements. Typing `rand(size(A))` produces an array of random entries that is the same size as `A`.

For example, the following script makes a random choice between two equally probable alternatives.

```
if rand < 0.5
   disp('heads')
else
   disp('tails')
end
```

In order to compare the results of two or more simulations, you sometimes will need to generate the same sequence of random numbers each time the simulation runs. To generate the same sequence, you must use the same state each time. The current state `s` of the uniform number generator can be obtained by typing `s = rand('twister')`. This returns a vector containing the current state of the uniform generator. To set the state of the generator to `s`, type `rand('twister',s)`. Typing `rand('twister',0)` resets the generator to its initial state. Typing `rand('twister',j)`, for integer `j`, resets the generator to state `j`. Typing `rand('twister',sum(100*clock))` resets the generator to a different state each time. Table 6.3–1 summarizes these functions.

The name `'twister'` refers to the specific algorithm used by MATLAB to generate random numbers. In MATLAB Version 4, `'seed'` was used instead of `'twister'`. In Versions 5 through 7.3, `'state'` was used. Use `'twister'` in Version 7.4 and later. The following session shows how to obtain the same sequence every time `rand` is called.

```
>>rand('twister',0)
>>rand
ans =
    0.5488
>>rand
ans =
    0.7152
>>rand('twister',0)
>>rand
```

Table 6.3–1 Random number functions

Command	Description
rand	Generates a single uniformly distributed random number between 0 and 1.
rand(n)	Generates an $n \times n$ matrix containing uniformly distributed random numbers between 0 and 1.
rand(m,n)	Generates an $m \times n$ matrix containing uniformly distributed random numbers between 0 and 1.
s = rand('state')	Returns a vector s containing the current state of the uniformly distributed generator.
rand('twister',s)	Sets the state of the uniformly distributed generator to s.
rand('twister',0)	Resets the uniformly distributed generator to its initial state.
rand('twister',j)	Resets the uniformly distributed generator to state j, for integer j.
rand('twister',sum(100*clock))	Resets the uniformly distributed generator to a different state each time it is executed.
randn	Generates a single normally distributed random number having a mean of 0 and a standard deviation of 1.
randn(n)	Generates an $n \times n$ matrix containing normally distributed random numbers having a mean of 0 and a standard deviation of 1.
randn(m,n)	Generates an $m \times n$ matrix containing normally distributed random numbers having a mean of 0 and a standard deviation of 1.
s = randn('state')	Like rand('state') but for the normally distributed generator.
randn('state',s)	Like rand('state',s) but for the normally distributed generator.
randn('state',0)	Like rand('state',0) but for the normally distributed generator.
randn('state',j)	Like rand('state',j) but for the normally distributed generator.
randn('state',sum(100*clock))	Like rand('state',sum(100*clock)) but for the normally distributed generator.
randperm(n)	Generates a random permutation of the integers from 1 to n.

```
ans =
    0.5488
>>rand
ans =
    0.7152
```

You need not start with the initial state in order to generate the same sequence. To show this, continue the above session as follows.

```
>>s = rand('twister');
>>rand('twister',s)
>>rand
ans =
    0.6028
>>rand('twister',s)
>>rand
ans =
    0.6028
```

You can use the `rand` function to generate random numbers in an interval other than [0, 1]. For example, to generate values in the interval [2, 10], first generate a random number between 0 and 1, multiply it by 8 (the difference between the upper and lower bounds), and then add the lower bound (2). The result is a value that is uniformly distributed in the interval [2, 10]. The general formula for generating a uniformly distributed random number y in the interval [a, b] is

$$y = (b - a)x + a \qquad (6.3-1)$$

where x is a random number uniformly distributed in the interval [0, 1]. For example, to generate a vector y containing 1000 uniformly distributed random numbers in the interval [2, 10], you type `y = 8*rand(1,1000)+2`. You can check the results with the `mean`, `min`, and `max` functions. You should obtain values close to 6, 2, and 10, respectively.

You can use `rand` to generate random results for games involving dice, for example, but you must use it to create integers. An easier way is to use the `randperm(n)` function, which generates a random permutation of the integers from 1 to n. For example, `randperm(6)` might generate the vector [3 2 6 4 1   5], or some other permutation of the numbers from 1 to 6. Note that `randperm` calls `rand` and therefore changes the state of the generator.

Test Your Understanding

T6.3–1 Use MATLAB to generate a vector y containing 1500 uniformly distributed random numbers in the interval [-5, 15]. Check your results with the `mean`, `min`, and `max` functions.

Normally Distributed Random Numbers

In a sequence of normally distributed random numbers, the values near the mean are more likely to occur. We have noted that the outcomes of many processes can be described by the normal distribution. Although a uniformly distributed random variable has definite upper and lower bounds, a normally distributed random variable does not.

The MATLAB function `randn` will generate a single number that is normally distributed with a mean equal to 0 and a standard deviation equal to 1. Type `randn(n)` to obtain an $n \times n$ matrix of such numbers. Type `randn(m,n)` to obtain an $m \times n$ matrix of random numbers.

The functions for retrieving and specifying the state of the normally distributed random number generator are identical to those for the uniformly distributed generator, except that `randn(...)` replaces `rand(...)` in the syntax and `'state'` is used instead of `'twister'`. These functions are summarized in Table 6.3–1.

You can generate a sequence of normally distributed numbers having a mean μ and standard deviation σ from a normally distributed sequence having a mean of 0 and a standard deviation of 1. You do this by multiplying the values by σ and adding μ to each result. Thus if x is a random number with a mean of 0 and a standard deviation of 1, use the following equation to generate a new random number y having a standard deviation of σ and a mean of μ.

$$y = \sigma x + \mu \qquad (6.3\text{--}2)$$

For example, to generate a vector `y` containing 2000 random numbers normally distributed with a mean of 5 and a standard deviation of 3, you type `y = 3*randn(1,2000) + 5`. You can check the results with the `mean` and `std` functions. You should obtain values close to 5 and 3, respectively.

Test Your Understanding

T6.3–2 Use MATLAB to generate a vector `y` containing 1800 random numbers normally distributed with a mean of 7 and a standard deviation of 10. Check your results with the `mean` and `std` functions. Why can't you use the `min` and `max` functions to check your results?

Functions of Random Variables   If y and x are linearly related, as

$$y = bx + c \qquad (6.3\text{--}3)$$

and if x is normally distributed with a mean μ_x and standard deviation σ_x, it can be shown that the mean and standard deviation of y are given by

$$\mu_y = b\mu_x + c \qquad (6.3\text{--}4)$$

$$\sigma_y = |b|\sigma_x \qquad (6.3\text{--}5)$$

However, it is easy to see that the means and standard deviations do not combine in a straightforward fashion when the variables are related by a nonlinear function. For example, if x is normally distributed with a mean of 0, and if $y = x^2$, it is easy to see that the mean of y is not 0, but is positive. In addition, y is not normally distributed.

Some advanced methods are available for deriving a formula for the mean and variance of $y = f(x)$, but for our purposes, the simplest way is to use random number simulation.

It was noted in the previous section that the mean of the sum (or difference) of two independent normally distributed random variables equals the sum (or difference) of their means, but the variance is always the sum of the two variances. However, if z is a nonlinear function of x and y, then the mean and variance of z cannot be found with a simple formula. In fact, the distribution of z will not even be normal. This outcome is illustrated by the following example.

Statistical Analysis and Manufacturing Tolerances   EXAMPLE 6.3–1

Suppose you must cut a triangular piece off the corner of a square plate by measuring the distances x and y from the corner (see Figure 6.3–1). The desired value of x is 10 in., and

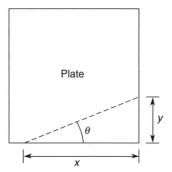

Figure 6.3–1 Dimensions of a triangular cut.

the desired value of θ is 20°. This requires that $y = 3.64$ in. We are told that measurements of x and y are normally distributed with means of 10 and 3.64, respectively, with a standard deviation equal to 0.05 in. Determine the standard deviation of θ and plot the relative frequency histogram for θ.

■ Solution

From Figure 6.3–1, we see that the angle θ is determined by $\theta = \tan^{-1}(y/x)$. We can find the statistical distribution of θ by creating random variables x and y that have means of 10 and 3.64, respectively, with a standard deviation of 0.05. The random variable θ is then found by calculating $\theta = \tan^{-1}(y/x)$ for each random pair (x, y). The following script file shows this procedure.

```
s = 0.05; % standard deviation of x and y
n = 8000; % number of random simulations
x = 10 + s*randn(1,n);
y = 3.64 + s*randn(1,n);
theta = (180/pi)*atan(y./x);
mean_theta = mean(theta)
sigma_theta = std(theta)
xp = [19:0.1:21];
z = hist(theta,xp);
yp = z/n;
bar(xp,yp),xlabel('Theta (degrees)'),ylabel('Relative Frequency')
```

The choice of 8000 simulations was a compromise between accuracy and the amount of time required to do the calculations. You should try different values of n and compare the results. The results gave a mean of 19.9993° for θ with a standard deviation of 0.2730°. The histogram is shown in Figure 6.3–2. Although the plot resembles the normal distribution, the values of θ are not distributed normally. From the histogram we can calculate that approximately 65 percent of the values of θ lie between 19.8 and 20.2. This range corresponds to a standard deviation of 0.2°, not 0.273° as calculated from the simulation data. Thus the curve is not a normal distribution.

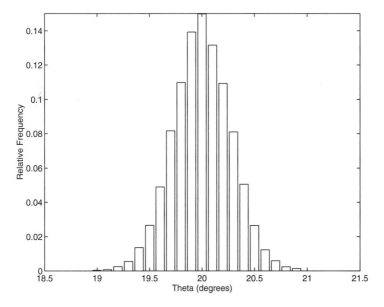

Figure 6.3–2 Scaled histogram of the angle θ.

This example shows that the interaction of two of more normally distributed variables does not produce a result that is normally distributed. In general, the result is normally distributed if and only if the result is a linear combination of the variables.

6.4 Interpolation

Paired data might represent a *cause and effect,* or *input-output relationship,* such as the current produced in a resistor as a result of an applied voltage, or a *time history,* such as the temperature of an object as a function of time. Another type of paired data represents a *profile,* such as a road profile (which shows the height of the road along its length). In some applications we want to estimate a variable's value between the data points. This process is called *interpolation*. In other cases we might need to estimate the variable's value outside of the given data range. This process is *extrapolation*. Interpolation and extrapolation are greatly aided by plotting the data. Such plots, some perhaps using logarithmic axes, often help to discover a functional description of the data.

INTERPOLATION

Suppose we have the following temperature measurements, taken once an hour starting at 7:00 A.M. The measurements at 8 A.M. and 10 A.M. are missing for some reason, perhaps because of equipment malfunction.

Time	7 A.M.	9 A.M.	11 A.M.	12 noon
Temperature (°F)	49	57	71	75

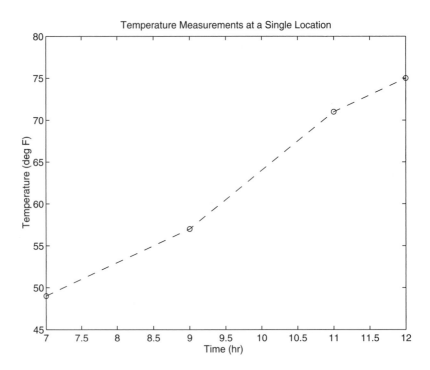

Figure 6.4–1 A plot of temperature data versus time.

A plot of this data is shown in Figure 6.4–1 with the data points connected by dashed lines. If we need to estimate the temperature at 10 A.M., we can read the value from the dashed line that connects the data points at 9 A.M. and 11 A.M. From the plot we thus estimate the temperature at 8 A.M. to be 53°F and at 10 A.M. to be 64°F. We have just performed *linear interpolation* on the data to obtain an *estimate* of the missing data. Linear interpolation is so named because it is equivalent to connecting the data points with a linear function (a straight line).

Of course we have no reason to believe that the temperature follows the straight lines shown in the plot, and our estimate of 64°F will most likely be incorrect, but it might be close enough to be useful. Using straight lines to connect the data points is the simplest form of interpolation. Another function could be used if we have a good reason to do so. Later in this section we use polynomial functions to do the interpolation.

Linear interpolation in MATLAB is obtained with the `interp1` and `interp2` functions. Suppose that x is a vector containing the independent variable data and that y is a vector containing the dependent variable data. If x_int is a vector containing the value or values of the independent variable at which we wish to estimate the dependent variable, then typing `interp1(x,y,x_int)` produces a vector the same size as x_int containing the interpolated values of *y* that correspond to x_int. For example, the following session produces an estimate of the temperatures at 8 A.M. and 10 A.M. from the preceding data. The vectors x and y contain the times and temperatures, respectively.

```
>>x = [7, 9, 11, 12];
>>y = [49, 57, 71, 75];
>>x_int = [8, 10];
>>interp1(x,y,x_int)
ans =
    53
    64
```

You must keep in mind two restrictions when using the interp1 function. The values of the independent variable in the vector x must be in ascending order, and the values in the interpolation vector x_int must lie within the range of the values in x. Thus we cannot use the interp1 function to estimate the temperature at 6 A.M., for example.

The interp1 function can be used to interpolate in a table of values by defining y to be a matrix instead of a vector. For example, suppose we now have temperature measurements at three locations and that the measurements at 8 A.M. and 10 A.M. are missing for all three locations. The data is

Time	Temperatures (°F)		
	Location 1	**Location 2**	**Location 3**
7 A.M.	49	52	54
9 A.M.	57	60	61
11 A.M.	71	73	75
12 noon	75	79	81

We define x as before, but now we define y to be a matrix whose three columns contain the second, third, and fourth columns of the preceding table. The following session produces an estimate of the temperatures at 8 A.M. and 10 A.M. at each location.

```
>>x = [7, 9, 11, 12]';
>>y(:,1) = [49, 57, 71, 75]';
>>y(:,2) = [52, 60, 73, 79]';
>>y(:,3) = [54, 61, 75, 81]';
>>x_int = [8, 10]';
>>interp1(x,y,x_int)
ans =
    53.0000    56.0000    57.5000
    64.0000    65.5000    68.0000
```

Thus the estimated temperatures at 8 A.M. at each location are 53, 56, and 57.5°F, respectively. At 10 A.M. the estimated temperatures are 64, 65.5, and 68. From this example we see that if the first argument x in the `interp1(x,y,x_int)` function is a *vector* and the second argument y is a *matrix*, the function interpolates between the rows of y and computes a matrix having the same number of columns as y and the same number of rows as the number of values in `x_int`.

Note that we need not define two separate vectors x and y. Rather, we can define a single matrix that contains the entire table. For example, by defining the matrix `temp` to be the preceding table, the session would look like this:

```
>>temp(:,1) = [7, 9, 11, 12]';
>>temp(:,2) = [49, 57, 71, 75]';
>>temp(:,3) = [52, 60, 73, 79]';
>>temp(:,4) = [54, 61, 75, 81]';
>>x_int = [8, 10]';
>>interp1(temp(:,1),temp(:,2:4),x_int)
ans =
      53.0000       56.0000       57.5000
      64.0000       65.5000       68.0000
```

Two-Dimensional Interpolation

Now suppose that we have temperature measurements at four locations at 7 A.M. These locations are at the corners of a rectangle 1 mi wide and 2 mi long. Assigning a coordinate system origin (0, 0) to the first location, the coordinates of the other locations are (1, 0), (1, 2), and (0, 2); see Figure 6.4–2. The temperature measurements are shown in the figure. The temperature is a function of two

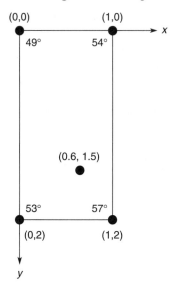

Figure 6.4–2 Temperature measurements at four locations.

variables, the coordinates x and y. MATLAB provides the `interp2` function to interpolate functions of two variables. If the function is written as $z = f(x, y)$ and we wish to estimate the value of z for $x = x_i$ and $y = y_i$, the syntax is `interp2(x,y,z,x_i,y_i)`.

Suppose we want to estimate the temperature at the point whose coordinates are (0.6, 1.5). Put the x coordinates in the vector `x` and the y coordinates in the vector `y`. Then put the temperature measurements in a matrix `z` such that going across a row represents an increase in x and going down a column represents an increase in y. The session to do this is as follows:

```
>>x = [0,1];
>>y = [0,2];
>>z = [49,54;53,57]
z =
    49    54
    53    57
>>interp2(x,y,z,0.6,1.5)
ans =
    54.5500
```

Thus the estimated temperature is 54.55°.

The syntax of the `interp1` and `interp2` functions is summarized in Table 6.4–1. MATLAB also provides the `interpn` function for interpolating multidimensional arrays.

Cubic-Spline Interpolation

High-order polynomials can exhibit undesired behavior between the data points, and this can make them unsuitable for interpolation. An alternative procedure that is widely used is to fit the data points using a lower-order polynomial between *each pair* of adjacent data points. This method is called *spline* interpolation and is so named for the splines used by illustrators to draw a smooth curve through a set of points.

Spline interpolation obtains an exact fit that is also smooth. The most common procedure uses cubic polynomials, called *cubic splines,* and thus is called

CUBIC SPLINES

Table 6.4–1 Linear interpolation functions

Command	Description
`y_int=interp1(x,y,x_int)`	Used to linearly interpolate a function of one variable: $y = f(x)$. Returns a linearly interpolated vector `y_int` at the specified value `x_int`, using data stored in `x` and `y`.
`z_int=interp2(x,y,z,x_,y_int)`	Used to linearly interpolate a function of two variables: $y = f(x, y)$. Returns a linearly interpolated vector `z_int` at the specified values `x_int` and `y_int`, using data stored in `x`, `y`, and `z`.

cubic-spline interpolation. If the data is given as n pairs of (x, y) values, then $n - 1$ cubic polynomials are used. Each has the form

$$y_i(x) = a_i(x - x_i)^3 + b_i(x - x_i)^2 + c_i(x - x_i) + d_i$$

for $x_i \leq x \leq x_{i+1}$ and $i = 1, 2, \ldots, n - 1$. The coefficients a_i, b_i, c_i, and d_i for each polynomial are determined so that the following three conditions are satisfied for each polynomial:

1. The polynomial must pass through the data points at its endpoints at x_i and x_{i+1}.
2. The slopes of adjacent polynomials must be equal at their common data point.
3. The curvatures of adjacent polynomials must be equal at their common data point.

For example, a set of cubic splines for the temperature data given earlier follows (y represents the temperature values, and x represents the hourly values). The data is repeated here.

x	7	9	11	12
y	49	57	71	75

We will shortly see how to use MATLAB to obtain these polynomials. For $7 \leq x \leq 9$,

$$y_1(x) = -0.35(x - 7)^3 + 2.85(x - 7)^2 - 0.3(x - 7) + 49$$

For $9 \leq x \leq 11$,

$$y_2(x) = -0.35(x - 9)^3 + 0.75(x - 9)^2 + 6.9(x - 9) + 57$$

For $11 \leq x \leq 12$,

$$y_3(x) = -0.35(x - 11)^3 - 1.35(x - 11)^2 + 5.7(x - 11) + 71$$

MATLAB provides the `spline` command to obtain a cubic-spline interpolation. Its syntax is `y_int = spline(x,y,x_int)`, where `x` and `y` are vectors containing the data and `x_int` is a vector containing the values of the independent variable x at which we wish to estimate the dependent variable y. The result `y_int` is a vector the same size as `x_int` containing the interpolated values of y that correspond to `x_int`. The spline fit can be plotted by plotting the vectors `x_int` and `y_int`. For example, the following session produces and plots a cubic-spline fit to the preceding data, using an increment of 0.01 in the x values.

```
>>x = [7,9,11,12];
>>y = [49,57,71,75];
>>x_int = [7:0.01:12];
>>y_int = spline(x,y,x_int);
>>plot(x,y,'o',x,y,'--',x_int,y_int),...
    xlabel('Time (hr)'),ylabel('Temperature (deg F)'), ...
```

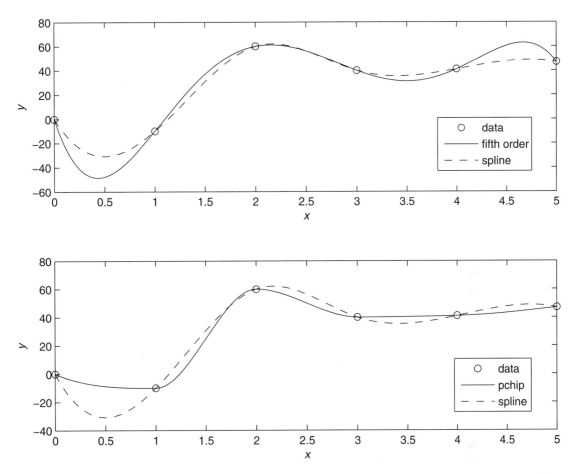

Figure 6.4–5 *Top graph:* Interpolation with a fifth-order polynomial and a cubic spline. *Bottom graph:* Interpolation with piecewise continuous Hermite polynomials (pchip) and a cubic spline.

Key Terms with Page References

Problems

You can find the answers to problems marked with an asterisk at the end of the text.

Section 6.1

1. The following list gives the measured gas mileage in miles per gallon for 22 cars of the same model. Plot the absolute frequency histogram and the relative frequency histogram.

23	25	26	25	27	25	24	22	23	25	26
26	24	24	22	25	26	24	24	24	27	23

2. Thirty pieces of structural timber of the same dimensions were subjected to an increasing lateral force until they broke. The measured force in pounds required to break them is given in the following list. Plot the absolute frequency histogram. Try bin widths of 50, 100, and 200 lb. Which gives the most meaningful histogram? Try to find a better value for the bin width.

243	236	389	628	143	417	205
404	464	605	137	123	372	439
497	500	535	577	441	231	675
132	196	217	660	569	865	725
457	347					

3. The following list gives the measured breaking force in newtons for a sample of 60 pieces of certain type of cord. Plot the absolute frequency histogram. Try bin widths of 10, 30, and 50 N. Which gives the most meaningful histogram? Try to find a better value for the bin width.

311	138	340	199	270	255	332	279	231	296	198	269
257	236	313	281	288	225	216	250	259	323	280	205
279	159	276	354	278	221	192	281	204	361	321	282
254	273	334	172	240	327	261	282	208	213	299	318
356	269	355	232	275	234	267	240	331	222	370	226

Section 6.2

4. For the data given in Problem 1,

    a. Plot the scaled frequency histogram.

    b. Compute the mean and standard deviation and use them to estimate the lower and upper limits of gas mileage corresponding to 68 percent of cars of this model. Compare these limits with those of the data.

5. For the data given in Problem 2,
   a. Plot the scaled frequency histogram.
   b. Compute the mean and standard deviation and use them to estimate the lower and upper limits of strength corresponding to 68 percent and to 96 percent of such timber pieces. Compare these limits with those of the data.

6. For the data given in Problem 3,
   a. Plot the scaled frequency histogram.
   b. Compute the mean and standard deviation, and use them to estimate the lower and upper limits of breaking force corresponding to 68 percent and 96 percent of cord pieces of this type. Compare these limits with those of the data.

7.* Data analysis of the breaking strength of a certain fabric shows that it is normally distributed with a mean of 200 lb and a variance of 9.
   a. Estimate the percentage of fabric samples that will have a breaking strength no less than 194 lb.
   b. Estimate the percentage of fabric samples that will have a breaking strength no less than 197 lb and no greater than 203 lb.

8. Data from service records shows that the time to repair a certain machine is normally distributed with a mean of 50 min and a standard deviation of 5 min. Estimate how often it will take more than 60 min to repair a machine.

9. Measurements of a number of fittings show that the pitch diameter of the thread is normally distributed with a mean of 5.007 mm and a standard deviation of 0.005 mm. The design specifications require that the pitch diameter must be 5 ± 0.01 mm. Estimate the percentage of fittings that will be within tolerance.

10. A certain product requires that a shaft be inserted into a bearing. Measurements show that the diameter d_1 of the cylindrical hole in the bearing is normally distributed with a mean of 3 cm with a variance of 0.0064. The diameter d_2 of the shaft is normally distributed with a mean of 2.96 cm and a variance of 0.0036.
   a. Compute the mean and the variance of the clearance $c = d_1 - d_2$.
   b. Find the probability that a given shaft will not fit into the bearing. (Hint: Find the probability that the clearance is negative.)

11.* A shipping pallet holds 10 boxes. Each box holds 300 parts of different types. The part weight is normally distributed with a mean of 1 lb and a standard deviation of 0.2 lb.
   a. Compute the mean and standard deviation of the pallet weight.
   b. Compute the probability that the pallet weight will exceed 3015 lb.

12. A certain product is assembled by placing three components end to end. The components' lengths are L_1, L_2, and L_3. Each component is manufactured on a different machine, so the random variation in their lengths is independent of each other. The lengths are normally distributed with means of 1, 2, and 1.5 ft and variances of 0.00014, 0.0002, and 0.0003, respectively.

a. Compute the mean and variance of the length of the assembled product.

b. Estimate what percentage of assembled products will be no less than 4.48 and no more than 4.52 ft in length.

Section 6.3

13. Use a random number generator to produce 1000 uniformly distributed numbers with a mean of 10, a minimum of 2, and a maximum of 18. Obtain the mean and the histogram of these numbers and discuss whether or not they appear uniformly distributed with the desired mean.

14. Use a random number generator to produce 1000 normally distributed numbers with a mean of 20 and a variance of 4. Obtain the mean, variance, and the histogram of these numbers and discuss whether or not they appear normally distributed with the desired mean and variance.

15. The mean of the sum (or difference) of two independent random variables equals the sum (or difference) of their means, but the variance is always the sum of the two variances. Use random number generation to verify this statement for the case where $z = x + y$, where x and y are independent and normally distributed random variables. The mean and variance of x are $\mu_x = 10$ and $\sigma_x^2 = 2$. The mean and variance of y are $\mu_y = 15$ and $\sigma_y^2 = 3$. Find the mean and variance of z by simulation and compare the results with the theoretical prediction. Do this for 100, 1000, and 5000 trials.

16. Suppose that $z = xy$, where x and y are independent and normally distributed random variables. The mean and variance of x are $\mu_x = 10$ and $\sigma_x^2 = 2$. The mean and variance of y are $\mu_y = 15$ and $\sigma_y^2 = 3$. Find the mean and variance of z by simulation. Does $\mu_z = \mu_x \mu_y$? Does $\sigma_z^2 = \sigma_x^2 \sigma_y^2$? Do this for 100, 1000, and 5000 trials.

17. Suppose that $y = x^2$, where x is a normally distributed random variable with a mean and variance of $\mu_x = 0$ and $\sigma_x^2 = 4$. Find the mean and variance of y by simulation. Does $\mu_y = \mu_x^2$? Does $\sigma_y = \sigma_x^2$? Do this for 100, 1000, and 5000 trials.

18.* Suppose you have analyzed the price behavior of a certain stock by plotting the scaled frequency histogram of the price over a number of months. Suppose that the histogram indicates that the price is normally distributed with a mean $100 and a standard deviation of $5. Write a MATLAB program to simulate the effects of buying 50 shares of this stock whenever the price is below the $100 mean, and selling all your shares whenever the price is above $105. Analyze the outcome of this strategy over 250 days (the approximate number of business days in a year). Define the profit as the yearly income from selling stock plus the value of the stocks you own at year's end, minus the yearly cost of buying stock. Compute the mean yearly profit you would expect to make, the minimum expected yearly profit, the maximum expected yearly profit, and the standard deviation of the yearly profit. The broker charges 6 cents per share bought or sold with a minimum fee of $40 per transaction. Assume you make only one transaction per day.

19. Suppose that data shows that a certain stock price is normally distributed with a mean of $150 and a variance of 100. Create a simulation to compare the results of the following two strategies over 250 days. You start the year with 1000 shares. With the first strategy, every day the price is below $140 you buy 100 shares, and every day the price is above $160 you sell all the shares you own. With the second strategy, every day the price is below $150 you buy 100 shares, and every day the price is above $160 you sell all the shares you own. The broker charges 5 cents per share traded with a minimum of $35 per transaction.

20. Write a script file to simulate 100 plays of a game in which you flip two coins. You win the game if you get two heads, lose if you get two tails, and you flip again if you get one head and one tail. Create three user-defined functions to use in the script. Function `flip` simulates the flip of one coin, with the state `s` of the random number generator as the input argument, and the new state `s` and the result of the flip (0 for a tail and 1 for a head) as the outputs. Function `flips` simulates the flipping of two coins, and calls `flip`. The input of `flips` is the state `s`, and the outputs are the new state `s` and the result (0 for two tails, 1 for a head and a tail, and 2 for two heads). Function `match` simulates a turn at the game. Its input is the state `s`, and its outputs are the result (1 for win, 0 for lose) and the new state `s`. The script should first reset the random number generator to its initial state, compute the state `s`, and then pass this state to the user-defined functions.

21. Write a script file to play a simple number guessing game as follows. The script should generate a random integer in the range 1, 2, 3, . . . , 14, 15. It should provide for the player to make repeated guesses of the number, and should indicate if the player has won or give the player a hint after each wrong guess. The responses and hints are:
- ■ "You won," and then stop the game.
- ■ "Very close," if the guess is within 1 of the correct number.
- ■ "Getting close," if the guess is within 2 or 3 of the correct number.
- ■ "Not close," if the guess is not within 3 of the correct number.

Section 6.4

22.* Interpolation is useful when one or more data points are missing. This situation often occurs with environmental measurements, such as temperature, because of the difficulty of making measurements around the clock. The following table of temperature versus time data is missing readings at

5 and 9 hours. Use linear interpolation with MATLAB to estimate the temperature at those times.

Time (hours, P.M.)	1	2	3	4	5	6	7	8	9	10	11	12
Temperature (°C)	10	12	18	24	?	21	20	18	?	15	13	8

23. The following table gives temperature data in °C as a function of time of day and day of the week at a specific location. Data is missing for the entries marked with a question mark (?). Use linear interpolation with MATLAB to estimate the temperature at the missing points.

Hour	Day				
	Mon	Tues	Wed	Thurs	Fri
1	17	15	12	16	16
2	13	?	8	11	12
3	14	14	9	?	15
4	17	15	14	15	19
5	23	18	17	20	24

24. Computer-controlled machines are used to cut and to form metal and other materials when manufacturing products. These machines often use cubic splines to specify the path to be cut or the contour of the part to be shaped. The following coodinates specify the shape of a certain car's front fender. Fit a series of cubic splines to the coordinates and plot the splines along with the coordinate points.

x (ft)	0	0.25	0.75	1.25	1.5	1.75	1.875	2	2.125	2.25
y (ft)	1.2	1.18	1.1	1	0.92	0.8	0.7	0.55	0.35	0

25. The following data is the measured temperature T of water flowing from a hot water faucet after it is turned on at time $t = 0$.

t (sec)	T (°F)	t (sec)	T (°F)
0	72.5	6	109.3
1	78.1	7	110.2
2	86.4	8	110.5
3	92.3	9	109.9
4	110.6	10	110.2
5	111.5		

  a. Plot the data first connecting them with straight lines, and then with a cubic spline.

  b. Estimate the temperature values at the following times using linear interpolation and then cubic-spline interpolation: $t = 0.6, 2.5, 4.7, 8.9$.

  c. Use both the linear and cubic-spline interpolations to estimate the time it will take for the temperature to equal the following values: $T = 75$, 85, 90, 105.

Numerical Methods for Calculus and Differential Equations

OUTLINE

This chapter covers numerical methods for computing integrals and derivatives and for solving ordinary differential equations. Some integrals cannot be evaluated analytically, and we need to compute them numerically with an approximate method (Section 7.1). In addition, it is often necessary to use data to estimate rates of change, and this requires a numerical estimate of the derivative (Section 7.2). Finally, many differential equations cannot be solved analytically, and so we must be able to solve them by using appropriate numerical techniques. Section 7.3 covers first-order differential equations, and Section 7.4 extends the methods to higher-order equations. More powerful methods are available for linear equations. Section 7.5 treats these methods.

When you have finished this chapter, you should be able to

■ Use MATLAB to numerically evaluate integrals.

■ Use numerical methods with MATLAB to estimate derivatives.

■ Use the MATLAB numerical differential equation solvers to obtain solutions.

7.1 Numerical Integration

The integral of a function $f(x)$ for $a \le x \le b$ can be interpreted as the area between the $f(x)$ curve and the x-axis, bounded by the limits $x = a$ and $x = b$. If we denote this area by A, then we can write A as

$$A = \int_a^b f(x)\,dx \qquad (7.1\text{--}1)$$

DEFINITE
INTEGRAL

INDEFINITE

IMPROPER
INTEGRAL

SINGULARITIES

An integral is called a *definite* integral if it has specified limits of integration. *Indefinite* integrals have specified limits. *Improper* integrals can have infinite values, depending on their integration limits. For example, the following integral can be found in most integral tables:

$$\int \frac{1}{x-1}\,dx = \ln|x-1|$$

However, it is an improper integral if the integration limits include the point $x = 1$. So, even though an integral can be found in an integral table, you should examine the integrand to check for *singularities,* which are points at which the integrand is undefined. The same warning applies when you are using numerical methods to evaluate integrals.

Trapezoidal Integration

The simplest way to find the area under a curve is to split the area into rectangles (Figure 7.1–1a). If the widths of the rectangles are small enough, the sum of their areas gives the approximate value of the integral. A more sophisticated method is to use trapezoidal elements (Figure 7.1–1b). Each trapezoid is called a *panel*. It is not necessary to use panels of the same width; to increase the method's

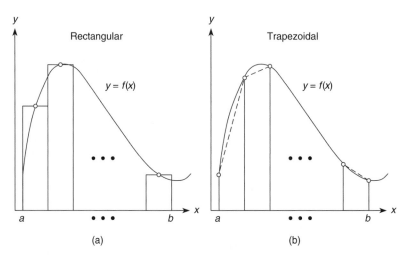

Figure 7.1–1 Illustration of (a) rectangular and (b) trapezoidal numerical integration.

accuracy, you can use narrow panels where the function is changing rapidly. When the widths are adjusted according to the function's behavior, the method is said to be *adaptive*. MATLAB implements trapezoidal integration with the `trapz` function. Its syntax is `trapz(x,y)`, where the array `y` contains the function values at the points contained in the array `x`. If you want the integral of a single function, then `y` is a vector. To integrate more than one function, place their values in a matrix `y`; `trapz(x,y)` will compute the integral of each column of `y`.

You cannot directly specify a function to integrate with the `trapz` function; you must first compute and store the function's values ahead of time in an array. Later we will discuss two other integration functions, the `quad` and `quadl` functions, that can accept functions directly. However, they cannot handle arrays of values. So the functions complement one another. The `trapz` function is summarized in Table 7.1–1.

As a simple example of the use of the `trapz` function, let us compute the integral

$$A = \int_0^\pi \sin x \, dx \qquad (7.1\text{–}2)$$

whose exact answer is $A = 2$. To investigate the effect of panel width, let us first use 10 panels with equal widths of $\pi/10$. The script file is

```
x = linspace(0,pi,10);
y = sin(x);
A = trapz(x,y)
```

Table 7.1–1 Basic syntax of numerical integration functions

Command	Description
`dblquad(fun,a,b,c,d)`	Computes the double integral of the function $f(x,y)$ between the limits $a \le x \le b$ and $c \le y \le d$. The input `fun` specifies the function that computes the integrand. It must accept a vector argument x and scalar y, and it must return a vector result.
`polyint(p,C)`	Computes the integral of the polynomial `p` using an optional user-specified constant of integration `C`.
`quad(fun,a,b)`	Uses an adaptive Simpson rule to compute the integral of the function `fun` between the limits a and b. The input `fun`, which represents the integrand $f(x)$, is a function handle for the integrand function. It must accept a vector argument x and return the vector result y.
`quadl(fun,a,b)`	Uses Lobatto integration. The syntax is identical to `quad`.
`trapz(x,y)`	Uses trapezoidal integration to compute the integral of `y` with respect to `x`, where the array `y` contains the function values at the points contained in the array `x`.
`triplequad(fun,a,b,c,d,e,f)`	Computes the triple integral of the function $f(x,y,z)$ between the limits $a \le x \le b$, $c \le y \le d$, and $e \le y \le f$. The input `fun` specifies the function that computes the integrand. It must accept a vector argument x, scalar y, and scalar z, and it must return a vector result.

The answer is $A = 1.9797$, which gives a relative error of $100(2 - 1.9797)$ /2 = 1%. Now try 100 panels of equal width; replace the array x with x = `linspace(0,pi,100)`. The answer is $A = 1.9998$ for a relative error of $100(2 - 1.9998)/2 = 0.01\%$. If we examine the plot of the integrand $\sin x$, we see that the function is changing faster near $x = 0$ and $x = \pi$ than near $x = \pi/2$. Thus we could achieve the same accuracy by using fewer panels if narrower panels are used near $x = 0$ and $x = \pi$.

We normally use the `trapz` function when the integrand is given as a table of values. Otherwise, if the integrand is given as a function, use the `quad` or `quadl` functions, to be introduced shortly.

EXAMPLE 7.1–1   Velocity from an Accelerometer

An *accelerometer* is used in aircraft, rockets, and other vehicles to estimate the vehicle's velocity and displacement. The accelerometer integrates the acceleration signal to produce an estimate of the velocity, and it integrates the velocity estimate to produce an estimate of displacement. Suppose the vehicle starts from rest at time $t = 0$, and its measured acceleration is given in the following table.

Time (s)	0	1	2	3	4	5	6	7	8	9	10
Acceleration (m/s²)	0	2	4	7	11	17	24	32	41	48	51

(a) Estimate the velocity v after 10 s.
(b) Estimate the velocity at times $t = 1, 2, \ldots , 10$ s.

■ Solution
(a) The initial velocity is zero, so $v(0) = 0$. The relation between the velocity and acceleration $a(t)$ is

$$v(10) = \int_0^{10} a(t)\,dt + v(0) = \int_0^{10} a(t)\,dt$$

The script file is shown below.

```
t = [0:10];
a = [0,2,4,7,11,17,24,32,41,48,51];
v10 = trapz(t,a)
```

The answer given for the velocity after 10 s is given by `v10`, and it is 211.5 m/s.
(b) The following script file uses the fact that the velocity can be expressed as

$$v(t_{k+1}) = \int_{t_k}^{t_{k+1}} a(t)\,dt + v(t_k) \qquad k = 1, 2, \ldots , 10$$

where $v(t_1) = 0$.

```
t = [0:10];
a = [0,2,4,7,11,17,24,32,41,48,51];
```

```
v(1) = 0;
for k = [1:10]
   v(k+1) = trapz(t(k:k+1), a(k:k+1))+v(k);
end
disp([t',v'])
```

The answers are given in the following table.

Time (s)	0	1	2	3	4	5	6	7	8	9	10
Velocity (m/s)	0	1	4	9.5	18.5	32.5	53	81	117	162	211.5

Test Your Understanding

T7.1–1 Modify the script file given in part (b) of Example 7.1-1 to estimate the displacement at times $t = 1, 2, \ldots, 10$ s. (Partial answer: The displacement after 10 s is 584.25 m.)

Quadrature Functions

Another approach to numerical integration is *Simpson's rule*, which divides the integration range $b - a$ into an even number of sections and uses a different quadratic function to represent the integrand for each panel. A quadratic function has three parameters, and Simpson's rule computes these parameters by requiring that the quadratic pass through the function's three points corresponding to the two adjacent panels. To obtain greater accuracy, we can use polynomials of degree higher than 2.

The MATLAB function quad implements an adaptive version of Simpson's rule. The quadl function is based on an adaptive Lobatto integration method, where the letter "l" in quadl stands for Lobatto. The term *quad* is an abbreviation of *quadrature,* which is an old term for the process of measuring areas. Some writers distinguish between the terms *quadrature* and *integration* and reserve *integration* to mean numerical integration of ordinary differential equations. We will not make that distinction.

QUADRATURE

The function quad(fun,a,b) computes the integral of the function fun between the limits a and b. The input fun, which represents the integrand $f(x)$, is either a function handle of the integrand function (the preferred method) or the name of the function as a character string (i.e., placed in single quotes). The function $y = f(x)$ must accept a vector argument x and must return the vector result y. The basic syntax of quadl is identical and is summarized in Table 7.1–1.

To illustrate, let us compute the integral given in (7.1–2). The session consists of one command: A = quad(@sin, 0, pi) or A = quad('sin', 0, pi). The answer given by MATLAB *is* $A = 2.0000$, which is correct to four decimal places. We use quadl the same way.

Because the quad and quadl functions call the integrand function using vector arguments, you must always use array operations when defining the function. The following example shows how this is done.

| EXAMPLE 7.1–2 |
                                                              Evaluation of Fresnel's Cosine Integral

Some simple-looking integrals cannot be evaluated in closed form. An example is Fresnel's cosine integral

$$A = \int_0^b \cos x^2 \, dx \tag{7.1–3}$$

a) Demonstrate two ways to compute the integral when the upper limit is $b = \sqrt{2\pi}$.

b) Demonstrate the use of a nested function to compute the more general integral

$$A = \int_0^b \cos x^n \, dx \tag{7.1–4}$$

for $n = 2$ and for $n = 3$.

Solution

a) The integrand $\cos x^2$ obviously does not contain any singularities that might cause problems for the integration function. We demonstrate two ways to use the `quad` function.

1.   With a function file: Define the integrand with a user-defined function as shown by the following function file.

```
function c2 = cossq(x)
c2 = cos(x.^2);
```

The `quad` function is called as follows: `A = quad(@cossq,0,sqrt(2*pi))`. The result is $A = 0.6119$.

2.   With an anonymous function (anonymous functions are discussed in Section 3.3): The session is

```
>>cossq = @(x)cos(x.^2);
>>A = quad(cossq,0,sqrt(2*pi))
  A =
     0.6119
```

The two lines can be combined into one as follows:

```
A = quad(@(x)cos(x.^2),0,sqrt(2*pi))
```

The advantage of using an anonymous function is that you need not create and save a function file. However, for complicated integrand functions, using a function file is preferable.

b) Because `quad` requires that the integrand function have only one argument, the following code will not work.

```
>>cossq = @(x)cos(x.^n);
>>n = 2;
>>A = quad(cossq,0,sqrt(2*pi))
??? Undefined function or variable 'n'.
```

Instead we will use parameter passing with a nested function (Nested functions are discussed in Section 3.3). First create and save the following function.

```
function A = integral_n(n)
A = quad(@cossq_n,0,sqrt(2*pi));

% Nested function
   function integrand = cossq_n(x)
       integrand = cos(x.^n);
   end
end
```

The session for $n = 2$ and $n = 3$ is as follows.

```
>>A = integral_n(2)
   A =
       0.6119
>>A = integral_n(3)
   A =
       0.7734
```

The quad functions have some optional arguments for analyzing and adjusting the algorithm's efficiency and accuracy. Type help quad for details.

Test Your Understanding

T7.1–2 Use both the quad and quadl functions to compute the integral

$$A = \int_2^5 \frac{1}{x}\, dx$$

and compare the answers with that obtained from the closed-form solution, which is $A = 0.9163$.

Polynomial Integration

MATLAB provides the polyint function to compute the integral of a polynomial. The syntax q = polyint(p, C) returns a polynomial q representing the integral of polynomial p with a user-specified scalar constant of integration C. The elements of the vector p are the coefficients of the polynomial, arranged in descending powers. The syntax polyint(p) assumes the constant of integration C is zero.

For example, the integral of $12x^3 + 9x^2 + 8x + 5$ is obtained from q = polyint([12,9,8,5], 10). The answer is q = [3, 3, 4, 5, 10], which corresponds to $3x^4 + 3x^3 + 4x^2 + 5x + 10$. Because polynomial integrals can be obtained from a symbolic formula, the polyint function is not a numerical integration operation.

Double Integrals

The function `dblquad` computes double integrals. Consider the integral

$$A = \int_c^d \int_a^b f(x,y)\,dx\,dy$$

The basic syntax is

```
A = dblquad(fun, a, b, c, d)
```

where `fun` is the handle to a user-defined function that defines the integrand $f(x,y)$. The function must accept a vector x and a scalar y, and it must return a vector result, so the appropriate array operations must be used. The extended syntax enables the user to adjust the accuracy and to use `quadl` or a user-defined quadrature routine. See the MATLAB help for details.

For example, using an anonymous function to compute the integral

$$A = \int_0^1 \int_1^3 xy^2\,dx\,dy$$

you type

```
>>fun = @(x,y)x.*y^2;
>>A = dblquad(fun, 1, 3, 0, 1)
```

The answer is $A = 1.3333$.

The preceding integral is carried out over the rectangular region specified by $1 \le x \le 3, 0 \le y \le 1$. Some double integrals are specified over a nonrectangular region. These problems can be handled by a transformation of variables. You can also use a rectangular region that encloses the nonrectangular region and force the integrand to be zero outside of the nonrectangular region, by using the MATLAB relational operators, for example. See Problem 15. The following example illustrates the former approach.

EXAMPLE 7.1–3   Double Integral over a Nonrectangular Region

Compute the integral

$$A = \int \int_R (x - y)^4 (2x + y)^2 \, dx\,dy$$

over the region R bounded by the lines

$$x - y = \pm 1 \qquad 2x + y = \pm 2$$

Solution

We must convert the integral into one that is specified over a rectangular region. To do this, let $u = x - y$ and $v = 2x + y$. Thus, using the Jacobian, we obtain

$$dx\,dy = \begin{vmatrix} \partial x/\partial u & \partial x/\partial v \\ \partial y/\partial u & \partial y/\partial v \end{vmatrix} du\,dv = \begin{vmatrix} 1/3 & 1/3 \\ -2/3 & 1/3 \end{vmatrix} du\,dv = \frac{1}{3}\,du\,dv$$

Then the region R is specified as a rectangular region in terms of u and v. Its boundaries are given by $u = \pm 1$ and $v = \pm 2$, and the integral becomes

$$A = \frac{1}{3} \int_{-2}^{2} \int_{-1}^{1} u^4 v^2 du\,dv$$

and the MATLAB session is

```
>>fun = @(u,v)u.^4*v^2;
>>A = (1/3)*dblquad(fun, -1, 1, -2, 2)
```

The answer is $A = 0.7111$.

Triple Integrals

The function `triplequad` computes triple integrals. Consider the integral

$$A = \int_{e}^{f} \int_{c}^{d} \int_{a}^{b} f(x,y,z)\,dx\,dy\,dz$$

The basic syntax is

```
A = triplequad(fun, a, b, c, d, e, f)
```

where `fun` is the handle to a user-defined function that defines the integrand $f(x,y,z)$. The function must accept a vector x, a scalar y, and a scalar z, and it must return a vector result, so the appropriate array operations must be used. The extended syntax enables the user to adjust the accuracy and to use `quadl` or a user-defined quadrature routine. See the MATLAB help for details. For example, to compute the integral

$$A = \int_{1}^{2} \int_{0}^{2} \int_{1}^{3} \left(\frac{xy - y^2}{z} \right) dx\,dy\,dz$$

You type

```
>>fun = @(x,y,z)(x*y-y^2)/z;
>>A = triplequad(fun, 1, 3, 0, 2, 1, 2)
```

The answer is $A = 1.8484$.

7.2 Numerical Differentiation

The derivative of a function can be interpreted graphically as the slope of the function. This interpretation leads to various methods for computing the derivative of a set of data. Figure 7.2–1 shows three data points that represent a function $y(x)$. Recall that the definition of the derivative is

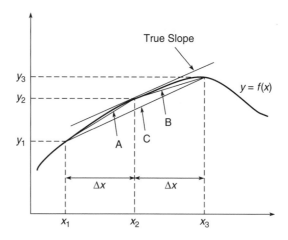

Figure 7.2–1 Illustration of methods for estimating the derivative dy/dx.

$$\frac{dy}{dx} = \lim_{\Delta x \to 0} \frac{\Delta y}{\Delta x} \tag{7.2–1}$$

The success of numerical differentiation depends heavily on two factors: the spacing of the data points and the scatter present in the data due to measurement error. The greater the spacing, the more difficult it is to estimate the derivative. We assume here that the spacing between the measurements is regular; that is, $x_3 - x_2 = x_2 - x_1 = \Delta x$. Suppose we want to estimate the derivative dy/dx at the point x_2. The correct answer is the slope of the straight line passing through the point (x_2, y_2); but we do not have a second point on that line, so we cannot find its slope. Therefore, we must estimate the slope by using nearby data points. One estimate can be obtained from the straight line labeled A in the figure. Its slope is

$$m_A = \frac{y_2 - y_1}{x_2 - x_1} = \frac{y_2 - y_1}{\Delta x} \tag{7.2–2}$$

BACKWARD DIFFERENCE

This estimate of the derivative is called the *backward difference* estimate, and it is actually a better estimate of the derivative at $x = x_1 + (\Delta x)/2$ than at $x = x_2$. Another estimate can be obtained from the straight line labeled B. Its slope is

$$m_B = \frac{y_3 - y_2}{x_3 - x_2} = \frac{y_3 - y_2}{\Delta x} \tag{7.2–3}$$

FORWARD DIFFERENCE

This estimate is called the *forward difference* estimate, and it is a better estimate of the derivative at $x = x_2 + (\Delta x)/2$ than at $x = x_2$. Examining the plot, you might think that the average of these two slopes would provide a better estimate of the derivative at $x = x_2$, because the average tends to cancel out the effects of measurement error. The average of m_A and m_B is

$$m_C = \frac{m_A + m_B}{2} = \frac{1}{2}\left(\frac{y_2 - y_1}{\Delta x} + \frac{y_3 - y_2}{\Delta x}\right) = \frac{y_3 - y_1}{2\,\Delta x} \qquad (7.2\text{-}4)$$

This is the slope of the line labeled C, which connects the first and third data points. This estimate of the derivative is called the *central difference* estimate.

CENTRAL DIFFERENCE

The diff Function

MATLAB provides the diff function to use for computing derivative estimates. Its syntax is d = diff(x), where x is a vector of values, and the result is a vector d containing the differences between adjacent elements in x. That is, if x has *n* elements, d will have *n* − 1 elements, where $d = [x(2) - x(1), x(3) - x(2), \ldots, x(n) - x(n-1)]$. For example, if x = [5, 7, 12, −20], then diff(x) returns the vector [2, 5, −32]. The derivative *dy/dx* can be estimated from diff(y)./diff(x).

The following script file implements the backward difference and central difference methods for artificial data generated from a sinusoidal signal that is measured 51 times during one half-period. The measurement error is uniformly distributed between −0.025 and 0.025.

```
x = [0:pi/50:pi];
n = length(x);
% Data-generation function with +/-0.025 random error.
y = sin(x)+.05*(rand(1,51)-0.5);
% Backward difference estimate of dy/dx.
d1 = diff(y)./diff(x);
subplot(2,1,1)
plot(x(2:n),d1,x(2:n),d1,'o')
% Central difference estimate of dy/dx.
d2 = (y(3:n)-y(1:n-2))./(x(3:n)-x(1:n-2));
subplot(2,1,2)
plot(x(2:n-1),d2,x(2:n-1),d2,'o')
```

Test Your Understanding

T7.2–1 Modify the previous program to use the forward difference method to estimate the derivative. Plot the results, and compare with the results from the backward and central difference methods.

Polynomial Derivatives

MATLAB provides the polyder function to compute the derivative of a polynomial. Its syntax has several forms. The basic form is d = polyder(p), where p is a vector whose elements are the coefficients of the polynomial, arranged in descending powers. The output d is a vector containing the coefficients of the derivative polynomial.

The second syntax form is d = polyder(p1,p2). This form computes the derivative of the *product* of the two polynomials p1 and p2. The third form is [num, den] = polyder(p2,p1). This form computes the derivative of the *quotient* p2/p1. The vector of coefficients of the numerator of the derivative is given by num. The denominator is given by den.

Here are some examples of the use of polyder. Let $p_1 = 5x + 2$ and $p_2 = 10x^2 + 4x - 3$. Then

$$\frac{dp_2}{dx} = 20x + 4$$

$$p_1p_2 = 50x^3 + 40x^2 - 7x - 6$$

$$\frac{d(p_1p_2)}{dx} = 150x^2 + 80x - 7$$

$$\frac{d(p_2/p_1)}{dx} = \frac{50x^2 + 40x + 23}{25x^2 + 20x + 4}$$

These results can be obtained with the following program.

```
p1 = [5, 2];p2 = [10, 4, -3];
% Derivative of p2.
der2 = polyder(p2)
% Derivative of p1*p2.
prod = polyder(p1,p2)
% Derivative of p2/p1.
[num, den] = polyder(p2,p1)
```

The results are der2 = [20, 4], prod = [150, 80, -7], num = [50, 40, 23], and den = [25, 20, 4].

Because polynomial derivatives can be obtained from a symbolic formula, the polyder function is not a numerical differentiation operation.

Gradients

The *gradient* ∇f of a function $f(x,y)$ is a vector pointing in the direction of increasing values of $f(x,y)$. It is defined by

$$\nabla f = \frac{\partial f}{\partial x}\mathbf{i} + \frac{\partial f}{\partial y}\mathbf{j}$$

where $\mathbf{i}$ and $\mathbf{j}$ are the unit vectors in the x and y directions, respectively. The concept can be extended to functions of three or more variables.

In MATLAB the gradient of a set of data representing a two-dimensional function $f(x,y)$ can be computed with the gradient function. Its syntax is [df_dx, df_dy] = gradient (f, dx, dy), where df_dx and df_dy represent $\partial f/\partial x$ and $\partial f/\partial y$, and dx and dy are the spacing in the x and y values

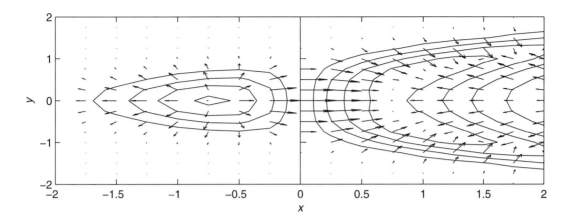

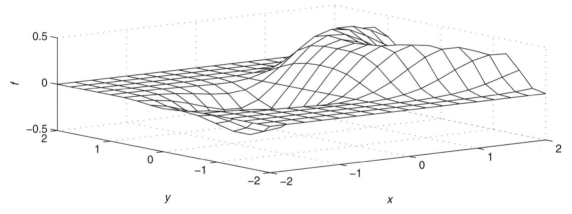

Figure 7.2–2 Gradient, contour, and surface plots of the function $f(x,y) = xe^{-(x^2+y^2)^2} + y^2$.

associated with the numerical values of f. The syntax can be extended to include functions of three or more variables.

The following program plots the contour plot and the gradient (shown by arrows) for the function

$$f(x,y) = xe^{-(x^2+y^2)^2} + y^2$$

The plots are shown in Figure 7.2–2. The arrows point in the direction of increasing f.

```
[x,y] = meshgrid(-2:0.25:2);
f = x.*exp(-((x-y.^2).^2+y.^2));
dx = x(1,2) - x(1,1); dy = y(2,1) - y(1,1);
[df_dx, df_dy] = gradient(f, dx, dy);
subplot(2,1,1)
contour(x,y,f), xlabel('x'), ylabel('y'), . . .
```

Table 7.2–1 Numerical differentiation functions

Command	Description
`d = diff(x)`	Returns a vector `d` containing the differences between adjacent elements in the vector `x`.
`[df_dx,df_dy] = gradient(f,dx,dy)`	Computes the gradient of the function $f(x, y)$, where `df_dx` and `df_dy` represent $\partial f/\partial x$ and $\partial f/\partial y$, and `dx` and `dy` are the spacing in the x and y values associated with the numerical values of f.
`d = polyder(p)`	Returns a vector `d` containing the coefficients of the derivative of the polynomial represented by the vector `p`.
`d = polyder(p1,p2)`	Returns a vector `d` containing the coefficients of the polynomial that is the derivative of the product of the polynomials represented by `p1` and `p2`.
`[num, den] = polyder(p2,p1)`	Returns the vectors `num` and `den` containing the coefficients of the numerator and denominator polynomials of the derivative of the quotient p^2/p^1, where `p1` and `p2` are polynomials.

```
hold on, quiver(x,y,df_dx, df_dy), hold off
subplot(2,1,2)
mesh(x,y,f),xlabel('x'),ylabel('y'),zlabel('f')
```

LAPLACIAN

The curvature is given by the second-order derivative expression called the *Laplacian*.

$$\nabla^2 f(x,y) = \frac{\partial^2 f}{\partial x^2} + \frac{\partial^2 f}{\partial y^2}$$

It can be computed with the `del2` function. See the MATLAB help for details.

The MATLAB differentiation functions discussed here are summarized in Table 7.2–1.

7.3 First-Order Differential Equations

In this section, we introduce numerical methods for solving first-order differential equations. In Section 7.4 we show how to extend the techniques to higher-order equations.

ORDINARY DIFFERENTIAL EQUATION

An *ordinary differential equation* (ODE) is an equation containing ordinary derivatives of the dependent variable. An equation containing partial derivatives with respect to two or more independent variables is a *partial differential equation* (PDE). Solution methods for PDEs are an advanced topic, and we will not treat them in this text. In this chapter we limit ourselves to *initial-value problems* (IVPs). These are problems where the ODE must be solved for a given set of

INITIAL-VALUE PROBLEM

values specified at some initial time, which is usually taken to be $t = 0$. Other types of ODE problems are discussed at the end of Section 7.6.

It will be convenient to use the following abbreviated "dot" notation for derivatives.

$$\dot{y}(t) = \frac{dy}{dt} \qquad \ddot{y}(t) = \frac{d^2y}{dt^2}$$

The *free response* of a differential equation, sometimes called the homogeneous solution or the initial response, is the solution for the case where there is no forcing function. The free response depends on the initial conditions. The *forced response* is the solution due to the forcing function when the initial conditions are zero. For linear differential equations, the complete or total response is the sum of the free and the forced responses. Nonlinear ODEs can be recognized by the fact that the dependent variable or its derivatives appear raised to a power or in a transcendental function. For example, the equations $\dot{y} = y^2$ and $\dot{y} = \cos y$ are nonlinear.

FREE RESPONSE

FORCED RESPONSE

The essence of a numerical method is to convert the differential equation into a difference equation that can be programmed. Numerical algorithms differ partly as a result of the specific procedure used to obtain the difference equations. It is important to understand the concept of "step size" and its effects on solution accuracy. To provide a simple introduction to these issues, we consider the simplest numerical methods, the *Euler method* and the *predictor-corrector method*.

The Euler Method

The *Euler method* is the simplest algorithm for numerical solution of a differential equation. Consider the equations

EULER METHOD

$$\frac{dy}{dt} = f(t,y) \qquad y(0) = y_0 \tag{7.3–1}$$

where $f(t, y)$ is a known function and y_0 is the initial condition, which is the given value of $y(t)$ at $t = 0$. From the definition of the derivative,

$$\frac{dy}{dt} = \lim_{\Delta t \to 0} \frac{y(t + \Delta t) - y(t)}{\Delta t}$$

If the time increment Δt is chosen small enough, the derivative can be replaced by the approximate expression

$$\frac{dy}{dt} \approx \frac{y(t + \Delta t) - y(t)}{\Delta t} \tag{7.3–2}$$

Assume that the function $f(t,y)$ in (7.3–1) remains constant over the time interval $(t, t + \Delta t)$, and replace (7.3–1) by the following approximation:

$$\frac{y(t + \Delta t) - y(t)}{\Delta t} = f(t, y)$$

or

$$y(t + \Delta t) = y(t) + f(t,y)\,\Delta t \tag{7.3–3}$$

The smaller Δt is, the more accurate are our two assumptions leading to (7.3–3). This technique for replacing a differential equation with a difference equation is

STEP SIZE

the *Euler method*. The increment Δt is called the *step size*.

Equation (7.3–3) can be written in more convenient form as

$$y(t_{k+1}) = y(t_k) + \Delta t\, f[t_k, y(t_k)] \tag{7.3–4}$$

where $t_{k+1} = t_k + \Delta t$. This equation can be applied successively at the times t_k by putting it in a `for` loop. The accuracy of the Euler method can be improved sometimes by using a smaller step size. However, very small step sizes require longer run times and can result in a large accumulated error due to roundoff effects.

PREDICTOR-
CORRECTOR
METHOD

The Predictor-Corrector Method

The Euler method can have a serious deficiency in problems where the variables are rapidly changing, because the method assumes the variables are constant over the time interval Δt. One way of improving the method is to use a better approximation to the right-hand side of (7.3–1). Suppose instead of the Euler approximation (7.3–4) we use the average of the right-hand side of (7.3–1) on the interval (t_k, t_{k+1}). This gives

$$y(t_{k+1}) = y(t_k) + \frac{\Delta t}{2}(f_k + f_{k+1}) \tag{7.3–5}$$

where

$$f_k = f[t_k, y(t_k)] \tag{7.3–6}$$

with a similar definition for f_{k+1}. Equation (7.3–5) is equivalent to integrating (7.3–1) with the trapezoidal rule.

The difficulty with (7.3–5) is that f_{k+1} cannot be evaluated until $y(t_{k+1})$ is known, but this is precisely the quantity being sought. A way out of this difficulty is to use the Euler formula (7.3–4) to obtain a preliminary estimate of $y(t_{k+1})$. This estimate is then used to compute f_{k+1} for use in (7.3–5) to obtain the required value of $y(t_{k+1})$.

The notation can be changed to clarify the method. Let $h = \Delta t$ and $y_k = y(t_k)$, and let x_{k+1} be the estimate of $y(t_{k+1})$ obtained from the Euler formula (7.3–4). Then, by omitting the t_k notation from the other equations, we obtain the following description of the predictor-corrector process.

$$\text{Euler predictor } x_{k+1} = y_k + hf(t_k, y_k) \tag{7.3–7}$$

$$\text{Trapezoidal corrector }\quad y_{k+1} = y_k + \frac{h}{2}[f(t_k, y_k) + f(t_{k+1}, x_{k+1})] \tag{7.3–8}$$

MODIFIED EULER
METHOD

This algorithm is sometimes called the *modified Euler method*. However, note that any algorithm can be tried as a predictor or a corrector. Thus many other methods can be classified as predictor-corrector.

Runge-Kutta Methods

The Taylor series representation forms the basis of several methods of solving differential equations, including the Runge-Kutta methods. The Taylor series may be used to represent the solution $y(t + h)$ in terms of $y(t)$ and its derivatives as follows.

$$y(t + h) = y(t) + h\dot{y}(t) + \frac{1}{2}h^2\ddot{y}(t) + \cdots \qquad (7.3-9)$$

The number of terms kept in the series determines its accuracy. The required derivatives are calculated from the differential equation. If these derivatives can be found, (7.3–9) can be used to march forward in time. In practice, the high-order derivatives can be difficult to calculate, and the series (7.3–9) is truncated at some term. The Runge-Kutta methods were developed because of the difficulty in computing the derivatives. These methods use several evaluations of the function $f(t,y)$ in a way that approximates the Taylor series. The number of terms in the series that is duplicated determines the order of the Runge-Kutta method. Thus, a fourth-order Runge-Kutta algorithm duplicates the Taylor series through the term involving h^4.

MATLAB ODE Solvers

In addition to the many variations of the predictor-corrector and Runge-Kutta algorithms that have been developed, there are more advanced algorithms that use a variable step size. These "adaptive" algorithms use larger step sizes when the solution is changing more slowly. MATLAB provides several functions, called *solvers*, that implement the Runge-Kutta and other methods with variable step size. Two of these are the `ode45` and `ode15s` functions. The `ode45` function uses a combination of fourth- and fifth-order Runge-Kutta methods. It is a general-purpose solver whereas `ode15s` is suitable for more difficult equations called "stiff" equations. These solvers are more than sufficient to solve the problems in this text. It is recommended that you try `ode45` first. If the equation proves difficult to solve (as indicated by a lengthy solution time or by a warning or error message), then use `ode15s`.

In this section we limit our coverage to first-order equations. Solution of higher-order equations is covered in Section 7.4. When used to solve the equation $\dot{y} = f(t,y)$, the basic syntax is (using `ode45` as the example)

```
[t,y] = ode45(@ydot, tspan, y0)
```

where `@ydot` is the handle of the function file whose inputs must be t and y, and whose output must be a column vector representing dy/dt, that is, $f(t, y)$. The number of rows in this column vector must equal the order of the equation. The syntax for `ode15s` is identical. The function file `ydot` may also be specified by a character string (i.e., its name placed in single quotes), but use of the function handle is now the preferred approach.

The vector `tspan` contains the starting and ending values of the independent variable t, and optionally any intermediate values of t where the solution is

desired. For example, if no intermediate values are specified, `tspan is [t0,
tf]`, where `t0` and `tf` are the desired starting and ending values of the independent parameter $t$. As another example, using `tspan = [0, 5, 10]` tells MATLAB
to find the solution at $t = 5$ and at $t = 10$. You can solve equation backward in
time by specifying `t0` to be greater than `tf`.

The parameter `y0` is the initial value y(0). The function file must have its first
two input arguments as `t` and `y` in that order, even for equations where $f(t, y)$ is not
a function of t. You need not use array operations in the function file because the
ODE solvers call the file with scalar values for the arguments. The solvers may
have an additional argument, `options`, which is discussed at the end of this
section.

First consider an equation whose solution is known in closed form, so that
we can make sure we are using the method correctly.

EXAMPLE 7.3–1   Response of an *RC* Circuit

The model of the *RC* circuit shown in Figure 7.3–1 can be found from Kirchhoff's voltage law and conservation of charge. It is $RC\dot{y} + y = v(t)$. Suppose the value of *RC* is
0.1 s. Use a numerical method to find the free response for the case where the applied voltage v is zero and the initial capacitor voltage is y(0) = 2 V. Compare the results with the
analytical solution, which is $y(t) = 2e^{-10t}$.

Solution
The equation for the circuit becomes $0.1\dot{y} + y = 0$. First solve this for y: $\dot{y} = -10y$.
Next define and save the following function file. Note that the order of the input
arguments must be t and y even though t does not appear on the right-hand side of the
equation.

```
function ydot = RC_circuit(t,y)
% Model of an RC circuit with no applied voltage.
ydot = -10*y;
```

The initial time is $t = 0$, so set `t0` to be 0. Here we know from the analytical solution that
y(t) will be close to 0 for $t \geq 0.5$ s, so we choose `tf` to be 0.5 s. In other problems we
generally do not have a good guess for `tf`, so we must try several increasing values of `tf`
until we see enough of the response on the plot.

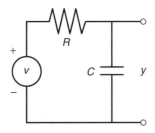

Figure 7.3–1 An *RC* circuit.

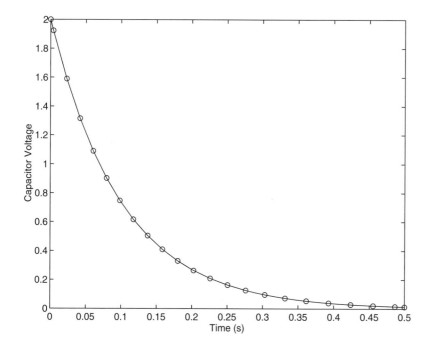

Figure 7.3–2 Free response of an *RC* circuit.

The function `ode45` is called as follows, and the solution plotted along with the analytical solution `y_true`.

```
[t, y] = ode45(@RC_circuit, [0, 0.5], 2);
y_true = 2*exp(-10*t);
plot(t,y,'o',t,y_true), xlabel('Time(secs)'),...
    ylabel('Capacitor Voltage')
```

Note that we need not generate the array `t` to evaluate `y_true` because `t` is generated by the `ode45` function. The plot is shown in Figure 7.3–2. The numerical solution is marked by the circles, and the analytical solution is indicated by the solid line. Clearly the numerical solution gives an accurate answer. Note that the step size has been automatically selected by the `ode45` function.

Earlier versions of MATLAB required that the function name, here `RC_circuit`, be enclosed within single quotes, but this might not be allowed in future versions. The use of function handles is now preferred, such as `@RC_circuit`. As we will see, additional capabilities are available with function handles.

Test Your Understanding

T7.3–1 Use MATLAB to compute and plot the solution of the following equation.

$$10\frac{dy}{dt} + y = 20 + 7\sin 2t \qquad y(0) = 15$$

When the differential equation is nonlinear, we often have no analytical solution to use for checking our numerical results. In such cases we can use our physical insight to guard against grossly incorrect results. We can also check the equation for singularities that might affect the numerical procedure. Finally, we can sometimes use an approximation to replace the nonlinear equation with a linear one that can be solved analytically. Although the linear approximation does not give the exact answer, it can be used to see if our numerical answer is "in the ballpark." The following example illustrates this approach.

EXAMPLE 7.3–3   Liquid Height in a Spherical Tank

Figure 7.3–3 shows a spherical tank for storing water. The tank is filled through a hole in the top and drained through a hole in the bottom. If the tank's radius is r, you can use integration to show that the volume of water in the tank as a function of its height h is given by

$$V(h) = \pi r h^2 - \pi \frac{h^3}{3} \tag{7.3–10}$$

Torricelli's principle states that the liquid flow rate through the hole is proportional to the square root of the height h. Further studies in fluid mechanics have identified the relation more precisely, and the result is that the volume flow rate through the hole is given by

$$q = C_d A \sqrt{2gh} \tag{7.3–11}$$

where A is the area of the hole, g is the acceleration due to gravity, and C_d is an experimentally determined value that depends partly on the type of liquid. For water, $C_d = 0.6$ is a common value. We can use the principle of conservation of mass to obtain a differential equation for the height h. Applied to this tank, the principle says that the rate of change of liquid volume in the tank must equal the flow rate out of the tank; that is,

$$\frac{dV}{dt} = -q \tag{7.3–12}$$

From (7.3–10),

$$\frac{dV}{dt} = 2\pi r h \frac{dh}{dt} - \pi h^2 \frac{dh}{dt} = \pi h (2r - h) \frac{dh}{dt}$$

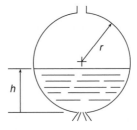

Figure 7.3–3 Draining of a spherical tank.

Substituting this and (7.3–11) into (7.3–12) gives the required equation for h.

$$\pi(2rh - h^2)\frac{dh}{dt} = -C_dA\sqrt{2gh} \tag{7.3–13}$$

Use MATLAB to solve this equation to determine how long it will take for the tank to empty if the initial height is 9 ft. The tank has a radius of $r = 5$ ft and has a 1-in., diameter hole in the bottom. Use $g = 32.2$ ft/sec^2. Discuss how to check the solution.

Solution:
With $C_d = 0.6$, $r = 5$, $g = 32.2$, and $A = \pi(1/24)^2$, (7.3–13) becomes

$$\frac{dh}{dt} = -\frac{0.0334\sqrt{h}}{10h - h^2} \tag{7.3–14}$$

We can first check the above expression for dh/dt for singularities. The denominator does not become zero unless $h = 0$ or $h = 10$, which correspond to a completely empty and a completely full tank. So we will avoid singularities if $0 < h < 10$.

Finally, we can use the following approximation to estimate the time to empty. Replace h on the right side of (7.3–14) with its average value, namely, $(9-0)/2 = 4.5$ ft. This gives $dh/dt = -0.00286$, whose solution is $h(t) = h(0) - 0.00286t = 9 - 0.00286t$. According to this equation the tank will be empty at $t = 9/0.00286 = 3147$ sec, or 52 min. We will use this value as a "reality check" on our answer.

The function file based on (7.3–14) is

```
function hdot = height(t,h)
hdot = -(0.0334*sqrt(h))/(10*h-h^2);
```

The file is called as follows, using the ode45 solver.

```
[t, h]=ode45(@height, [0, 2475], 9);
plot(t,h),xlabel('Time (sec)'), ylabel('Height (ft)')
```

The resulting plot is shown in Figure 7.3–4. Note how the height changes more rapidly when the tank is nearly full or nearly empty. This is to be expected because of the effects of the tank's curvature. The tank empties in 2475 sec, or 41 min. This value is not grossly different from our rough estimate of 52 min, so we should feel comfortable accepting the numerical results. The value of the final time of 2475 sec was found by increasing the final time until the plot showed that the height became 0.

7.4 Higher-Order Differential Equations

To use the ODE solvers to solve an equation higher than order 1, you must first write the equation as a set of first-order equations. This is easily done. Consider the second-order equation

$$5\ddot{y} + 7\dot{y} + 4y = f(t) \tag{7.4–1}$$

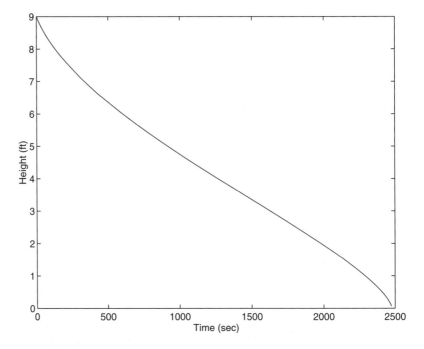

Figure 7.3–4 Plot of water height in a spherical tank.

Solve it for the highest derivative:

$$\ddot{y} = \frac{1}{5}f(t) - \frac{4}{5}y - \frac{7}{5}\dot{y} \qquad (7.4\text{–}2)$$

Define two new variables, x_1 and x_2, to be y and its derivative $\dot{y}$. That is, define $x_1 = y$ and $x_2 = \dot{y}$. This implies that

$$\dot{x}_1 = x_2$$

$$\dot{x}_2 = \frac{1}{5}f(t) - \frac{4}{5}x_1 - \frac{7}{5}x_2$$

**CAUCHY OR
STATE-VARIABLE
FORM**

This form is sometimes called the *Cauchy* form or the *state-variable* form.

Now write a function file that computes the values of $\dot{x}_1$ and $\dot{x}_2$ and stores them in a *column* vector. To do this, we must first have a function specified for $f(t)$. Suppose that $f(t) = \sin t$. Then the required file is

```
function xdot = example_1(t,x)
% Computes derivatives of two equations
xdot(1) = x(2);
xdot(2) = (1/5)*(sin(t)-4*x(1)-7*x(2));
xdot = [xdot(1); xdot(2)];
```

Note that xdot(1) represents $\dot{x}_1$, xdot(2) represents $\dot{x}_2$, x(1) represents x_1, and x(2) represents x_2. Once you become familiar with the notation for the state-variable form, you will see that the previous code could be replaced with the following shorter form.

```
function xdot = example_1(t,x)
% Computes derivatives of two equations
xdot = [x(2); (1/5)*(sin(t)-4*x(1)-7*x(2))];
```

Suppose we want to solve (7.4–1) for $0 \le t \le 6$ with the initial conditions $x(0) = 3$, $\dot{x}(0) = 9$. Then the initial condition for the *vector* x is [3, 9]. To use ode45, you type

```
[t, x] = ode45(@example_1, [0, 6], [3, 9]);
```

Each row in the vector x corresponds to a time returned in the column vector t. If you type plot(t,x), you will obtain a plot of both x_1 and x_2 versus t. Note x is a matrix with two columns. The first column contains the values of x_1 at the various times generated by the solver; the second column contains the values of x_2. Thus, to plot only x_1, type plot(t,x(:,1)). To plot only x_2, type plot(t,x(:,2)).

When we are solving nonlinear equations, sometimes it is possible to check the numerical results by using an approximation that reduces the equation to a linear one. The following example illustrates such an approach with a second-order equation.

A Nonlinear Pendulum Model

EXAMPLE 7.4–1

The pendulum shown in Figure 7.4–1 consists of a concentrated mass m attached to a rod whose mass is small compared to m. The rod's length is L. The equation of motion for this pendulum is

$$\ddot{\theta} + \frac{g}{L}\sin\theta = 0 \qquad (7.4\text{–}3)$$

Suppose that $L = 1$ m and $g = 9.81$ m/s². Use MATLAB to solve this equation for $\theta(t)$ for two cases: $\theta(0) = 0.5$ rad and $\theta(0) = 0.8\pi$ rad. In both cases $\dot{\theta}(0) = 0$. Discuss how to check the accuracy of the results.

Solution

If we use the small-angle approximation $\sin \approx \theta$, the equation becomes

$$\ddot{\theta} + \frac{g}{L}\theta = 0 \qquad (7.4\text{–}4)$$

which is linear and has the solution

$$\theta(t) = \theta(0)\cos\sqrt{\frac{g}{L}}\,t \qquad (7.4\text{–}5)$$

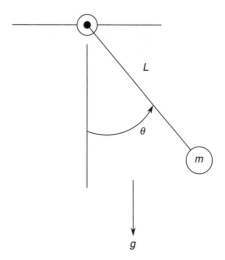

Figure 7.4–1 A pendulum.

if $\dot{\theta}(0) = 0$. Thus the amplitude of oscillation is $\theta(0)$, and the period is $P = 2\pi$ $\sqrt{L/g} = 2.006$ s. We can use this information to select a final time and to check our numerical results.

First rewrite the pendulum equation (7.4–3) as two first-order equations. To do this, let $x_1 = \theta$ and $x_2 = \dot{\theta}$. Thus

$$\dot{x}_1 = \dot{\theta} = x_2$$

$$\dot{x}_2 = \ddot{\theta} = -\frac{g}{L} \sin x_1$$

The following function file is based on the last two equations. Remember that the output `xdot` must be a *column* vector.

```
function xdot = pendulum(t,x)
g = 9.81; L = 1;
xdot = [x(2); -(g/L)*sin(x(1))];
```

This file is called as follows. The vectors `ta` and `xa` contain the results for the case where $\theta(0)$ = 0.5. In both cases, $\dot{\theta}(0)$ = 0. The vectors `tb` and `xb` contain the results for $\theta(0)$ = 0.8π.

```
[ta, xa] = ode45(@pendulum, [0,5], [0.5, 0];
[tb, xb] = ode45(@pendulum, [0,5], [0.8*pi, 0];
plot(ta, xa(:,1), tb,xb(:,1)), xlabel ('Time (s)'), ...
    ylabel('Angle (rad)'), gtext('Case 1'), gtext('Case 2')
```

The results are shown in Figure 7.4–2. The amplitude remains constant, as predicted by the small-angle analysis, and the period for the case where $\theta(0)$ = 0.5 is a little larger than 2 s, the value predicted by the small-angle analysis. So we can place some confidence in the numerical procedure. For the case where $\theta(0)$ = 0.8π, the period of the numerical

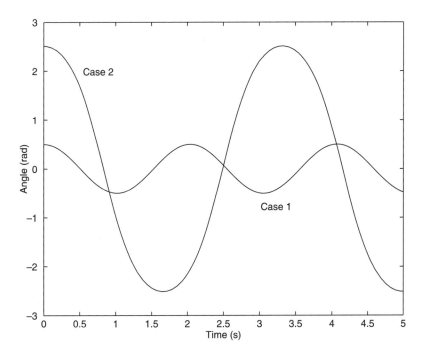

Figure 7.4–2 The pendulum angle as a function of time for two starting positions.

solution is about 3.3 s. This illustrates an important property of nonlinear differential equations. The free response of a linear equation has the same period for any initial conditions; however, the form and therefore the period of the free response of a nonlinear equation often depends on the particular values of the initial conditions.

In this example, the values of *g* and *L* were encoded in the function `pendulum(t,x)`. Now suppose you want to obtain the pendulum response for different lengths *L* or different gravitational accelerations *g*. You could use the `global` command to declare *g* and *L* as global variables, or you could pass parameter values through an argument list in the `ode45` function; but starting with MATLAB 7, the preferred method is to use a nested function. Nested functions are discussed in Section 3.3. The following program shows how this is done.

```
function pendula
g = 9.81; L = 0.75; % First case.
tf = 6*pi*sqrt(L/g); % Approximately 3 periods.
[t1, x1] = ode45(@pendulum, [0,tf], [0.4, 0];
%
g = 1.63; L = 2.5; % Second case.
tf = 6*pi*sqrt(L/g); % Approximately 3 periods.
[t2, x2] = ode45(@pendulum, [0,tf], [0.2, 0];
plot(t1, x1(:,1), t2, x2(:,1)), ...
```

```
   xlabel ('time (s)'), ylabel ('\theta (rad)')
   % Nested function.
      function xdot = pendulum(t,x)
      xdot = [x(2);-(g/L)*sin(x(1))];
   end
end
```

Advanced Solver Capabilities

The complete preferred ODE solver syntax in MATLAB 7, using `ode45` as an example, is

```
[t, y] = ode45(@ydot, tspan, y0, options)
```

where the `options` argument is created with the `odeset` function.

The `odeset` Function The `odeset` function creates an options structure to be supplied to the solver. Its syntax is

```
options = odeset('name1', 'value1' 'name2', 'value2', ...)
```

where `name` is the name of a *property* and `value` is the value to be assigned to the property.

A simple example will clarify things. The `Refine` property is used to increase the number of output points from the solver by an integer factor n. For `ode45` the default value of n is 4 because of the solver's large step sizes. Suppose we want to solve the following equation: $\dot{y} = \sin^2 t$ for $0 \le t \le 4\pi$ with $y(0) = 0$. Define the following function file.

```
function ydot = sinefn(t,y)
ydot = sin(t)^2;
```

Then use the `odeset` function to set the value of `Refine` to $n = 8$, and call the `ode45` solver, as shown in the following code. This will produce twice as many points to plot to obtain a smoother curve.

```
options = odeset('Refine',8);
[t, y] = ode45(@sinefn, [0, 4*pi], 0, options);
```

Another property is the `Events` property, which has two possible values: `on` and `off`. It can be used to locate transitions to, from, or through zeros of a user-defined function. This can be used to detect in the ODE solution when a variable makes a transition to, from, or through a certain value, such as zero. This feature can be used to simulate a dropped ball bouncing up from the floor. A program to do this is given in Chapter 8 of [Palm, 2005a]. See the MATLAB help for other examples.

There are many properties that can be set with the `odeset` function. To see a list of these, type `odeset`. Table 7.4–1 summarizes the syntax of the ODE solvers using `ode45` as an example.

Table 7.4–1 Syntax of the ODE solver `ode45`

Command	Description
`[t, y] = ode45(@ydot, tspan, y0, options)`	Solves the vector differential equation $\dot{\mathbf{y}} = \mathbf{f}(t,\mathbf{y})$ specified by the function file whose handle is `@ydot` and whose inputs must be t and $\mathbf{y}$, and whose output must be a *column* vector representing $d\mathbf{y}/dt$; that is, $\mathbf{f}(t, \mathbf{y})$. The number of rows in this column vector must equal the order of the equation. The vector `tspan` contains the starting and ending values of the independent variable t, and optionally any intermediate values of t where the solution is desired. The vector `y0` contains the initial values. The function file must have two input arguments, `t` and `y`, even for equations where $f(t, y)$ is not a function of t. The `options` argument is created with the `odeset` function. The syntax is identical for the solver `ode15s`.
`options = odeset ('name1', 'value1' 'name2', 'value2', . . .)`	Creates an integrator options structure `options` to be used with the ODE solver, in which the named properties have the specified values, where `name` is the name of a *property* and `value` is the value to be assigned to the property. Any unspecified properties have default values. Typing `odeset` with no input arguments displays all property names and their possible values.

7.5 Special Methods for Linear Equations

MATLAB provides some convenient tools to use if the differential equation model is linear. Even though there are general methods available for finding the analytical solutions of linear differential equations, it is sometimes more convenient to use a numerical method to find the solution. Examples of such situations are when the forcing function is a complicated function or when the order of the differential equation is higher than 2. In such cases the algebra involved in obtaining the analytical solution might not be worth the effort, especially if the main objective is to obtain a plot of the solution.

Matrix Methods

We can use matrix operations to reduce the number of lines to be typed in the derivative function file. For example, the following equation describes the motion of a mass connected to a spring, with viscous friction acting between the mass and the surface. Another force $u(t)$ also acts on the mass.

$$m\ddot{y} + c\dot{y} + ky = u(t) \qquad (7.5-1)$$

This can be put into Cauchy form by letting $x_1 = y$ and $x_2 = \dot{y}$. This gives

$$\dot{x}_1 = x_2$$

$$\dot{x}_2 = \frac{1}{m}u(t) - \frac{k}{m}x_1 - \frac{c}{m}x_2$$

This can be written as one matrix equation as follows.

$$\begin{bmatrix} \dot{x}_1 \\ \dot{x}_2 \end{bmatrix} = \begin{bmatrix} 0 & 1 \\ -\dfrac{k}{m} & -\dfrac{c}{m} \end{bmatrix} \begin{bmatrix} x_1 \\ x_2 \end{bmatrix} + \begin{bmatrix} 0 \\ \dfrac{1}{m} \end{bmatrix} u(t)$$

In compact form this is

$$\dot{\mathbf{x}} = \mathbf{A}\mathbf{x} + \mathbf{B}u(t) \tag{7.5–2}$$

where

$$\mathbf{A} = \begin{bmatrix} 0 & 1 \\ -\dfrac{k}{m} & -\dfrac{c}{m} \end{bmatrix} \qquad \mathbf{B} = \begin{bmatrix} 0 \\ \dfrac{1}{m} \end{bmatrix} \qquad \mathbf{x} = \begin{bmatrix} x_1 \\ x_2 \end{bmatrix}$$

The following function file shows how to use matrix operations. In this example, $m = 1$, $c = 2$, $k = 5$, and the applied force is $u(t) = 10$.

```
function xdot = msd(t,x)
% Function file for mass with spring and damping.
% Position is first variable, velocity is second variable.
u = 10;
m = 1;c = 2;k = 5;
A = [0, 1;-k/m, -c/m];
B = [0; 1/m];
xdot = A*x+B*u;
```

Note that the output `xdot` will be a column vector because of the definition of matrix-vector multiplication. We try different values of the final time until we see the entire response. Using a final time of 5 and the initial conditions $x_1(0) = 0$ and $x_2(0) = 0$, we call the solver and plot the solution as follows:

```
[t, x] = ode45(@msd, [0,5], [0,0];
plot(t,x(:,1),t,x(:,2),'--')
```

Figure 7.5–1 shows the edited plot. Note that we could have avoided embedding the values of the parameters m, c, k, and u by making `msd` a nested function as was done with the functions `pendulum` and `pendula` in Section 7.4.

Test Your Understanding

T7.5–1 Plot the position and velocity of a mass with a spring and damping, having the parameter values $m = 2$, $c = 3$, and $k = 7$. The applied force is $u = 35$, the initial position is $y(0) = 2$, and the initial velocity is $\dot{y}(0) = -3$.

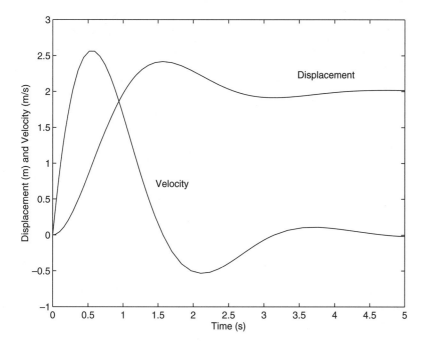

Figure 7.5–1 Displacement and velocity of the mass as a function of time.

Characteristic Roots from the `eig` Function

The characteristic roots of a linear differential equation give information about the speed of response and the oscillation frequency, if any.

MATLAB provides the `eig` function to compute the characteristic roots when the model is given in the state-variable form (7.5–2). Its syntax is `eig(A)`, where A is the matrix that appears in (7.5–2). (The function's name is an abbreviation of *eigenvalue*, which is another name for characteristic root.) For example, consider the equations

EIGENVALUE

$$\dot{x}_1 = -3x_1 + x_2 \tag{7.5–3}$$

$$\dot{x}_2 = -x_1 - 7x_2 \tag{7.5–4}$$

The matrix **A** for these equations is

$$\mathbf{A} = \begin{bmatrix} -3 & 1 \\ -1 & -7 \end{bmatrix}$$

To find the characteristic roots, type

```
>>A = [-3, 1;-1, -7];
>>r = eig(A)
```

The answer so obtained is `r = [-6.7321, -3.2679]`. To find the time constants, which are the negative reciprocals of the real parts of the roots, you type `tau = -1./real(r)`. The time constants are 0.1485 and 0.3060. Four times the dominant time constant, or 4(0.3060) = 1.224, gives the time it takes for the free response to become approximately zero.

ODE Solvers in the Control System Toolbox

Many of the functions from the Control System Toolbox are available in the Student Edition of MATLAB. Some of these can be used to solve linear, time-invariant (constant-coefficient) differential equations. They are sometimes more convenient to use and more powerful than the ODE solvers discussed thus far, because general solutions can be found for linear, time-invariant equations. Here we discuss several of these functions. These are summarized in Table 7.5–1. The other features of the Control System Toolbox require advanced methods, and will not be covered here. See [Palm, 2005b] for coverage of these methods.

LTI OBJECT

An *LTI object* describes a linear, time-invariant equation, or sets of equations, here referred to as the *system*. An LTI object can be created from different descriptions of the system, it can be analyzed with several functions, and it can be accessed to provide alternate descriptions of the system. For example, the equation

$$2\ddot{x} + 3\dot{x} + 5x = u(t) \tag{7.5--5}$$

is one description of a particular system. This description is called the *reduced form*. The following is a state-model description of the same system:

$$\dot{\mathbf{x}} = \mathbf{A}\mathbf{x} + \mathbf{B}u \tag{7.5--6}$$

where $x_1 = x$, $x_2 = \dot{x}$, and

$$\mathbf{A} = \begin{bmatrix} 0 & 1 \\ -\frac{5}{2} & -\frac{3}{2} \end{bmatrix} \qquad \mathbf{B} = \begin{bmatrix} 0 \\ \frac{1}{2} \end{bmatrix} \qquad \mathbf{x} = \begin{bmatrix} x_1 \\ x_2 \end{bmatrix} \tag{7.5--7}$$

Both model forms contain the same information. However, each form has its own advantages, depending on the purpose of the analysis.

Because there are two or more state variables in a state model, we need to be able to specify which state variable, or combination of variables, constitutes the output of the simulation. For example, models (7.5–6) and (7.5–7) can represent the motion of a mass, with x_1 the position and x_2 the velocity of the mass. We need to be able to specify whether we want to see a plot of the position, or the velocity, or both. This specification of the output, denoted by the vector **y**, is done in general with the matrices **C** and **D,** which must be compatible with the following equation.

$$\mathbf{y} = \mathbf{C}\mathbf{x} + \mathbf{D}u(t) \tag{7.5--8}$$

where the vector $\mathbf{u}(t)$ allows for multiple inputs. To continue the previous example, if we want the output to be the position $x = x_1$, then $y = x_1$, and we would select $\mathbf{C} = [1,0]$ and $\mathbf{D} = 0$. Thus, in this case, (7.5–8) reduces to $y = x_1$.

To create an LTI object from the reduced form (7.5–5), use the tf(right, left) function, and type

```
>>sys1 = tf(1, [2, 3, 5]);
```

where the vector right is the vector of coefficients of the right-hand side of the equation, arranged in descending derivative order, and left is the vector of coefficients of the left-hand side of the equation, also arranged in descending derivative order. The result, sys1, is the LTI object that describes the system in reduced form, also called the *transfer function form*. (The name of the function, tf, stands for *transfer function*, which is an equivalent way of describing the coefficients on the left- and right-hand sides of the equation.)

The LTI object sys2 in transfer function form for the equation

$$6\frac{d^3x}{dt^3} - 4\frac{d^2x}{dt^2} + 7\frac{dx}{dt} + 5x = 3\frac{d^2u}{dt^2} + 9\frac{du}{dt} + 2u \qquad (7.5\text{–}9)$$

is created by typing

```
>>sys2 = tf([3, 9, 2], [6, -4, 7, 5]);
```

To create an LTI object from a state model, you use the ss(A, B, C, D) function, where ss stands for *state space*. For example, to create an LTI object in state-model form for the system described by (7.5–6) through (7.5–8), you type

```
>>A = [0, 1; -5/2, -3/2]; B = [0; 1/2];
>>C = [1, 0]; D = 0;
>>sys3 = ss(A,B,C,D);
```

An LTI object defined using the tf function can be used to obtain an equivalent state-model description of the system. To create a state model for the system described by the LTI object sys1 created previously in transfer function form, you type ss(sys1). You will then see the resulting **A**, **B**, **C**, and **D** matrices on the screen. To extract and save the matrices, use the ssdata function as follows.

```
>>[A1, B1, C1, D1] = ssdata(sys1);
```

The results are

$$\mathbf{A1} = \begin{bmatrix} -1.5 & -1.25 \\ 2 & 0 \end{bmatrix} \qquad \mathbf{B1} = \begin{bmatrix} 0.5 \\ 0 \end{bmatrix} \qquad \mathbf{C1} = [0 \quad 0.5] \qquad \mathbf{D1} = [0]$$

When using ssdata to convert a transfer function form to a state model, note that the output y will be a scalar that is identical to the solution variable of the reduced form; in this case the solution variable of (7.5–1) is the variable y. To interpret the state model, we need to relate its state variables x_1 and x_2 to y. The values of the matrices **C1** and **D1** tell us that the output variable $y = 0.5x_2$. Thus we then see that $x_2 = 2y$. The other state variable x_1 is related to x_2 by $\dot{x}_2 = 2x_1$. Thus $x_1 = \dot{y}$.

Table 7.5–1 LTI object functions

Command	Description
sys = ss(A, B, C, D)	Creates an LTI object in state-space form, where the matrices A, B, C, and D correspond to those in the model $\dot{\mathbf{x}} = \mathbf{Ax} + \mathbf{Bu}$, $\mathbf{y} = \mathbf{Cx} + \mathbf{Du}$.
[A, B, C, D] = ssdata(sys)	Extracts the matrices A, B, C and D corresponding to those in the model $\dot{\mathbf{x}} = \mathbf{Ax} + \mathbf{Bu}$, $\mathbf{y} = \mathbf{Cx} + \mathbf{Du}$.
sys = tf(right,left)	Creates an LTI object in transfer function form, where the vector right is the vector of coefficients of the right-hand side of the equation, arranged in descending derivative order, and left is the vector of coefficients of the left-hand side of the equation, also arranged in descending derivative order.
sys2 = tf(sys1)	Creates the transfer function model sys2 from the state model sys1.
sys1 = ss(sys2)	Creates the state model sys1 from the transfer function model sys2.
[right, left] = tfdata(sys,'v')	Extracts the coefficients on the right- and left-hand sides of the reduced-form model specified in the transfer function model sys. When the optional parameter 'v' is used, the coefficients are returned as vectors rather than as cell arrays.

To create a transfer function description of the system sys3, previously created from the state model, you type tfsys3 = tf(sys3); To extract and save the coefficients of the reduced form, use the tfdata function as follows:

```
[right, left] = tfdata(sys3, 'v')
```

For this example, the vectors returned are right = 1 and left = [1, 1.5, 2.5]. The optional parameter 'v' tells MATLAB to return the coefficients as vectors; otherwise, they are returned as cell arrays. These functions are summarized in Table 7.5–1.

Test Your Understanding

T7.5–2 Obtain the state model for the reduced-form model

$$5\ddot{x} + 7\dot{x} + 4x = u(t)$$

Then convert the state model back to reduced form, and see if you get the original reduced-form model.

Linear ODE Solvers

The Control System Toolbox provides several solvers for linear models. These solvers are categorized by the type of input function they can accept: zero input, impulse input, step input, and a general input function. These are summarized in Table 7.5–2.

The initial Function The initial function computes and plots the free response of a state model. This is sometimes called the *initial condition response* or the *undriven response* in the MATLAB documentation. The basic

Table 7.5–2 Basic syntax of the LTI ODE solvers

Command	Description
impulse(sys)	Computes and plots the impulse response of the LTI object sys.
initial(sys,x0)	Computes and plots the free response of the LTI object sys given in state-model form, for the initial conditions specified in the vector x0.
lsim(sys,u,t)	Computes and plots the response of the LTI object sys to the input specified by the vector u, at the times specified by the vector t.
step(sys)	Computes and plots the step response of the LTI object sys.

See the text for description of extended syntax.

syntax is initial(sys,x0), where sys is the LTI object in state-model form and x0 is the initial-condition vector. The time span and number of solution points are chosen automatically. For example, to find the free response of the state model (7.5–5) through (7.5–8), for $x_1(0) = 5$ and $x_2(0) = -2$, first define it in state-model form. This was done previously to obtain the system sys3. Then use the initial function as follows.

```
>>initial(sys3, [5, -2])
```

The plot shown in Figure 7.5–2 will be displayed on the screen. Note that MATLAB automatically labels the plot, computes the steady-state response, and displays it with a dotted line.

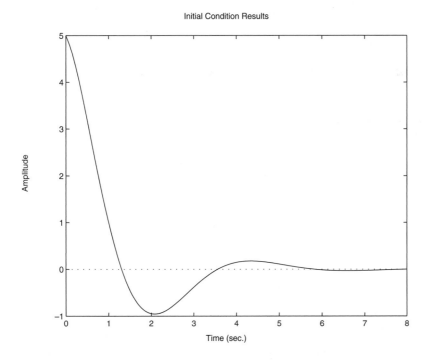

Figure 7.5–2 Free response of the model given by (7.5–5) through (7.5–8) for $x_1(0) = 5$ and $x_2(0) = -2$.

To specify the final time $\texttt{tf}$, use the syntax $\texttt{initial(sys,x0,tf)}$. To specify a vector of times of the form $\texttt{t = [0:dt:tf]}$, at which to obtain the solution, use the syntax $\texttt{initial(sys,x0,t)}$.

When called with left-hand arguments, as $\texttt{[y, t, x] = initial}$ $\texttt{(sys,x0, . . .)}$, the function returns the output response $\texttt{y}$, the time vector $\texttt{t}$ used for the simulation, and the state vector $\texttt{x}$ evaluated at those times. The columns of the matrices $\texttt{y}$ and $\texttt{x}$ are the outputs and the states, respectively. The number of rows in $\texttt{y}$ and $\texttt{x}$ equals $\texttt{length(t)}$. No plot is drawn. The syntax $\texttt{initial(sys1, sys2, . . .,x0,t)}$ plots the free response of multiple LTI systems on a single plot. The time vector $\texttt{t}$ is optional. You can specify line color, line style, and marker for each system; for example, $\texttt{initial(sys1,'r', sys2,'y--',sys3,'gx',x0)}$.

The $\texttt{impulse}$ Function   The $\texttt{impulse}$ function plots the unit-impulse response for each input-output pair of the system, assuming that the initial conditions are zero. (The unit impulse is also called the Dirac delta function.) The basic syntax is $\texttt{impulse(sys)}$, where $\texttt{sys}$ is the LTI object. Unlike the $\texttt{initial}$ function, the $\texttt{impulse}$ function can be used with either a state model or a transfer function model. The time span and number of solution points are chosen automatically. For example, the impulse response of (7.5–5) is found as follows:

```
>>sys1 = tf(1, [2, 3, 5]);
>>impulse(sys1)
```

The extended syntax of the $\texttt{impulse}$ function is similar to that of the $\texttt{initial}$ function.

The $\texttt{step}$ Function   The $\texttt{step}$ function plots the unit-step response for each input-output pair of the system, assuming that the initial conditions are zero. [The unit step function $u(t)$ is 0 for $t < 0$ and 1 for $t > 0$]. The basic syntax is $\texttt{step(sys)}$, where $\texttt{sys}$ is the LTI object. The $\texttt{step}$ function can be used with either a state model or a transfer function model. The time span and number of solution points are chosen automatically. The extended syntax of the $\texttt{step}$ function is similar to that of the $\texttt{initial}$ and the $\texttt{impulse}$ functions.

To find the unit-step response, for zero initial conditions, of the state model (7.5–6) through (7.5–8), and the reduced-form model

$$5\ddot{x} + 7\dot{x} + 5x = 5\dot{f} + f \tag{7.5–10}$$

the session is (assuming $\texttt{sys3}$ is still available in the workspace)

```
>>sys4 = tf([5, 1], [5, 7, 5]);
>>step(sys3,'b',sys4,'--')
```

The result is shown in Figure 7.5–3. The steady-state response is indicated by the horizontal dotted line. Note how the steady-state response and the time to reach that state are automatically determined.

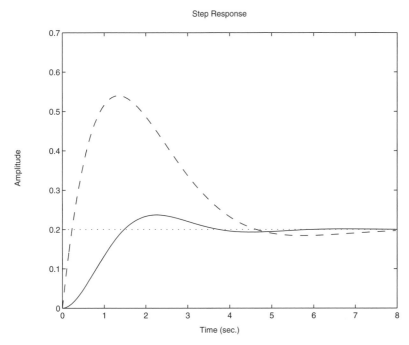

Figure 7.5–3 Step response of the model given by (7.5–6) through (7.5–8) and the model (7.5–10), for zero initial conditions.

Step response can be characterized by the following parameters.

- *Steady-state value*: The limit of the response as $t \rightarrow \infty$.
- *Settling time*: The time for the response to reach and stay within a certain percentage (usually 2%) of its steady-state value.
- *Rise time*: The time required for the response to rise from 10% to 90% of its steady-state value.
- *Peak response*: The largest value of the response.
- *Peak time*: The time at which the peak response occurs.

When the `step(sys)` function puts a plot on the screen, you may use the plot to calculate these parameters by right-clicking anywhere within the plot area. This brings up a menu. Choose "Characteristics" to obtain a submenu that contains the response characteristics. When you select a specific characteristic, for example "peak response," MATLAB puts a large dot on the peak and displays dashed lines indicating the value of the peak response and the peak time. Move the cursor over this dot to see a display of the values. You can use the other solvers in the same way, although the menu choices may be different. For example, peak response and settling time are available when you use the `impulse(sys)` function, but not the rise time. If instead of choosing "Characteristics," you choose "Properties" and select the "Options" tab, you can change the defaults for the settling time and rise time, which are 2% and 10% to 90%.

Using this method, we find that the solid curve in Figure 7.5–3 has the following characteristics:

- Steady-state value: 0.2
- 2% Settling time: 5.22
- 10–90% Rise time: 1.01
- Peak response: 0.237
- Peak time: 2.26

You can also read values off any part of the curve by placing the cursor on the curve at the desired point. You can also move the cursor along the curve and read the values as they change. Using this method, we find that the solid curve in Figure 7.5–3 crosses the steady-state value of 0.2 for the second time at $t = 3.74$.

You can suppress the plot generated by `step` and create your own plot as follows, assuming `sys3` is still available in the workspace.

```
[x,t] = step(sys3);
plot(t,x)
```

You can then use the Plot Editor tools to edit the plot. However, with this approach, right-clicking on the plot will no longer give you information about the step response characteristics.

Suppose the step input is not a *unit* step but instead is 0 for $t < 0$ and 10 for $t > 0$. There are two ways to obtain the solution with the factor 10. Using `sys3` as the example, these are `step(10*sys3)`; and

```
[x,t] = step(sys3);
plot(t,10*x)
```

The `lsim` Function   The `lsim` function plots the response of the system to an arbitrary input. The basic syntax for zero-initial conditions is `lsim(sys,u,t)`, where `sys` is the LTI object, `t` is a time vector having regular spacing, as `t = [0:dt:tf]`, and `u` is a matrix with as many columns as inputs, and whose *i*th row specifies the value of the input at time `t(i)`. To specify nonzero initial conditions for a state-space model, use the syntax `lsim(sys,u,t,x0)`. This computes and plots the total response (the free plus forced response). Right-clicking on the plot brings up the menu containing the "Characteristics" choice, although the only characteristic available is the peak response.

When called with left-hand arguments, as `[y, t] = lsim(sys, u, . . .)`, the function returns the output response `y` and the time vector `t` used for the simulation. The columns of the matrix `y` are the outputs, and the number of its rows equals `length(t)`. No plot is drawn. To obtain the state vector solution for state-space models, use the syntax `[y, t, x] = lsim(sys,u,. . .)`. The syntax `lsim(sys1,sys2,. . .,u,t,x0)` plots the responses of multiple LTI systems on a single plot. The initial condition vector `xo` is needed only if the initial conditions are nonzero. You can specify line color, line style, and marker for each system; for example, `lsim(sys1,'r',sys2,'y--',sys3,'gx',u,t)`.

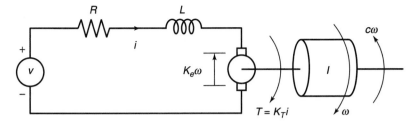

Figure 7.5–4 An armature-controlled dc motor.

We will see an example of the `lsim` function shortly.

Programming Detailed Forcing Functions

As a final example of higher-order equations, we now show how to program a detailed forcing function for use with the `lsim` function. We use a dc motor as the application. The equations for an armature-controlled dc motor (such as a permanent magnet motor) shown in Figure 7.5–4 are the following. They result from Kirchhoff's voltage law and Newton's law applied to a rotating inertia. The motor's current is i and its rotational velocity is ω.

$$L\frac{di}{dt} = -Ri - K_e\omega + v(t) \tag{7.5–11}$$

$$I\frac{d\omega}{dt} = K_Ti - c\omega \tag{7.5–12}$$

where L, R, and I are the motor's inductance, resistance, and inertia; K_T and K_e are the torque constant and back emf constant; c is a viscous damping constant; and $v(t)$ is the applied voltage. These equations can be put into matrix form as follows, where $x_1 = i$ and $x_2 = \omega$.

$$\begin{bmatrix} \dot{x}_1 \\ \dot{x}_2 \end{bmatrix} = \begin{bmatrix} -\frac{R}{L} & -\frac{K_e}{L} \\ \frac{K_T}{I} & -\frac{c}{I} \end{bmatrix} \begin{bmatrix} x_1 \\ x_2 \end{bmatrix} + \begin{bmatrix} \frac{1}{L} \\ 0 \end{bmatrix} v(t)$$

Trapezoidal Profile for a DC Motor   EXAMPLE 7.5–1

In many applications we want to accelerate the motor to a desired speed, and allow it to run at that speed for some time before decelerating to a stop. Investigate whether an applied voltage having a trapezoidal profile will accomplish this. Use the values $R = 0.6\ \Omega$, $L = 0.002$ H, $K_T = 0.04$ N · m/A, $K_e = 0.04$ V· s/rad, $c = 0$, and $I = 6 \times 10^{-5}$ kg · m^2. The applied voltage in volts is given by

$$v(t) = \begin{cases} 100t & 0 \le t < 0.1 \\ 10 & 0.1 \le t \le 0.4 \\ -100(t - 0.4) + 10 & 0.4 < t \le 0.5 \\ 0 & t > 0.5 \end{cases}$$

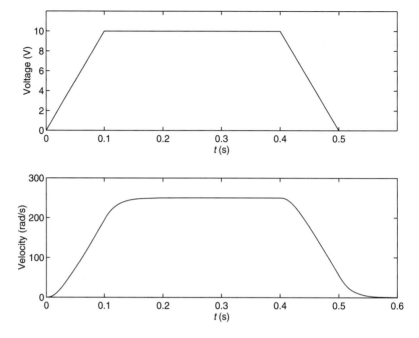

Figure 7.5–5 Voltage input and resulting velocity response of a dc motor.

This is shown in the top graph in Figure 7.5–5.

■ Solution

The following program first creates the model sys from the matrices **A, B, C,** and **D.** We choose **C** and **D** to obtain the speed x_2 as the only output. (To obtain both the speed and the current as outputs, we would choose C = [1, 0; 0, 1] and D = [0; 0].) The program then computes the time constants using the eig function, and then creates time, the array of time values to be used by lsim. We choose the time increment 0.0001 to be a very small fraction of the total time, 0.6 s.

The trapezoidal voltage function is then created with a for loop. This is perhaps the easiest way because the if-elseif-else structure mimics the equations that define $v(t)$. The initial conditions $x_1(0)$ and $x_2(0)$ are assumed to be zero, and so they need not be specified in the lsim function.

```
% File dcmotor.m
R = 0.6; L = 0.002; c = 0;
K_T = 0.04; K_e = 0.04; I = 6e-5;
A = [-R/L, -K_e/L; K_T/I, -c/I];
B = [1/L; 0]; C = [0,1]; D = [0];
sys = ss(A,B,C,D);
Time_constants = -1./real(eig(A))
time = [0:0.0001:0.6];
k = 0;
for t = [0:0.0001:0.6]
```

```
   k = k + 1;
   if t < 0.1
      v(k) = 100*t;
   elseif t <= 0.4
      v(k) = 10;
   elseif t <= 0.5
      v(k) = -100*(t-0.4) + 10;
    else
      v(k) = 0;
    end
end
[y,t] = lsim(sys, v, time);
subplot(2,1,1), plot(time,v)
subplot(2,1,2), plot(time,y)
```

The time constants are computed to be 0.0041 and 0.0184 s. The largest time constant indicates that the motor's response time is approximately 4(0.0184) = 0.0736 s. Because this time is less than the time needed for the applied voltage to reach 10 V, the motor should be able to follow the desired trapezoidal profile reasonably well. To know for certain, we must solve the motor's differential equations. The results are plotted in the bottom graph of Figure 7.5–5. The motor's velocity follows a trapezoidal profile as expected, although there is some slight deviation because of its electrical resistance and mechanical inertia.

LTI Viewer The Control System Toolbox contains the LTI Viewer, which assists in the analysis of LTI systems. It provides an interactive user interface that allows you to switch between different types of response plots and between the analysis of different systems. The viewer is invoked by typing `ltiview`. See the MATLAB help for more information.

Predefined Input Functions

You can always create any complicated input function to use with the ODE solver `ode45` or `lsim` by defining a vector containing the input function's values at specified times, as was done in Example 7.5–1 for the trapezoidal profile. However, MATLAB provides the `gensig` function that makes it easy to construct periodic input functions.

The syntax `[u, t] = gensig(type, period)` generates a periodic input of a specified type `type`, having a period `period`. The following types are available: sine wave (`type` = 'sin'), square wave (`type` = 'square'), and narrow-width periodic pulse (`type` = 'pulse'). The vector `t` contains the times, and the vector `u` contains the input values at those times. All generated inputs have unit amplitudes. The syntax `[u, t] = gensig(type, period, tf,dt)` specifies the time duration `tf` of the input and the spacing `dt` between the time instants.

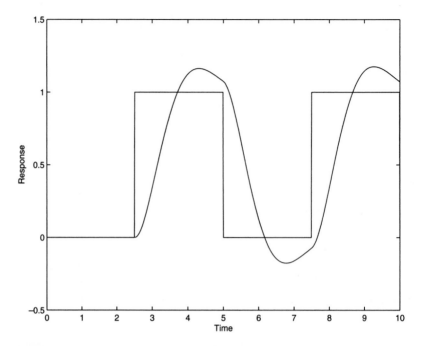

Figure 7.5–6 Square-wave response of the model $\ddot{x} + 2\dot{x} + 4x = 4f$.

For example, suppose a square wave with period 5 is applied to the following reduced-form model.

$$\ddot{x} + 2\dot{x} + 4x = 4f \tag{7.5–13}$$

To find the response for zero initial conditions, over the interval $0 \leq t \leq 10$, using a step size of 0.01, the session is

```
>>sys5 = tf(4,[1,2,4]);
>>[u, t] = gensig('square',5,10,0.01);
>>[y, t] = lsim (sys5,u,t);plot(t,y,u), . . .
   axis([0 10 -0.5 1.5]), . . .
   xlabel('Time'),ylabel('Response')
```

The result is shown in Figure 7.5–6.

7.6 Summary

This chapter covered numerical methods for computing integrals and derivatives, and for solving ordinary differential equations. Now that you have finished this chapter, you should be able to do the following.

■ Numerically evaluate single, double, and triple integrals whose integrands are given functions.

- Numerically evaluate single integrals whose integrands are given as numerical values.
- Numerically estimate the derivative of a set of data.
- Compute the gradient and Laplacian of a given function.
- Obtain in closed form the integral and derivative of a polynomial function.
- Use the MATLAB ODE solvers to solve single first-order ordinary differential equations whose initial conditions are specified.
- Convert higher-order ordinary differential equations into a set of first-order equations.
- Use the MATLAB ODE solvers to solve sets of higher-order ordinary differential equations whose initial conditions are specified.
- Use MATLAB to convert a model from transfer function form to state-variable form, and vice versa.
- Use the MATLAB linear solvers to solve linear differential equations to obtain the free response and the step response for arbitrary forcing functions.

We have not covered all the differential equation solvers provided in MATLAB, but limited our coverage to ordinary differential equations whose initial conditions are specified. MATLAB provides algorithms for solving *boundary-value problems* (BVPs) such as

$$\ddot{x} + 7\dot{x} + 10x = 0 \qquad x(0) = 2 \qquad x(5) = 8 \qquad 0 \le t \le 5$$

See the help for the function `bvp4c`. Some differential equations are specified *implicitly* as $f(t, y, \dot{y}) = 0$. The solver `ode15i` can be used for such problems. MATLAB can also solve *delay-differential equations* (*DDE*) such as

$$\ddot{x} + 7\dot{x} + 10x + 5x(t - 3) = 0$$

See the help for the functions `dde23`, `ddesd`, and `deval`. The function `pdepe` can solve *partial* differential equations. See also `pdeval`. In addition, MATLAB provides support for analyzing and plotting the solver's output. See the functions `odeplot`, `odephas2`, `odephas3`, and `odeprint`.

Key Terms with Page References

Backward difference, 314
Cauchy form, 326
Central difference, 315
Definite integral, 306
Eigenvalue, 333
Euler method, 319
Forced response, 319
Forward difference, 314
Free response, 319
Improper integral, 306
Indefinite integral, 306

Initial value problem (IVP), 318
Laplacian, 318
LTI object, 334
Modified Euler method, 320
Ordinary differential equation
(ODE), 318
Predictor-corrector method, 320
Quadrature, 309
Singularities, 306
State-variable form, 326
Step size, 320

Problems

You can find answers to problems marked with an asterisk at the end of the text.

Section 7.1

1.* An object moves at a velocity $v(t) = 5 + 7t^2$ m/s starting from the position $x(2) = 5$ m at $t = 2$ s. Determine its position at $t = 10$ s.

2. The total distance traveled by an object moving at velocity $v(t)$ from the time $t = a$ to the time $t = b$ is

$$x(b) = \int_a^b |v(t)|\, dt + x(a)$$

The absolute value $|v(t)|$ is used to account for the possibility that $v(t)$ might be negative. Suppose an object starts at time $t = 0$ and moves with a velocity of $v(t) = \cos(\pi t)$ m. Find the object's location at $t = 1$ s if $x(0) = 2$ m.

3. An object starts with an initial velocity of 3 m/s at $t = 0$, and it accelerates with an acceleration of $a(t) = 5t$ m/s^2. Find the total distance the object travels in 5 s.

4. The equation for the voltage $v(t)$ across a capacitor as a function of time is

$$v(t) = \frac{1}{C}\left(\int_0^t i(t)\, dt + Q_0 \right)$$

where $i(t)$ is the applied current and Q_0 is the initial charge. A certain capacitor initially holds no charge. Its capacitance is $C = 10^{-7}$ F. If a current $i(t) = 0.2\,[1 + \sin(0.2t)]$ A is applied to the capacitor, compute the voltage $v(t)$ at $t = 1.2$ s.

5. A certain object's position as a function of time is given by $x(t) = 6t \sin 5t$ m. Compute its velocity and acceleration at $t = 5$ s.

6. A certain object moves with the velocity $v(t)$ given in the table below: Determine the object's position $x(t)$ at $t = 10$ s if $x(0) = 3$.

Time (s)	0	1	2	3	4	5	6	7	8	9	10
Velocity (m/s)	0	2	5	7	9	12	15	18	22	20	17

7.* A tank having vertical sides and a bottom area of 100 ft^2 is used to store water. The tank is initially empty. To fill the tank, water is pumped into the top at the rate given in the following table. Determine the water height $h(t)$ at $t = 10$ min.

Time (min)	0	1	2	3	4	5	6	7	8	9	10
Flow rate (ft^3/min)	0	80	130	150	150	160	165	170	160	140	120

8. A cone-shaped paper drinking cup (like the kind supplied at water fountains) has a radius R and a height H. If the water height in the cup is h, the water volume is given by

$$V = \frac{1}{3}\pi\left(\frac{R}{H}\right)^2 h^3$$

Suppose that the cup's dimensions are $R = 1.5$ in. and $H = 4$ in.
   (a) If the flow rate from the fountain into the cup is 2 in.³/s, how long will it take to fill the cup to the brim?
   (b) If the flow rate from the fountain into the cup is given by $2(1 - e^{-2t})$ in.³/s, how long will it take to fill the cup to the brim?

9. A certain object has a mass of 100 kg and is acted on by a force $f(t) = 500[2 - e^{-t}\sin(5\pi t)]$ N. The mass is at rest at $t = 0$. Determine the object's velocity at $t = 5$ s.

10.* A rocket's mass decreases as it burns fuel. The equation of motion for a rocket in vertical flight can be obtained from Newton's law, and it is

$$m(t)\frac{dv}{dt} = T - m(t)g$$

where T is the rocket's thrust and its mass as a function of time is given by $m(t) = m_0(1 - rt/b)$. The rocket's initial mass is m_0, the burn time is b, and r is the fraction of the total mass accounted for by the fuel.
   Use the values $T = 48{,}000$ N, $m_0 = 2200$ kg, $r = 0.8$, $g = 9.81$ m/s², and $b = 40$ s. Determine the rocket's velocity at burnout.

11. The equation for the voltage $v(t)$ across a capacitor as a function of time is

$$v(t) = \frac{1}{C}\left(\int_0^t i(t)\,dt + Q_0\right)$$

where $i(t)$ is the applied current and Q_0 is the initial charge. Suppose that $C = 10^{-7}$ F and that $Q_0 = 0$. Suppose the applied current is $i(t) = 0.3 + 0.1e^{-5t}\sin(25\pi t)$ A. Plot the voltage $v(t)$ for $0 \le t \le 7$ s.

12. Compute the indefinite integral of $p(x) = 5x^2 - 6x + 8$.

13. Compute the double integral

$$A = \int_0^3\int_1^3 (x^2 + 3xy)\,dx\,dy$$

14. Compute the double integral

$$A = \int_0^4\int_0^\pi x^2\sin y\,dx\,dy$$

15. Compute the double integral

$$A = \int_0^1 \int_y^3 x^2(x + y)\,dx\,dy$$

Note that the region of integration lies to the right of the line $y = x$. Use this fact and a MATLAB relational operator to eliminate values for which $y > x$.

16. Compute the triple integral

$$A = \int_1^2 \int_0^1 \int_1^3 xe^{yz}\,dx\,dy\,dz$$

Section 7.2

17. Plot the estimate of the derivative dy/dx from the following data. Do this by using forward, backward, and central differences. Compare the results.

x	0	1	2	3	4	5	6	7	8	9	10
y	0	2	5	7	9	12	15	18	22	20	17

18. At a relative maximum of a curve $y(x)$, the slope dy/dx is zero. Use the following data to estimate the values of x and y that correspond to a maximum point.

x	0	1	2	3	4	5	6	7	8	9	10
y	0	2	5	7	9	10	8	7	6	8	10

19. Compare the performance of the forward, backward, and central difference methods for estimating the derivative of the following function: $y(x) = e^{-x} \sin(3x)$. Use 101 points from $x = 0$ to $x = 4$. Use a random additive error of ± 0.01.

20. Compute the expressions for dp_2/dx, $d(p_1 p_2)/dx$, and $d(p_2/p_1)/dx$ for $p_1 = 3x^2 + 7$ and $p_2 = 5x^2 - 6x + 8$.

21. Plot the contour plot and the gradient (shown by arrows) for the function

$$f(x,y) = -x^2 + 2xy + 3y^2$$

Section 7.3

22. Plot the solution of the equation

$$5\dot{y} + y = f(t)$$

if $f(t) = 0$ for $t < 0$ and $f(t) = 10$ for $t \geq 0$. The initial condition is $y(0) = 5$.

23. The equation for the voltage y across the capacitor of an RC circuit is

$$RC\frac{dy}{dt} + y = v(t)$$

where $v(t)$ is the applied voltage. Suppose that $RC = 0.2$ s and that the capacitor voltage is initially 2 V. Suppose also that the applied voltage goes from 0 to 10 V at $t = 0$. Plot the voltage $y(t)$ for $0 \le t \le 1$ s.

24. The following equation describes the temperature $T(t)$ of a certain object immersed in a liquid bath of constant temperature T_b.

$$10\frac{dT}{dt} + T = T_b$$

Suppose the object's temperature is initially $T(0) = 70°F$ and the bath temperature is $T_b = 170°F$.
  (a) How long will it take for the object's temperature T to reach the bath temperature?
  (b) How long will it take for the object's temperature T to reach 168°F?
  (c) Plot the object's temperature $T(t)$ as a function of time.

25.* The equation of motion of a rocket-propelled sled is, from Newton's law,

$$m\dot{v} = f - cv$$

where m is the sled mass, f is the rocket thrust, and c is an air resistance coefficient. Suppose that $m = 1000$ kg and $c = 500$ N · s/m. Suppose also that $v(0) = 0$ and $f = 75,000$ N for $t \ge 0$. Determine the speed of the sled at $t = 10$ s.

26.* The following equation describes the motion of a mass connected to a spring, with viscous friction on the surface.

$$m\ddot{y} + c\dot{y} + ky = 0$$

Plot $y(t)$ for $y(0) = 10$, $\dot{y}(0) = 5$ if
  (a) $m = 3$, $c = 18$, and $k = 102$
  (b) $m = 3$, $c = 39$ and $k = 120$

27. The equation for the voltage y across the capacitor of an RC circuit is

$$RC\frac{dy}{dt} + y = v(t)$$

where $v(t)$ is the applied voltage. Suppose that $RC = 0.2$ s and that the capacitor voltage is initially 2 V. Suppose also that the applied voltage is $v(t) = 10[2 - e^{-t}\sin(5\pi t)]$ V. Plot the voltage $y(t)$ for $0 \le t \le 5$ s.

28. The equation describing the water height h in a spherical tank with a drain at the bottom is

$$\pi(2rh - h^2)\frac{dh}{dt} = -C_d A\sqrt{2gh}$$

Suppose the tank's radius is $r = 3$ m and that the circular drain hole has a radius of 2 cm. Assume that $C_d = 0.5$ and that the initial water height is $h(0) = 5$ m. Use $g = 9.81$ m/s^2.

(a) Use an approximation to estimate how long it takes for the tank to empty.

(b) Plot the water height as a function of time until $h(t) = 0$.

29. The following equation describes a certain dilution process, where $y(t)$ is the concentration of salt in a tank of fresh water to which salt brine is being added.

$$\frac{dy}{dt} + \frac{2}{10 + 2t}y = 4$$

Suppose that $y(0) = 0$. Plot $y(t)$ for $0 \le t \le 10$.

Section 7.4

30. The following equation describes the motion of a certain mass connected to a spring, with viscous friction on the surface,

$$3\ddot{y} + 18\dot{y} + 102y = f(t)$$

where $f(t)$ is an applied force. Suppose that $f(t) = 0$ for $t < 0$ and $f(t) = 10$ for $t \ge 0$.

(a) Plot $y(t)$ for $y(0) = \dot{y}(0) = 0$.

(b) Plot $y(t)$ for $y(0) = 0$ and $\dot{y}(0) = 10$. Discuss the effect of the nonzero initial velocity.

31. The following equation describes the motion of a certain mass connected to a spring, with viscous friction on the surface,

$$3\ddot{y} + 39\dot{y} + 120y = f(t)$$

where $f(t)$ is an applied force. Suppose that $f(t) = 0$ for $t < 0$ and $f(t) = 10$ for $t \ge 0$.

(a) Plot $y(t)$ for $y(0) = \dot{y}(0) = 0$.

(b) Plot $y(t)$ for $y(0) = 0$ and $\dot{y}(0) = 10$. Discuss the effect of the nonzero initial velocity.

32.* The following equation describes the motion of a certain mass connected to a spring, with no friction,

$$3\ddot{y} + 75y = f(t)$$

where $f(t)$ is an applied force. Suppose the applied force is sinusoidal with a frequency of ω rad/s and an amplitude of 10 N: $f(t) = 10 \sin(\omega t)$. Suppose that the initial conditions are $y(0) = \dot{y}(0) = 0$. Plot $y(t)$ for $0 \le t \le 20$ s. Do this for the following three cases. Compare the results of each case.

(a) $\omega = 1$ rad/s
(b) $\omega = 5$ rad/s
(c) $\omega = 10$ rad/s

33. Van der Pol's equation has been used to describe many oscillatory processes. It is

$$\ddot{y} - \mu(1 - y^2)\dot{y} + y = 0$$

Plot $y(t)$ for $\mu = 1$ and $0 \le t \le 20$, using the initial conditions $y(0) = 2$, $\dot{y}(0) = 0$.

34. The equation of motion for a pendulum whose base is accelerating horizontally with an acceleration $a(t)$ is

$$L\ddot{\theta} + g \sin \theta = a(t) \cos \theta$$

Suppose that $g = 9.81$ m/s^2, $L = 1$ m, and $\dot{\theta}(0) = 0$. Plot $\theta(t)$ for $0 \le t \le 10$ s for the following three cases.

(a) The acceleration is constant: $a = 5$ m/s^2, and $\theta(0) = 0.5$ rad.
(b) The acceleration is constant: $a = 5$ m/s^2, and $\theta(0) = 3$ rad.
(c) The acceleration is linear with time: $a = 0.5t$ m/s^2, and $\theta(0) = 3$ rad.

35. Van der Pol's equation is

$$\ddot{y} - \mu(1 - y^2)\dot{y} + y = 0$$

This equation is stiff for large values of the parameter μ. Compare the performance of `ode45` and `ode15s` for this equation. Use $\mu = 1000$ and $0 \le t \le 3000$, with the initial conditions $y(0) = 2$, $\dot{y}(0) = 0$. Plot $y(t)$ versus t.

Section 7.5

36. The equations for an armature-controlled dc motor are the following. The motor's current is i and its rotational velocity is ω.

$$L\frac{di}{dt} = -Ri - K_e\omega + v(t) \tag{7.6-1}$$

$$I\frac{d\omega}{dt} = K_Ti - c\omega \tag{7.6-2}$$

where L, R, and I are the motor's inductance, resistance, and inertia; K_T and K_e are the torque constant and back emf constant; c is a viscous damping constant; and $v(t)$ is the applied voltage.

Use the values $R = 0.8\ \Omega$, $L = 0.003$ H, $K_T = 0.05$ N $\cdot$ m/A, $K_e = 0.05$ V $\cdot$ s/rad, $c = 0$, and $I = 8 \times 10^{-5}$ kg $\cdot$ m^2.

(a) Suppose the applied voltage is 20 V. Plot the motor's speed and current versus time. Choose a final time large enough to show the motor's speed becoming constant.

(b) Suppose the applied voltage is trapezoidal as given below.

$$v(t) = \begin{cases} 400t & 0 \le t < 0.05 \\ 20 & 0.05 \le t \le 0.2 \\ -400(t - 0.2) + 20 & 0.2 < t \le 0.25 \\ 0 & t > 0.25 \end{cases}$$

Plot the motor's speed versus time for $0 \le t \le 0.3$ s. Also plot the applied voltage versus time. How well does the motor speed follow a trapezoidal profile?

37.* Compute and plot the unit-impulse response of the following model.

$$10\ddot{y} + 3\dot{y} + 7y = f(t)$$

38. Compute and plot the unit-step response of the following model.

$$10\ddot{y} + 6\dot{y} + 2y = f + 3\dot{f}$$

39.* Find the reduced form of the following state model.

$$\begin{bmatrix} \dot{x}_1 \\ \dot{x}_2 \end{bmatrix} = \begin{bmatrix} -4 & -1 \\ 2 & -3 \end{bmatrix} \begin{bmatrix} x_1 \\ x_2 \end{bmatrix} + \begin{bmatrix} 2 \\ 5 \end{bmatrix} u(t)$$

40. The following state model describes the motion of a certain mass connected to a spring, with viscous friction on the surface, where $m = 1$, $c = 2$, and $k = 5$.

$$\begin{bmatrix} \dot{x}_1 \\ \dot{x}_2 \end{bmatrix} = \begin{bmatrix} 0 & 1 \\ -5 & -2 \end{bmatrix} \begin{bmatrix} x_1 \\ x_2 \end{bmatrix} + \begin{bmatrix} 0 \\ 1 \end{bmatrix} f(t)$$

(a) Use the `initial` function to plot the position x_1 of the mass, if the initial position is 5 and the initial velocity is 3.

(b) Use the `step` function to plot the step response of the position and velocity for zero initial conditions, where the magnitude of the step input is 10. Compare your plot with that shown in Figure 7.5–1.

41. Consider the following equation.

$$5\ddot{y} + 2\dot{y} + 10y = f(t)$$

(a) Plot the free response for the initial conditions $y(0) = 10$, $\dot{y}(0) = -5$.

(b) Plot the unit-step response (for zero initial conditions).

(c) The *total response* to a step input is the sum of the free response and the step response. Demonstrate this fact for this equation by plotting the sum of the solutions found in parts (a) and (b), and comparing the plot with that generated by solving for the total response with $y(0) = 10$, $\dot{y}(0) = -5$.

42. The model for the RC circuit shown in Figure P42 is

$$RC\frac{dv_o}{dt} + v_o = v_i$$

For $RC = 0.1$ s, plot the voltage response $v_o(t)$ for the case where the applied voltage is a single square pulse of height 10 V and duration 0.2 s, starting at $t = 0$. The initial capacitor voltage is zero.

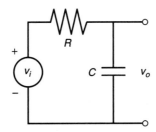

Figure P42

8

CHAPTER

Symbolic Processing

OUTLINE

Symbolic processing is the term used to describe how a computer performs operations on mathematical expressions in the way, for example, that humans do algebra with pencil and paper. Whenever possible, we wish to obtain solutions in closed form because they give us greater insight into the problem. Symbolic processing is a way to obtain closed-form solutions.

SYMBOLIC EXPRESSION

This chapter explains how to define a *symbolic expression* such as $y = \sin x/\cos x$ in MATLAB and how to use MATLAB to simplify expressions wherever possible. For example, the previous function simplifies to $y = \sin x/\cos x = \tan x$. MATLAB can perform such simplifications and operations such as addition and multiplication on mathematical expressions. We can use MATLAB to obtain symbolic solutions to algebraic equations such as $x^2 + 2x + a = 0$ (the solution for x is $x = -1 \pm \sqrt{1-a}$). MATLAB can also perform symbolic differentiation and integration and can solve ordinary differential equations in closed form.

To use the methods of this chapter, you must have either the Symbolic Math toolbox or the Student Edition of MATLAB, which contains all the functions of the Symbolic Math toolbox but has limited access to the Maple® kernel.

The programs in this chapter are compatible with versions 2 through 3.2 of the toolbox, *although different versions might give slightly different error messages and slightly different displays of expressions.*

The symbolic processing capabilities in MATLAB are based on the Maple V software package, which was developed by Waterloo Maple Software, Inc. The MathWorks has licensed the Maple "engine," that is, the core of Maple. If you have used Maple before, however, or plan to use it in the future, you should be aware that the syntax used by MATLAB differs from that used by the commercially available Maple package.

When you have finished this chapter, you should be able to use MATLAB to

- Create symbolic expressions and manipulate them algebraically.
- Obtain symbolic solutions to algebraic and transcendental equations.
- Perform symbolic differentiation and integration.
- Evaluate limits and series symbolically.
- Obtain symbolic solutions to ordinary differential equations.
- Obtain Laplace transforms.
- Perform symbolic linear algebra operations, including obtaining expressions for determinants, matrix inverses, and eigenvalues.

8.1 Symbolic Expressions and Algebra

The `sym` function can be used to create "symbolic objects" in MATLAB. If the input argument to `sym` is a string, the result is a symbolic number or variable. If the input argument is a numeric scalar or matrix, the result is a symbolic representation of the given numeric values. For example, typing x = sym('x') creates the symbolic variable with name x. Typing x = sym('x','real') tells MATLAB to assume that x is real. Typing x = sym('x','unreal') tells MATLAB to assume that x is not real. Typing x = sym('x', 'positive') makes x a positive (real) variable.

In the syntax object_name = sym('string'), string may be a single letter or combination of letters, a combination of letters and digits starting with a letter, or a number. The object names are case-sensitive.

The `syms` function enables you to combine more than one such statement into a single statement. For example, typing syms x is equivalent to typing x = sym('x'), and typing syms x y u v creates the four symbolic variables x, y, u, and v. When used without arguments, syms lists the symbolic objects in the workspace. The syms command, however, cannot be used to create symbolic constants; you must use sym for this purpose.

The `syms` command enables you to specify that certain variables are real. For example,

```
>>syms x y real
```

You can use the `sym` function to create *symbolic constants* by using a numerical value for the argument. For example, typing pi = sym('pi'), fraction = sym('1/3'), and sqroot2 = sym('sqrt(2)') create symbolic constants that avoid the floating-point approximations inherent in the

SYMBOLIC CONSTANT

values of π, 1/3, and $\sqrt{2}$. If you create the symbolic constant π this way, it temporarily replaces the built-in numeric constant, and you no longer obtain a numerical value when you type its name. For example,

```
>>pi = sym('pi')
pi =
pi
>>sqroot2 = sym('sqrt(2)')
sqroot2 =
sqrt(2)
>>a = 3*sqrt(2)     % This gives a numeric result.
a =
    4.2426
>>b = 3*sqroot2     % This gives a symbolic result.
b =
3*2^(1/2)
>>c = sym (2);
>>d = 5/3
d =
    1.6667
>>f = c*d
f =
10/3
```

Note that MATLAB indents numeric results but not symbolic results. Note also that when a numerical variable (such as d in the previous session) is used in a symbolic expression (such as f), the exact value (5/3 here) is used.

Symbolic constants can look like numbers but are actually symbolic expressions. Symbolic expressions can look like character strings but are a different sort of quantity. You can use the class function to determine whether or not a quantity is symbolic, numeric, or a character string.

Symbolic Expressions

You can use symbolic variables in expressions and as arguments of functions. You use the operators $+ - * / \wedge$ and the built-in functions just as you use them with numerical calculations. For example, typing

```
>>syms x y
>>s = x + y;
>>r = sqrt(x^2 + y^2);
```

creates the symbolic variables s and r. The terms s = x + y and r = sqrt (x^2 +y^2) are examples of symbolic *expressions*. The variables s and r created this way are *not* the same as user-defined function files. That is, if you later assign x and y numeric values, typing r will not cause MATLAB to evaluate the equation $r = \sqrt{x^2 + y^2}$. We will see later how to evaluate symbolic expressions numerically. Note that execution of the line syms x y does not display the variables even though there is no semicolon at the end of the line.

The vector and matrix notation used in MATLAB also applies to symbolic variables. For example, you can create a symbolic matrix A as follows:

```
>>n = 3; syms x;
>>A = x.^((0:n)'*(0:n))
A =
[ 1, 1, 1, 1]
[ 1, x, x^2, x^3]
[ 1, x^2, x^4, x^6]
[ 1, x^3, x^6, x^9]
```

In MATLAB the variable x is the *default* independent variable, but other variables can be specified to be the independent variable. It is important to know which variable is the independent variable in an expression. The function findsym(E) can be used to determine the symbolic variable used by MATLAB in a particular expression E.

DEFAULT VARIABLE

The function findsym(E) finds the symbolic variables in a symbolic expression or matrix, where E is a scalar or matrix symbolic expression, and returns a string containing all of the symbolic variables appearing in E. The variables are returned in alphabetical order and are separated by commas. If no symbolic variables are found, findsym returns the empty string.

By contrast, the function findsym(E,n) returns the n symbolic variables in E closest to x, with the tie breaker going to the variable closer to z. The following session shows some examples of its use:

```
>>syms b x1 y
>>findsym(6*b+y)
ans =
b,y
>>findsym(6*b+y+x) %Note: x has not been declared symbolic.
??? Undefined function or variable 'x'.
>>findsym(6*b+y,1) %Find the one variable closest to x.
ans =
y
>>findsym(6*b+y+x1,1) %Find the one variable closest to x.
ans =
x1
>>findsym(6*b+y*i) %i is not symbolic.
ans =
b, y
```

Manipulating Expressions

The function collect(E) collects coefficients of like powers in the expression E. If there is more than one variable, you can use the optional form collect(E,v), which collects all the coefficients with the same power of v.

```
>>syms x y
>>E = (x-5)^2+(y-3)^2;
>>collect(E)
ans =
x^2-10*x+25+(y-3)^2
>>collect(E,y)
ans =
y^2-6*y+(x-5)^2+9
```

The function `expand(E)` expands the expression `E` by carrying out powers. For example,

```
>>syms x y
>>expand((x+y)^2) % Applies algebra rules.
ans =
x^2+2*x*y+y^2
>>expand(sin(x+y)) % Applies trig identities.
ans =
sin(x)*cos(y)+cos(x)*sin(y)
>>simplify(6*((sin(x))^2+(cos(x))^2)) % Another identity.
ans =
     6
```

The function `factor(E)` factors the expression `E`. For example,

```
>>syms x y
>>factor(x^2-1)
ans =
(x-1)*(x+1)
```

The function `simplify(E)` simplifies the expression `E`, using Maple's simplification rules. For example,

```
>>syms x y
>>simplify(x*sqrt(x^8*y^2))
ans =
x*(x^8*y^2)^(1/2)
```

The function `simple(E)` searches for the shortest form of the expression `E` in terms of number of characters. When called, the function displays the results of each step of its search. When called without the argument, `simple` acts on the previous expression and displays the simplification steps. The form `[r, how] = simple(E)` does not display intermediate steps, but saves those steps in the string `how`. The shortest form found is stored in `r`. Typing `r = simple(E)` stores the shortest form in `r` without displaying the simplification steps.

The function `[num den] = numden(E)` returns two symbolic expressions that represent the numerator `num` and denominator `den` for the rational representation of the expression `E`.

The function `double(E)` converts the expression `E` to floating-point, double-precision numeric form. The expression `E` must not contain any symbolic variables. For example,

```
>>sqroot2 = sym('sqrt(2)');
>>y = 6*sqroot2
y =
6*2^(1/2)
z = double(y)
z =
    8.4853
```

The function `poly2sym(p)` converts a coefficient vector `p` to a symbolic polynomial in terms of `x`. The form `poly2sym(p, 'v')` generates the polynomial in terms of the variable `v`. For example,

```
>>poly2sym([5,-3,7],'y')
ans =
5*y^2-3*y+7
```

The function `sym2poly(E)` converts the expression `E` to a polynomial coefficient vector.

```
>>syms x
>>sym2poly(9*x^2+4*x+6)
ans =
9 4 6
```

The function `pretty(E)` displays the expression `E` on the screen in a form that resembles typeset mathematics.

The function `subs(E,old,new)` substitutes `new` for `old` in the expression `E`, where `old` can be a symbolic variable or expression and `new` can be a symbolic variable, expression, or matrix, or a numeric value or matrix. For example,

```
>>syms x y
>>E = x^2+6*x+7;
>>F = subs(E,x,y)
F =
y^2+6*y+7
```

If `old` and `new` are cell arrays of the same size, each element of `old` is replaced by the corresponding element of `new`. If `E` and `old` are scalars and `new` is an array or cell array, the scalars are expanded to produce an array result.

If you want to tell MATLAB that f is a function of the variable t, type `f = sym('f(t)')`. Thereafter, `f` behaves like a function of `t`, and you can manipulate it with the toolbox commands. For example, to create a new function $g(t) = f(t + 2) - f(t)$, the session is

```
>>syms t
>>f = sym('f(t)');
```

```
>>g = subs(f,t,t+2)-f
g =
f(t+2)-f(t)
```

Once a specific function is defined for *f*(*t*), the function *g*(*t*) will be available. We will use this technique with the Laplace transform in Section 8.5.

To perform multiple substitutions, enclose the new and old elements in braces. For example, to substitute *a* = *x* and *b* = 2 into the expression *E* = *a* sin *b*, the session is

```
>>syms a b x
>>E = a*sin(b);
>>F = subs(E,{a, b}, {x, 2})
F =
x*sin(2)
```

Evaluating Expressions

In most applications we eventually want to obtain numerical values or a plot from the symbolic expression. Use the `subs` and `double` functions to evaluate an expression numerically. Use `subs(E,old,new)` to replace `old` with a numeric value `new` in the expression E. The result is of class double. For example,

```
>>syms x y
>>E = x^2+6*x+7; F = y*x^5;
>>G = subs(E,x,2)
G =
   23
>>class(G)
ans =
double
>>H = subs (E, x, [1:3])
H=
   14 23 34
>>J = subs (F, y, [1:3])
J =
[x^5, 2*x^5, 3*x^5]
```

The MATLAB function `ezplot(E, [xmin xmax])` generates a plot of a symbolic expression E, a function of one variable over the range from `xmin` to `xmax`.

Order of Precedence

MATLAB does not always arrange expressions in a form that we normally would use. For example, MATLAB might provide an answer in the form −c+b, whereas we would normally write b−c. The order of precedence used by MATLAB must be constantly kept in mind to avoid misinterpreting the MATLAB output (see Table 1.1–2 for the order of precedence).MATLAB frequently expresses results in

Table 8.1–1 Functions for creating and evaluating symbolic expressions

Command	Description
class(E)	Returns the class of the expression E.
digits(d)	Sets the number of decimal digits used to do variable precision arithmetic. The default is 32 digits.
double(E)	Converts the expression E to double precision numeric form.
ezplot(E)	Generates a plot of a symbolic expression E, which is a function of one variable. The default range of the independent variable is the interval $[-2\pi, 2\pi]$ unless this interval contains a singularity. The optional form ezplot(E,[xmin xmax]) generates a plot over the range from xmin to xmax.
findsym(E)	Finds the symbolic variables in a symbolic expression or matrix, where E is a scalar or matrix symbolic expression, and returns a string containing all the symbolic variables appearing in E. The variables are returned in alphabetical order and are separated by commas. If no symbolic variables are found, findsym returns the empty string.
findsym(E,n)	Returns the n symbolic variables in E closest to x, with the tie breaker going to the variable closer to z.
Latex(E)	Converts the expression E to a LaTeX typeset expression.
[num den] = numden(E)	Returns two symbolic expressions that represent the numerator expression num and denominator expression den for the rational representation of the expression E.
x = sym('x')	Creates the symbolic variable with name x. Typing x = sym('x','real') makes x real. Typing x = sym('x','unreal') makes x not real. Typing x = sym ('x', 'positive') makes x a positive (real) variable.
syms x y u v	Creates the symbolic variables x, y, u, and v. When used without arguments, syms lists the symbolic objects in the workspace.
vpa(E,d)	Sets the number of digits used to evaluate the expression E to d. Typing vpa(E) causes E to be evaluated to the number of digits specified by the default value of 32 or by the current setting of digits.

the form 1/a*b, whereas we would normally write b/a. MATLAB sometimes writes x^(1/2)*y^(1/2) instead of grouping the terms as (x*y)^(1/2), and it often fails to cancel negative signs where possible, as in -a/(-b*c-d), instead of a/(b*c+d).

The function latex(E) converts the expression E to a LaTeX typeset expression. Tables 8.1–1 and 8.1–2 summarize the functions for creating, evaluating, and manipulating symbolic expressions.

Test Your Understanding

T8.1–1 Given the expressions: $E_1 = x^3 - 15x^2 + 75x - 125$ and $E_2 = (x + 5)^2 - 20x$, use MATLAB to

a. Find the product E_1E_2 and express it in its simplest form.
b. Find the quotient E_1/E_2 and express it in its simplest form.
c. Evaluate the sum $E_1 + E_2$ at $x = 7.1$ in symbolic form and in numeric form.

(Answers: *a.* $(x - 5)^5$; *b.* $x - 5$; *c.* 13,671/1000 in symbolic form, 13.6710 in numeric form.)

Table 8.1–2 Functions for manipulating symbolic expressions

Command	Description
collect(E)	Collects coefficients of like powers in the expression E.
expand(E)	Expands the expression E by carrying out powers.
factor(E)	Factors the expression E.
poly2sym(p)	Converts a polynomial coefficient vector p to a symbolic polynomial. The form poly2sym(p,'v') generates the polynomial in terms of the variable v.
pretty(E)	Displays the expression E on the screen in a form that resembles typeset mathematics.
simple(E)	Searches for the shortest form of the expression E in terms of number of characters. When called, the function displays the results of each step of its search. When called without the argument, simple acts on the previous expression. The form [r, how] = simple(E) does not display intermediate steps, but saves those steps in the string how. The shortest form found is stored in r.
simplify(E)	Simplifies the expression E using Maple's simplification rules.
subs(E,old,new)	Substitutes new for old in the expression E, where old can be a symbolic variable or expression, new can be a symbolic variable, expression, or matrix, or a numeric value or matrix.
sym2poly(E)	Converts the expression E to a polynomial coefficient vector.

8.2 Algebraic and Transcendental Equations

The function solve(E) solves a symbolic expression or equation represented by the expression E. If E represents an *equation,* the equation's expression must be enclosed in single quotes. If E represents an expression, then the solution obtained will be the roots of the expression E; that is, the solution of the equation $E = 0$. Multiple expressions or equations can be solved by separating them with a comma, as solve(E1,E2, ..., En). Note that you need not declare the symbolic variable with the sym or syms function before using solve.

For example, to solve the equation $x + 5 = 0$, the session is

```
>>syms x
>>x = solve('x+5 = 0')
x =
-5
```

To solve the equation $e^{2x} + 3e^x = 54$, the session is

```
>>solve('exp(2*x)+3*exp(x)=54')
ans =
[2*log(3)+i*pi]
[ log(6)]
```

When more than one variable occurs in the expression, MATLAB assumes that the variable closest to x in the alphabet is the variable to be found. You can specify the solution variable using the syntax solve(E,'v'), where v is the solution variable. For example,

```
>>solve('b^2+8*c+2*b=0') %Solves for c because it is closer to x.
ans =
```

```
-1/8*b^2-1/4*b
>>solve('b^2+8*c+2*b=0','b') % Solves for b.
ans =
[-1+(1-8*c)^(1/2)]
[-1-(1-8*c)^(1/2)]
```

Thus the solution of $b^2 + 8c + 2b = 0$ for c is $c = -(b^2 + 2b)/8$. The solution for b is $b = -1 \pm \sqrt{1 - 8c}$.

You can solve simultaneous equations and save the solutions as vectors by using the form `[x, y] = solve(eq1, eq2)`. Note the difference in the output formats in the following example. In the first form the result `ans` is a structure, with the answers for `x` and `y` stored in the fields `ans.x` and `ans.y`. (See Section 2.8 for a discussion of structures and fields.) In the second form the solution is saved in a cell array. (See Section 2.7 for discussion of cell arrays.)

SOLUTION STRUCTURE

```
>>eq1 = '6*x+2*y=14';
>>eq2 = '3*x+7*y=31';
>>solve(eq1,eq2)
ans =
x: [1x1 sym]
y: [1x1 sym]
>>x = ans.x
x =
1
>>y = ans.y =
4
>>[x, y] = solve(eq1,eq2)
x =
1
y =
4
```

Test Your Understanding

T8.2–1 Use MATLAB to solve the equation $\sqrt{1 - x^2} = x$.
(Answer: $x = \sqrt{2}/2$.)

T8.2–2 Use MATLAB to solve the equation set $x + 6y = a$, $2x - 3y = 9$ for x and y in terms of the parameter a.
(Answer: $x = (a + 18)/5$, $y = (2a - 9)/15$.)

Intersection of Two Circles

EXAMPLE 8.2–1

We want to find the intersection points of two circles. The first circle has a radius of 2 and is centered at $x = 3$, $y = 5$. The second circle has a radius b and is centered at $x = 5$, $y = 3$. See Figure 8.2–1.

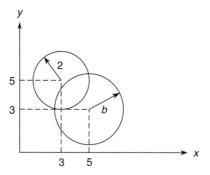

Figure 8.2–1 Intersection points of two circles.

(a) Find the (x, y) coordinates of the intersection points in terms of the parameter b.
(b) Evaluate the solution for the case where $b = \sqrt{3}$.

■ Solution

(a) The intersection points are found from the solutions of the two equations for the circles. These equations are $(x - 3)^2 + (y - 5)^2 = 4$ for the first circle and $(x - 5)^2 + (y - 3)^2 = b^2$. The session to solve these equations follows. Note that the result $\mathtt{x\!:\ [2x1\ sym]}$ indicates that there are two solutions for x. Similarly, there are two solutions for y.

```
>>syms x y b
>>S = solve((x-3)^2+(y-5)^2-4,(x-5)^2+(y-3)^2-b^2)
S =
x: [2x1 sym]
y: [2x1 sym]
>>S.x
ans =
[ 9/2-1/8*b^2+1/8*(-16+24*b^2-b^4)^(1/2)]
[ 9/2-1/8*b^2-1/8*(-16+24*b^2-b^4)^(1/2)]
```

The solution for the x coordinates of the intersection points is

$$x = \tfrac{9}{2} - \tfrac{1}{8}b^2 \pm \tfrac{1}{8}\sqrt{-16 + 24b^2 - b^4}$$

The solution for the y coordinates can be found in a similar way by typing $\mathtt{S.y}$.
(b) Continue the session by substituting $b = \sqrt{3}$ into the expression for x.

```
>>subs(S.x,b,sqrt(3))
ans =
    4.9820
    3.2680
```

Thus the x coordinates of the two intersection points are $x = 4.982$ and $x = 3.268$. The y coordinates can be found in a similar way.

Table 8.2–1 Functions for solving algebraic and transcendental equations

Command	Description
solve(E)	Solves a symbolic expression or equation represented by the expression E. If E represents an *equation,* the equation's expression must be enclosed in single quotes. If E represents an expression, then the solution obtained will be the roots of the expression E; that is, the solution of the equation $E = 0$. You need not declare the symbolic variable with the sym or syms function before using solve.
solve(E1,. . .,En)	Solves multiple expressions or equations.
S = solve(E)	Saves the solution in the structure S.

Test Your Understanding

T8.2–3 Find the y coordinates of the intersection points in Example 8.2–1. Use $b = \sqrt{3}$.

(Answer: $y = 4.7320, 3.0180$.)

Equations containing periodic functions can have an infinite number of solutions. In such cases the solve function restricts the solution search to solutions near 0. For example, to solve the equation $\sin(2x) - \cos x = 0$, the session is

```
>>solve('sin(2*x)-cos(x)=0')
ans =
1/2*pi
-1/2*pi
1/6*pi
5/6*pi
```

Table 8.2–1 summarizes the solve function.

8.3 Calculus

The diff function is used to obtain the symbolic derivative. Although this function has the same name as the function used to compute numerical differences (see Section 7.2), MATLAB detects whether a symbolic expression is used in the argument and directs the calculation accordingly. The basic syntax is diff(E), which returns the derivative of the expression E with respect to the default independent variable.

For example,

```
>>syms n x y
>>diff(x^n)
ans =
x^n*n/x
>>simplify(ans)
ans =
```

```
x^(n-1)*n
>>diff(log(x))
ans =
1/x
>>diff((sin(x))^2)
ans =
2*sin(x)*cos(x)
>>diff(sin(y))
ans =
cos(y)
```

If the expression contains more than one variable, the `diff` function operates on the variable *x*, or the variable closest to *x*, unless told to do otherwise. When there is more than one variable, the `diff` function computes the *partial* derivative. For example,

```
>>diff(sin(x*y))
ans =
cos(x*y)*y
```

The function `diff(E,v,n)` returns the *n*th derivative of the expression E with respect to the variable v. Both v and n are optional. For example,

$$\frac{\partial^2 [x \sin(xy)]}{\partial y^2} = -x^3 \sin(xy)$$

is given by

```
>>syms x y
>>diff(x*sin(x*y),y,2)
ans =
-x^3*sin(x*y)
```

Table 8.3–1 summarizes the differentiation functions.

Table 8.3–1 Symbolic calculus functions

Command	Description
`diff(E,v,n)`	Returns the *n*th derivative of the expression E with respect to the variable v. Both v and n are optional.
`int(E,v,a,b)`	Returns the integral of the expression E with respect to the optional variable v over the interval [*a*, *b*].
`limit(E,v,a)`	Returns the limit of the expression E as the variable v goes to a. Both v and a are optional. If a is omitted, the limit is taken as v goes to 0.
`limit(E,v,a,'d')`	Returns the limit of the expression E as the variable v goes to a from the direction specified by d, which may be `right` or `left`.
`symsum(E)`	Returns the symbolic summation of the expression E.
`taylor(f,n,a)`	Gives the first n−1 terms in the Taylor series for the function defined in the expression f, evaluated at the point $x = a$. If the parameter a is omitted, the function returns the series evaluated at $x = 0$.

Test Your Understanding

T8.3–1 Given that $y = \sinh(3x)\cosh(5x)$, use MATLAB to find dy/dx at $x = 0.2$. (Answer: 9.2288.)

T8.3–2 Given that $z = 5\cos(2x)\ln(4y)$, use MATLAB to find $\partial z/\partial y$. (Answer: $5\cos(2x)/y$.)

Integration

The `int(E)` function is used to integrate a symbolic expression E. It attempts to find the symbolic expression I such that `diff(I) = E`. If the integral does not exist in closed form or MATLAB cannot find the integral even if it exists, the function will return the expression unevaluated. The function `int(E)` returns the integral of the expression E with respect to the default independent variable. For example,

```
>>syms n x y
>>int(x^n)
ans =
x^(n+1)/(n+1)
>>int(1/x)
ans =
log(x)
>>int(cos(x))
ans =
sin(x)
>>int(sin(y))
ans =
-cos(y)
```

The form `int(E,v,a,b)` returns the integral of the expression E with respect to the variable v evaluated over the interval [a, b]. The argument v is optional. For example,

```
>>syms x y
>>int(x*y^2,y,0,5)
ans =
125/3*x
```

Another example is

```
>>syms a b x
>>int(x^2,a,b)
ans =
1/3*b^3-1/3*a^3
```

The following session gives an example for which no integral can be found. The indefinite integral exists, but the definite integral does not exist if the limits of integration include the singularity at $x = 1$. The integral is

$$\int \frac{1}{x - 1} dx = \ln |x - 1|$$

The session is

```
>>syms x
>>int(1/(x-1))
ans =
log(x-1)
>>int(1/(x-1),0,2)
ans =
NaN
```

Table 8.3–1 summarizes the integration functions.

Test Your Understanding

T8.3–3 Given that $y = x \sin(3x)$, use MATLAB to find $\int y\, dx$.
(Answer: $[\sin(3x) - 3x \cos(3x)]/9$.)

T8.3–4 Given that $z = 6y^2 \tan(8x)$, use MATLAB to find $\int z\, dy$.
(Answer: $2y^3 \tan(8x)$.)

T8.3–5 Use MATLAB to evaluate

$$\int_{-2}^{5} x \sin(3x)\, dx$$

(Answer: 0.6672.)

Taylor Series

The `taylor(f,n,a)` function gives the first $n-1$ terms in the Taylor series for the function defined in the expression `f`, evaluated at the point $x = a$. If the parameter `a` is omitted the function returns the series evaluated at $x = 0$. Here are some examples:

```
>>syms x
>>f = exp(x);
>>taylor(f,4)
ans =
1+x+1/2*x^2+1/6*x^3
>>taylor(f,3,2)
ans =
exp(2)+exp(2)*(x-2)+1/2*exp(2)*(x-2)^2
```

The latter expression corresponds to

$$e^2[1 + (x - 2) + \tfrac{1}{2}(x - 2)^2]$$

The function `taylortool` opens a graphical interface that plots a function and the nth partial sum of its Taylor series. See the MATLAB help for details.

Sums

The `symsum(E,a,b)` function returns the sum of the expression `E` as the default symbolic variable varies from `a` to `b`. That is, if the symbolic variable is `x`, then `S = symsum(E,a,b)` returns

$$\sum_{x=a}^{b} E(x) = E(a) + E(a + 1) + E(a + 2) + \cdots + E(b)$$

For example, the summation

$$\sum_{k=0}^{n-1} k = 0 + 1 + 2 + 3 + \cdots + n - 1 = \frac{1}{2}n^2 - \frac{1}{2}n$$

is given by

```
>>syms k n
>>symsum(k,0,n-1)
ans =
1/2*n^2-1/2*n
```

Limits

The function `limit(E,v,a)` finds the limit of the expression `E` as $v \rightarrow a$. If `v` is omitted, the default symbolic variable is used. If `a` is omitted, the limit is taken to 0. For example,

$$\lim_{x \to 3} \frac{x - 3}{x^2 - 9} = \frac{1}{6}$$

$$\lim_{h \to 0} \frac{\sin(x + h) - \sin(x)}{h}$$

are given by

```
>>syms h x
>>limit((x-3)/(x^2-9),3)
ans =
1/6
>>limit((sin(x+h)-sin(x))/h,h,0)
ans =
cos(x)
```

The forms `limit(E,v,a,'right')` and `limit(E,v,a,'left')` specify the direction of the limit. For example,

$$\lim_{x \to 0-} \frac{1}{x} = -\infty$$

$$\lim_{x \to 0+} \frac{1}{x} = \infty$$

are given by

```
>>syms x
>>limit(1/x,x,0,'left')
ans =
-Inf
>>limit(1/x,x,0,'right')
ans =
Inf
```

Table 8.3–1 summarizes the series and limit functions.

Test Your Understanding

T8.3–6 Use MATLAB to find the first three nonzero terms in the Taylor series for cos x.
(Answer: $1 - x^2/2 + x^4/24$.)

T8.3–7 Use MATLAB to find a formula for the sum

$$\sum_{m=0}^{m-1} m^3$$

(Answer: $m^4/4 - m^3/2 + m^2/4$.)

T8.3–8 Use MATLAB to evaluate

$$\sum_{n=0}^{7} \cos(\pi n)$$

(Answer: 0.)

T8.3–9 Use MATLAB to evaluate

$$\lim_{x \to 5} \frac{2x - 10}{x^3 - 125}$$

(Answer: 2/75.)

8.4 Differential Equations

MATLAB provides the `dsolve` function for solving ordinary differential equations. Its various forms differ according to whether they are used to solve single equations or sets of equations, whether or not boundary conditions are specified, and whether or not the default independent variable t is acceptable. Note that t is the default independent variable and not x as with the other symbolic functions.

Solving a Single Differential Equation

The dsolve function's syntax for solving a single equation is dsolve('eqn'). The function returns a symbolic solution of the ODE specified by the symbolic expression eqn. Use the uppercase letter D to represent the first derivative, use D2 to represent the second derivative, and so on. Any character immediately following the differentiation operator is taken to be the dependent variable. Thus Dw represents dw/dt. Because of this syntax, you cannot use uppercase D as a symbolic variable when using the dsolve function.

The arbitrary constants in the solution are denoted by C1, C2, and so on. The number of such constants is the same as the order of the ODE. For example, the equation $\dot{y} + 2y = 12$ has the solution $y(t) = 6 + C_1 e^{-2t}$. The solution can be found with the following session. Note that you need not declare y to be symbolic prior to using dsolve.

```
>>dsolve('Dy+2*y=12')
ans =
6+C1*exp(-2*t)
```

There can be symbolic constants in the equation. The second-order equation $\ddot{y} = c^2 y$ has the solution $y(t) = C_1 e^{ct} + C_2 e^{-ct}$, which can be found with the following session:

```
>>dsolve('D2y=c^2*y')
ans =
C1*exp(c*t) + C2*exp(-c*t)
```

Solving Sets of Equations

Sets of equations can be solved with dsolve. The appropriate syntax is dsolve('eqn1','eqn2',...). The function returns a symbolic solution of the set of equations specified by the symbolic expressions eqn1 and eqn2.

For example, the set

$$\frac{dx}{dt} = 3x + 4y$$

$$\frac{dy}{dt} = -4x + 3y$$

has the solution $x(t) = C_1 e^{3t} \cos 4t + C_2 e^{3t} \sin 4t$, $y(t) = -C_1 e^{3t} \sin 4t + C_2 e^{3t} \cos 4t$. The session is

```
>>[x, y] = dsolve('Dx=3*x+4*y','Dy=-4*x+3*y')
x = C1*exp(3*t)*cos(4*t)+C2*exp(3*t)*sin(4*t)
y = -C1*exp(3*t)*sin(4*t)+C2*exp(3*t)*cos(4*t)
```

Specifying Initial and Boundary Conditions

Conditions on the solutions at specified values of the independent variable can be handled as follows. The form `dsolve('eqn', 'cond1', 'cond2',...)` returns a symbolic solution of the ODE specified by the symbolic expression `eqn`, subject to the conditions specified in the expressions `cond1`, `cond2`, and so on. If `y` is the dependent variable, these conditions are specified as follows: `y(a) = b`, `Dy(a) = c`, `D2y(a) = d`, and so on. These correspond to $y(a)$, $\dot{y}(a)$, $\ddot{y}(a)$, and so on. If the number of conditions is less than the order of the equation, the returned solution will contain arbitrary constants `C1`, `C2`, and so on.

For example, the problem

$$\frac{dy}{dt} = \sin(bt), \qquad y(0) = 0$$

has the solution $y(t) = [1 - \cos(bt)]/b$. It can be found as follows:

```
>>dsolve('Dy=sin(b*t)','y(0)=0')
ans =
-cos(b*t)/b+1/b
```

The problem

$$\frac{d^2y}{dt^2} = c^2y, \qquad y(0) = 1, \qquad \dot{y}(0) = 0$$

has the solution $y(t) = (e^{ct} + e^{-ct})/2$. The session is

```
>>dsolve('D2y=c^2*y','y(0)=1','Dy(0)=0')
ans =
1/2*exp(c*t)+1/2*exp(-c*t)
```

Arbitrary boundary conditions, such as $y(0) = c$, can be used. For example, the solution of the problem

$$\frac{dy}{dt} + ay = b, \qquad y(0) = c$$

is

$$y(t) = \frac{b}{a} + \left(c - \frac{b}{a}\right)e^{-at}$$

The session is

```
>>dsolve('Dy+a*y=b','y(0)=c')
ans =
1/a*b+exp(-a*t)*(-1/a*b+c)
```

Sets of equations with specified boundary conditions can be solved as follows. The function `dsolve('eqn1','eqn2',...,'cond1','cond2',...)` returns a symbolic solution of a set of equations specified by the symbolic expressions eqn1, eqn2, and so on, subject to the initial conditions specified in the expressions cond1, cond2, and so on.

It is not necessary to specify only initial conditions. The conditions can be specified at different values of t. For example, to solve the problem

$$\frac{d^2y}{dt^2} + 9y = 0, \quad y(0) = 1, \quad \dot{y}(\pi) = 2$$

the session is

```
>>dsolve('D2y+9*y=0','y(0)=1','Dy(pi)=2')
ans =
-2/3*sin(3*t)+cos(3*t)
```

Although the default independent variable is t, you can use the following syntax to specify a different independent variable. The function dsolve('eqn1', 'eqn2', ..., 'cond1','cond2', ..., 'x') returns a symbolic solution of a set of equations where the independent variable is x. The ezplot function can be used to plot the solution, just as with any other symbolic expression, provided no undetermined constants are present.

Test Your Understanding

T8.4–1 Use MATLAB to solve the equation

$$\frac{d^2y}{dt^2} + b^2y = 0$$

Check the answer by hand or with MATLAB.
(Answer: $y(t) = C_1 \cos(bt) + C_2 \sin(bt)$.)

T8.4–2 Use MATLAB to solve the problem

$$\frac{d^2y}{dt^2} + b^2y = 0, \quad y(0) = 1, \quad \dot{y}(0) = 0$$

Check the answer by hand or with MATLAB.
(Answer: $y(t) = \cos(bt)$.)

Solving Nonlinear Equations

MATLAB can solve some nonlinear differential equations. For example, the problem

$$\frac{dy}{dt} = 4 - y^2, \quad y(0) = 1 \tag{8.4–1}$$

Table 8.4–1 The `dsolve` function

Command	Description
`dsolve('eqn')`	Returns a symbolic solution of the ODE specified by the symbolic expression `eqn`. Use the uppercase letter `D` to represent the first derivative; use `D2` to represent the second derivative, and so on. Any character immediately following the differentiation operator is taken to be the dependent variable.
`dsolve('eqn1','eqn2', . . .)`	Returns a symbolic solution of the set of equations specified by the symbolic expressions `eqn1`, `eqn2`, and so on.
`dsolve('eqn','cond1','cond2', . . .)`	Returns a symbolic solution of the ODE specified by the symbolic expression `eqn`, subject to the conditions specified in the expressions `cond1`, `cond2`, and so on. If `y` is the dependent variable, these conditions are specified as follows: `y(a) = b, Dy(a) = c, D2(a) = d`, and so on.
`dsolve('eqn1','eqn2', . . ., 'cond1', 'cond2', . . .)`	Returns a symbolic solution of a set of equations specified by the symbolic expressions `eqn1`, `eqn2`, and so on, subject to the initial conditions specified in the expressions `cond1`, `cond2`, and so on.

can be solved with the following session

```
>>dsolve('Dy=4-y^2', 'y(0)=1')
ans =
2*(exp(4*t-log(-1/3))+1)/(-1+exp(4*t-log(-1/3)))
>>simple(ans)
ans =
2*(3*exp(4*t)-1)/(1+3*exp(4*t))
```

which is equivalent to

$$y(t) = 2\frac{3e^{4t} - 1}{1 + 3e^{4t}}$$

Not all nonlinear equations can be solved in closed form. If so, you will get a message that a solution could not be found.

Table 8.4–1 summarizes the functions for solving differential equations.

8.5 Laplace Transforms

The Laplace transform $\mathcal{L}[y(t)]$ of a function $y(t)$ is defined to be

$$\mathcal{L}[y(t)] = Y(s) = \int_0^\infty y(t)e^{-st}\,dt \qquad (8.5–1)$$

and can be obtained by typing `laplace(function)`, where `function` is a symbolic expression representing the function $y(t)$ in (8.5–1). The default independent variable is `t`, and the default result is a function of `s`.

Here is a session with some examples. The functions are t^3, e^{-at}, and $\sin bt$.

```
>>syms b t
>>laplace(t^3)
ans =
6/s^4
>>laplace(exp(-b*t))
ans =
1/(s+b)
>>laplace(sin(b*t))
ans =
b/(s^2+b^2)
```

The *inverse Laplace transform* $\mathcal{L}^{-1}[Y(s)]$ is that time function $y(t)$ whose transform is $Y(s)$; that is, $y(t) = \mathcal{L}^{-1}[Y(s)]$. Inverse transforms can be found using the `ilaplace` function. For example,

```
>>syms b s
>>ilaplace(1/s^4)
ans =
1/6*t^3
>>ilaplace(1/(s+b))
ans =
exp(-b*t)
>>ilaplace(b/(s^2+b^2)
ans =
sin(b*t)
```

The transforms of derivatives are useful for solving differential equations. Applying integration by parts to the definition of the transform, we obtain

$$\mathcal{L}\left(\frac{dy}{dt}\right) = s\mathcal{L}[y(t)] - y(0) = sY(s) - y(0) \qquad (8.5\text{–}2)$$

This procedure can be extended to higher derivatives. For example, the result for the second derivative is

$$\mathcal{L}\left(\frac{d^2y}{dt^2}\right) = s^2Y(s) - sy(0) - \dot{y}(0) \qquad (8.5\text{–}3)$$

The general result for any order derivative is

$$\mathcal{L}\left(\frac{d^ny}{dt^n}\right) = s^nY(s) - \sum_{k=1}^{n} s^{n-k}g_{k-1} \qquad (8.5\text{–}4)$$

where

$$g_{k-1} = \left.\frac{d^{k-1}y}{dt^{k-1}}\right|_{t=0} \tag{8.5-5}$$

Application to Differential Equations

The derivative and linearity properties of the transform can be used to solve differential equations such as

$$a\dot{y} + y = bv(t) \tag{8.5-6}$$

Application of the Laplace transform gives

$$Y(s) = \frac{ay(0)}{as + 1} + \frac{b}{as + 1}V(s) \tag{8.5-7}$$

The free response is given by

$$\mathcal{L}^{-1}\left[\frac{ay(0)}{as + 1}\right] = \mathcal{L}^{-1}\left[\frac{y(0)}{s + 1/a}\right] = y(0)e^{-t/a}$$

The forced response is given by

$$\mathcal{L}^{-1}\left[\frac{b}{as + 1}V(s)\right] \tag{8.5-8}$$

This inverse transform cannot be evaluated until $V(s)$ is specified. Suppose $v(t)$ is a unit-step function. Then $V(s) = 1/s$, and (8.5–8) becomes

$$\mathcal{L}^{-1}\left[\frac{b}{s(as + 1)}\right]$$

To find the inverse transform, enter

```
>>syms a b s
>>ilaplace(b/(s*(a*s+1)))
ans =
b*(1-exp(-t/a))
```

Thus the forced response of (8.5–6) to a unit-step input is $b(1 - e^{-t/a})$.
　　　Consider the second-order model

$$\ddot{x} + 1.4\dot{x} + x = f(t) \tag{8.5-9}$$

Transforming this equation gives

$$X(s) = \frac{x(0)s + \dot{x}(0) + 1.4x(0)}{s^2 + 1.4s + 1} + \frac{F(s)}{s^2 + 1.4s + 1}$$

The free response is obtained from

$$x(t) = \mathcal{L}^{-1}\left[\frac{x(0)s + \dot{x}(0) + 1.4x(0)}{s^2 + 1.4s + 1}\right]$$

Suppose the initial conditions are $x(0) = 2$ and $\dot{x}(0) = -3$. Then the free response is obtained from

$$x(t) = \mathcal{L}^{-1}\left[\frac{2s - 0.2}{s^2 + 1.4s + 1}\right] \tag{8.5–10}$$

It can be found by typing

```
>>syms s
>>ilaplace((2*s-0.2)/(s^2+1.4*s+1))
```

The free response thus found is

$$x(t) = e^{-0.7t}\left[2\cos\left(\frac{\sqrt{51}}{10}t\right) - \frac{16\sqrt{51}}{51}\sin\left(\frac{\sqrt{51}}{10}t\right)\right]$$

The forced response is obtained from

$$x(t) = \mathcal{L}^{-1}\left[\frac{F(s)}{s^2 + 1.4s + 1}\right]$$

If $f(t)$ is a unit-step function, $F(s) = 1/s$ and the forced response is

$$x(t) = \mathcal{L}^{-1}\left[\frac{1}{s(s^2 + 1.4s + 1)}\right]$$

To find the forced response, enter

```
>>ilaplace(1/(s*(s^2+1.4*s+1)))
```

The answer obtained is

$$x(t) = 1 - e^{-0.7t}\left[2\cos\left(\frac{\sqrt{51}}{10}t\right) + \frac{7\sqrt{51}}{51}\sin\left(\frac{\sqrt{51}}{10}t\right)\right] \tag{8.5–11}$$

Input Derivatives

Consider a differential equation containing the derivative of the input function $y(t)$.

$$m\ddot{x} + c\dot{x} + kx = d y + g\dot{y} \tag{8.5–12}$$

Suppose $y(t)$ is a unit-step function, which is obtained by the function heaviside(t) in MATLAB. Its derivative is the Dirac delta function, $\delta(t)$, which is given by dirac(t) in MATLAB. Use of the heaviside and dirac functions with the `dsolve` function to find the step response of equations containing derivatives of the input is not recommended. We now demonstrate how to use the Laplace transform to find the step response of equations containing derivatives of the input. Suppose the initial conditions are zero. Then transforming (8.5–12) gives

$$X(s) = \frac{d + gs}{ms^2 + cs + k} Y(s) \tag{8.5–13}$$

Suppose that $d = 1$, $g = 5$, $m = 1$, $c = 1.4$, and $k = 1$, with zero initial conditions. If $y(t)$ is a unit-step function, then $Y(s) = 1/s$, and (8.5–13) gives

$$X(s) = \frac{1 + 5s}{s(s^2 + 1.4s + 1)} \tag{8.5–14}$$

The response for the case $g = 0$ was found earlier. It is given by (8.5–11). The response for $g = 5$ is found by typing

```
>>syms s
>>ilaplace((1+5*s)/(s*(s^2+1.4*s+1)))
```

The response obtained is

$$x(t) = 1 - e^{0.7t}\left[\cos\left(\frac{\sqrt{51}}{10}t\right) + \frac{43\sqrt{51}}{51}\sin\left(\frac{\sqrt{51}}{10}t\right)\right] \tag{8.5–15}$$

Figure 8.5–1 shows the responses given by (8.5–11) and (8.5–15). The effect of differentiating the input is an increase in the response's peak value.

Direct Method

Instead of performing by hand the algebra required to find the response transform, we could use MATLAB to do the algebra for us. We now demonstrate the most direct way of using MATLAB to solve an equation with the Laplace transform. One advantage of this method is that we are not required to use the transform identities (8.5–2) through (8.5–5) for the derivatives. Let us solve the equation

$$a\frac{dy}{dt} + y = f(t) \tag{8.5–16}$$

with $f(t) = \sin t$, in terms of an unspecified value for $y(0)$. Here is the session:

```
>>syms a L s t
>>y = sym('y(t)');
>>dydt = sym('diff(y(t),t)');
>>f = sin(t);
>>eq = a*dydt+y-f;
>>E = laplace(eq,t,s)
```

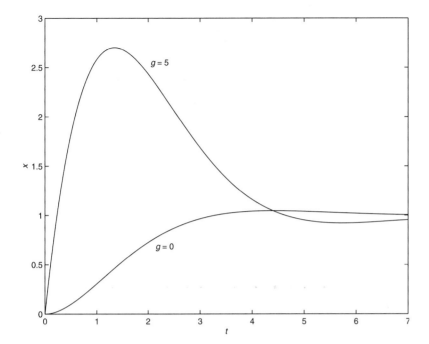

Figure 8.5–1 The unit-step response of the model $\ddot{x} + 1.4\dot{x} + x = y + g\dot{y}$ for $g = 0$ and $g = 5$.

```
E =
a*(s*laplace(y(t),t,s)-y(0)) + laplace(y(t),t,s)- 1/(s^2+1)
>>E = subs(E,'laplace(y(t),t,s)',L)

E =
a*(s*L-y(0))+L-1/(s^2+1)
>>L = solve(E,L)

L =
(a*y(0)*s^2+a*y(0)+1)/(a*s^3+a*s+s^2+1)
>>I = simplify(ilaplace(L))

I =
(-a*cos(t)+sin(t)+exp(-t/a)*y(0)+exp(-t/a) *a^2*y(0)+exp(-t/a)*a)/(1+a^2)
>>I = collect(I,exp(-t/a))

I =
1/(1+a^2)*(-a*cos(t)+sin(t))+(a+y(0)+y(0)*a^2)/(1+a^2)*exp(-t/a)
```

Table 8.5–1 Laplace transform functions

Command	Description
`ilaplace(function)`	Returns the inverse Laplace transform of `function`.
`laplace(function)`	Returns the Laplace transform of `function`.
`laplace(function,x,y)`	Returns the Laplace transform of `function`, which is a function of `x`, in terms of the Laplace variable `y`.

The answer is

$$y(t) = \frac{1}{1 + a^2}\{\sin t - a\cos t + e^{-t/a}[y(0) + a^2 y(0) + a]\}$$

Note that this session consists of the following steps:

1. Define the symbolic variables, including the derivatives that appear in the equation. Note that $y(t)$ is explicitly expressed as a function of t in these definitions.
2. Move all terms to the left side of the equation and define the left side as a symbolic expression.
3. Apply the Laplace transformation to the differential equation to obtain an algebraic equation.
4. Substitute a symbolic variable, here `L`, for the expression `laplace(y(t),t,s)` in the algebraic equation. Then solve the equation for the variable `L`, which is the transform of the solution.
5. Invert `L` to find the solution as a function of t.

Note that this procedure can also be used to solve sets of equations.

Test Your Understanding

T8.5–1 Find the Laplace transform of the following functions: $1 - e^{-at}$ and $\cos bt$. Use the `ilaplace` function to check your answers.

T8.5–2 Use the Laplace transform to solve the problem $5\ddot{y} + 20\dot{y} + 15y = 30u - 4\dot{u}$, where $u(t)$ is a unit-step function and $y(0) = 5$, $\dot{y}(0) = 1$. (Answer: $y(t) = -1.6e^{-3t} + 4.6e^{-t} + 2$.)

Table 8.5–1 summarizes the Laplace transform functions.

8.6 Symbolic Linear Algebra

You can perform operations with symbolic matrices in much the same way as with numeric matrices. Here we give examples of finding matrix products, the matrix inverse, eigenvalues, and the characteristic polynomial of a matrix.

Remember that using symbolic matrices avoids numerical imprecision in subsequent operations. You can create a symbolic matrix from a numeric matrix in several ways, as shown in the following session:

```
>>A = sym([3, 5; 2, 7]); % The most direct method.
>>B = [3, 5; 2, 7];
>>C = sym(B);% B is preserved as a numeric matrix.
>>D = subs(A,[3, 5; 2, 7]);
```

The first method is the most direct. Use the second method when you want to keep a numeric version of the matrix. The matrices A and C are symbolic and identical. The matrices B and D look like A and C but are numeric of class double.

You can create a symbolic matrix consisting of functions. For example, the relationship between the coordinates (x_2, y_2) of a coordinate system rotated counterclockwise through an angle a relative to the (x_1, y_1) coodinate system is

$$\begin{bmatrix} x_2 \\ y_2 \end{bmatrix} = \begin{bmatrix} \cos a & \sin a \\ -\sin a & \cos a \end{bmatrix} \begin{bmatrix} x_1 \\ y_1 \end{bmatrix} = \mathbf{R} \begin{bmatrix} x_1 \\ y_1 \end{bmatrix}$$

where the rotation matrix $\mathbf{R}(a)$ is defined as

$$\mathbf{R}(a) = \begin{bmatrix} \cos a & \sin a \\ -\sin a & \cos a \end{bmatrix} \tag{8.6–1}$$

The symbolic matrix $\mathbf{R}$ can be defined in MATLAB as follows:

```
>>syms a
>>R = [cos(a), sin(a); -sin(a), cos(a)]
R =
[cos(a), sin(a)]
[-sin(a), cos(a)]
```

If we rotate the coordinate system twice by the same angle to produce a third coordinate system (x_3, y_3), the result is the same as a single rotation with twice the angle. Let us see if MATLAB gives that result. The vector-matrix equation is

$$\begin{bmatrix} x_3 \\ y_3 \end{bmatrix} = \mathbf{R} \begin{bmatrix} x_2 \\ y_2 \end{bmatrix} = \mathbf{RR} \begin{bmatrix} x_1 \\ y_1 \end{bmatrix}$$

Thus $\mathbf{R}(a)\mathbf{R}(a)$ should be the same as $\mathbf{R}(2a)$. Continue the previous session as follows:

```
>>Q = R*R
Q =
[cos(a)^2-sin(a)^2, 2*cos(a)*sin(a)]
[-2*cos(a)*sin(a), cos(a)^2-sin(a)^2]
>>Q = simple(Q)
Q =
[cos(2*a), sin(2*a)]
[-sin(2*a), cos(2*a)]
```

The matrix $\mathbf{Q}$ is the same as $\mathbf{R}(2a)$, as we suspected.

To evaluate a matrix numerically, use the `subs` and `double` functions. For example, for a rotation of $a = \pi/4$ rad,

```
>>R = subs(R,a,pi/4);
```

Characteristic Polynomial and Roots

Sets of first-order linear differential equations can be expressed in vector-matrix notation as

$$\dot{\mathbf{x}} = \mathbf{Ax} + \mathbf{Bf}(t)$$

where $\mathbf{x}$ is the vector of dependent variables and $\mathbf{f}(t)$ is a vector containing the forcing functions. For example, for the equation set

$$\dot{x}_1 = x_2$$
$$\dot{x}_2 = -kx_1 - 2x_2 + f(t)$$

the vector $\mathbf{x}$ and the matrices $\mathbf{A}$ and $\mathbf{B}$ are

$$\mathbf{x} = \begin{bmatrix} x_1 \\ x_2 \end{bmatrix} \qquad \mathbf{A} = \begin{bmatrix} 0 & 1 \\ -k & -2 \end{bmatrix} \qquad \mathbf{B} = \begin{bmatrix} 0 \\ 1 \end{bmatrix}$$

EIGENVALUE

The equation $|s\mathbf{I} - \mathbf{A}| = 0$ is the characteristic equation of the model, where s represents the characteristic roots or "eigenvalues" of the model. Use the `poly(A)` function to find this polynomial, and note that MATLAB uses the default symbolic variable x to represent the roots. For example, to find the characteristic equation and solve for the roots in terms of the constant k, use the following session:

```
>>syms k
>>A = [0,1;-k,-2];
>>poly(A)
ans =
x^2+2*x+k
>>solve(ans)
ans =
-1+(1-k)^(1/2)
-1-(1-k)^(1/2)
```

Thus the roots are $s = -1 \pm \sqrt{1 - k}$.

You can use the `eig(A)` function to find the roots directly without finding the characteristic equation.

You can use the `inv(A)` and `det(A)` functions to invert and find the determinant of a matrix symbolically. For example, using the same matrix A from the previous session,

```
>>inv(A)
ans =
[-2/k, -1/k]
[1,  0]
```

```
>>det(A)
ans =
k
```

Solving Linear Algebraic Equations

You can use matrix methods in MATLAB to solve linear algebraic equations symbolically. You can use the matrix inverse method, if the inverse exists, or the left-division method (see Section 2.5 for a discussion of these methods). For example, to solve the set

$$2x - 3y = 3$$

$$5x + cy = 19$$

using both methods, the session is

```
>>syms c
>>A = sym([2, -3; 5, c]);
>>b = sym([3; 19]);
>>x = inv(A)*b % The matrix inverse method.
x =
3*c/(2*c+15)+57/(2*c+15)
23/(2*c+15)
>>x = A\b % The left-division method.
x =
3*(19+c)/(2*c+15)
23/(2*c+15)
```

Although the results appear to be different, they both reduce to the same solution: $x = 3(19 + c)/(2c + 15)$, $y = 23/(2c + 15)$.

Table 8.6–1 summarizes the functions used in this section. Note that their syntax is identical to the numeric versions used in earlier chapters.

Table 8.6–1 Linear algebra functions

Command	Description
det(A)	Returns the determinant of the matrix A in symbolic form.
eig(A)	Returns the eigenvalues (characteristic roots) of the matrix A in symbolic form.
inv(A)	Returns the inverse of the matrix A in symbolic form.
poly(A)	Returns the characteristic polynomial of the matrix A in symbolic form.

Test Your Understanding

T8.6–1 Consider three successive coordinate rotations using the same angle a. Show that the product **RRR** of the rotation matrix **R**(a) given by (8.6–1) equals **R**($3a$).

T8.6–2 Find the characteristic polynomial and roots of the following matrix.

$$\mathbf{A} = \begin{bmatrix} -2 & 1 \\ -3k & -5 \end{bmatrix}$$

(Answers: $s^2 + 7s + 10 + 3k$ and $s = (-7 \pm \sqrt{9 - 12k})/2$.)

T8.6–3 Use the matrix inverse and the left-division method to solve the following set.

$$-4x + 6y = -2c$$
$$7x - 4y = 23$$

(Answer: $x = (69 - 4c)/13$, $y = (46 - 7c)/13$.)

8.7 Summary

This chapter covers a subset of the capabilities of the Symbolic Math toolbox. Now that you have finished this chapter, you should be able to use MATLAB to

- Create symbolic expressions and manipulate them algebraically.
- Obtain symbolic solutions to algebraic and transcendental equations.
- Perform symbolic differentiation and integration.
- Evaluate limits and series symbolically.
- Obtain symbolic solutions to ordinary differential equations.
- Obtain Laplace transforms.
- Perform symbolic linear algebra operations, including obtaining expressions for determinants, matrix inverses, and eigenvalues.

Key Terms with Page References

Boundary condition, 372
Default variable, 357
Eigenvalue, 382
Initial condition, 373

Solution structure, 363
Symbolic constant, 355
Symbolic expression, 354

Problems

You can find the answers to problems marked with an asterisk at the end of the text.

Section 8.1

1. Use MATLAB to prove the following identities:
   a. $\sin^2 x + \cos^2 x = 1$
   b. $\sin(x + y) = \sin x \cos y + \cos x \sin y$
   c. $\sin 2x = 2 \sin x \cos x$
   d. $\cosh^2 x - \sinh^2 x = 1$

2. Use MATLAB to express $\cos 5\theta$ as a polynomial in x, where $x = \cos \theta$.

3.* Two polynomials in the variable x are represented by the coefficient vectors `p1 = [6,2,7,-3]` and `p2 = [10,-5,8]`.

   a. Use MATLAB to find the product of these two polynomials; express the product in its simplest form.

   b. Use MATLAB to find the numeric value of the product if $x = 2$.

4.* The equation of a circle of radius r centered at $x = 0$, $y = 0$ is

$$x^2 + y^2 = r^2$$

   Use the `subs` and other MATLAB functions to find the equation of a circle of radius r centered at the point $x = a$, $y = b$. Rearrange the equation into the form $Ax^2 + Bx + Cxy + Dy + Ey^2 = F$ and find the expressions for the coefficients in terms of a, b, and r.

5. The equation for a curve called the "lemniscate" in polar coordinates (r, θ) is

$$r^2 = a^2 \cos(2\theta)$$

   Use MATLAB to find the equation for the curve in terms of Cartesian coordinates (x, y), where $x = r \cos \theta$ and $y = r \sin \tau$.

Section 8.2

6.* The law of cosines for a triangle states that $a^2 = b^2 + c^2 - 2bc \cos A$, where a is the length of the side opposite the angle A, and b and c are the lengths of the other sides.

   a. Use MATLAB to solve for b.

   b. Suppose that $A = 60°$, $a = 5$ m, and $c = 2$ m. Determine b.

7. Use MATLAB to solve the polynomial equation $x^3 + 8x^2 + ax + 10 = 0$ for x in terms of the parameter a, and evaluate your solution for the case $a = 17$. Use MATLAB to check the answer.

8.* The equation for an ellipse centered at the origin of the Cartesian coordinates (x,y) is

$$\frac{x^2}{a^2} + \frac{y^2}{b^2} = 1$$

   where a and b are constants that determine the shape of the ellipse.

   a. In terms of the parameter b, use MATLAB to find the points of intersection of the two ellipses described by

$$x^2 + \frac{y^2}{b^2} = 1$$

and

$$\frac{x^2}{100} + 4y^2 = 1$$

b. Evaluate the solution obtained in part *a* for the case $b = 2$.

9. The equation

$$r = \frac{p}{1 - \epsilon \cos \theta}$$

describes the polar coordinates of an orbit with the coordinate origin at the sun. If $\epsilon = 0$, the orbit is circular; if $0 < \epsilon < 1$, the orbit is elliptical. The planets have orbits that are nearly circular; comets have orbits that are highly elongated with ϵ nearer to 1. It is of obvious interest to determine whether or not a comet's or an asteroid's orbit will intersect that of a planet. For each of the following two cases, use MATLAB to determine whether or not orbits A and B intersect. If they do, determine the polar coordinates of the intersection point. The units of distance are AU, where 1 AU is the mean distance of the Earth from the sun.
a. Orbit A: $p = 1$, $\epsilon = 0.01$. Orbit B: $p = 0.1$, $\epsilon = 0.9$.
b. Orbit A: $p = 1$, $\epsilon = 0.01$. Orbit B: $p = 1.1$, $\epsilon = 0.5$.

Section 8.3

10. Use MATLAB to find all the values of x where the graph of $y = 3^x - 2x$ has a horizontal tangent line.

11.* Use MATLAB to determine all the local minima and local maxima and all the inflection points where $dy/dx = 0$ of the following function:

$$y = x^4 - \tfrac{16}{3}x^3 + 8x^2 - 4$$

12. The surface area of a sphere of radius r is $S = 4\pi r^2$. Its volume is $V = 4\pi r^3/3$.
a. Use MATLAB to find the expression for dS/dV.
b. A spherical balloon expands as air is pumped into it. What is the rate of increase in the balloon's surface area with volume when its volume is 30 in.^{3}?

13. Use MATLAB to find the point on the line $y = 2 - x/3$ that is closest to the point $x = -3$, $y = 1$.

14. A particular circle is centered at the origin and has a radius of 5. Use MATLAB to find the equation of the line that is tangent to the circle at the point $x = 3$, $y = 4$.

15. Ship A is traveling north at 6 mi/hr, and ship B is traveling west at 12 mi/hr. When ship A was dead ahead of ship B, it was 6 mi away. Use MATLAB to determine how close the ships come to each other.

16. Suppose you have a wire of length L. You cut a length x to make a square, and use the remaining length $L - x$ to make a circle. Use MATLAB to find the length x that maximizes the sum of the areas enclosed by the square and the circle.

17.* A certain spherical street lamp emits light in all directions. It is mounted on a pole of height h (see Figure P17). The brightness B at point P on the sidewalk is directly proportional to sin θ and inversely proportional to the square of the distance d from the light to the point. Thus

$$B = \frac{c}{d^2}\sin\theta$$

where c is a constant. Use MATLAB to determine how high h should be to maximize the brightness at point P, which is 30 ft from the base of the pole.

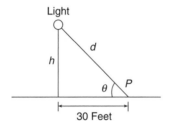

Figure P17

18.* A certain object has a mass $m = 100$ kg and is acted on by a force $f(t) = 500[2 - e^{-t}\sin(5\pi t)]$ N. The mass is at rest at $t = 0$. Use MATLAB to compute the object's velocity v at $t = 5$ s. The equation of motion is $m\dot{v} = f(t)$.

19. A rocket's mass decreases as it burns fuel. The equation of motion for a rocket in vertical flight can be obtained from Newton's law and is

$$m(t)\frac{dv}{dt} = T - m(t)g$$

where T is the rocket's thrust and its mass as a function of time is given by $m(t) = m_0(1 - rt/b)$. The rocket's initial mass is m_0, the burn time is b, and r is the fraction of the total mass accounted for by the fuel. Use the values $T = 48,000$ N; $m_0 = 2200$ kg; $r = 0.8$; $g = 9.81$ m/s²; and $b = 40$ s.
a. Use MATLAB to compute the rocket's velocity as a function of time for $t \leq b$.
b. Use MATLAB to compute the rocket's velocity at burnout.

20. The equation for the voltage $v(t)$ across a capacitor as a function of time is

$$v(t) = \frac{1}{C}\left(\int_0^t i(t)dt + Q_0\right)$$

where $i(t)$ is the applied current and Q_0 is the initial charge. Suppose that $C = 10^{-6}$ F and that $Q_0 = 0$. If the applied current is $i(t) = [0.01 + 0.3e^{-5t} \sin(25\pi t)]10^{-3}$ A, use MATLAB to compute and plot the voltage $v(t)$ for $0 \le t \le 0.3$ s.

21. The power P dissipated as heat in a resistor R as a function of the current $i(t)$ passing through it is $P = i^2 R$. The energy $E(t)$ lost as a function of time is the time integral of the power. Thus

$$E(t) = \int_0^t P(t)dt = R\int_0^t i^2(t)\,dt$$

If the current is measured in amperes, the power is in watts and the energy is in joules (1 W = 1 J/s). Suppose that a current $i(t) = 0.2[1 + \sin(0.2t)]$ A is applied to the resistor.

a. Determine the energy $E(t)$ dissipated as a function of time.

b. Determine the energy dissipated in 1 min if $R = 1000$ Ω.

22. The *RLC* circuit shown in Figure P22 can be used as a *narrowband filter*. If the input voltage $v_i(t)$ consists of a sum of sinusoidally varying voltages with different frequencies, the narrowband filter will allow to pass only those voltages whose frequencies lie within a narrow range. These filters are used in tuning circuits, such as those used in AM radios, to allow reception only of the carrier signal of the desired radio station. The magnification ratio M of a circuit is the ratio of the amplitude of the output voltage $v_o(t)$ to the amplitude of the input voltage $v_i(t)$. It is a function of the radian frequency ω of the input voltage. Formulas for M are derived in elementary electrical circuits courses. For this particular circuit, M is given by

$$M = \frac{RC\omega}{\sqrt{(1 - LC\omega^2)^2 + (RC\omega)^2}}$$

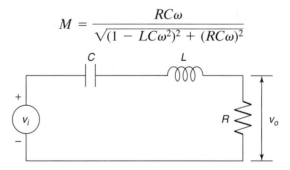

Figure P22

The frequency at which M is a maximum is the frequency of the desired carrier signal.

    a. Determine this frequency as a function of R, C, and L.

    b. Plot M versus ω for two cases where $C = 10^{-5}$ F and $L = 5 \times 10^{-3}$ H. For the first case, $R = 1000$ Ω. For the second case, $R = 10$ Ω. Comment on the filtering capability of each case.

23. The shape of a cable hanging with no load other than its own weight is a *catenary* curve. A particular bridge cable is described by the catenary $y(x) = 10 \cosh((x - 20)/10)$ for $0 \leq x \leq 50$, where x and y are the horizontal and vertical coordinates measured in feet. (See Figure P23.) It is desired to hang plastic sheeting from the cable to protect passersby while the bridge is being repainted. Use MATLAB to determine how many square feet of sheeting are required. Assume that the bottom edge of the sheeting is located along the x-axis at $y = 0$.

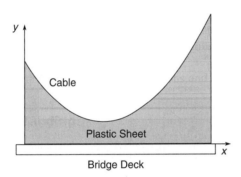

Figure P23

24. The shape of a cable hanging with no load other than its own weight is a *catenary* curve. A particular bridge cable is described by the catenary $y(x) = 10 \cosh((x - 20)/10)$ for $0 \leq x \leq 50$, where x and y are the horizontal and vertical coordinates measured in feet.

    The length L of a curve described by $y(x)$ for $a \leq x \leq b$ can be found from the following integral:

$$L = \int_a^b \sqrt{1 + \left(\frac{dy}{dx}\right)^2}\, dx$$

Determine the length of the cable.

25. Use the first five nonzero terms in the Taylor series for e^{ix}, $\sin x$, and $\cos x$ about $x = 0$ to demonstrate the validity of Euler's formula $e^{ix} = \cos x + i \sin x$.

26. Find the Taylor series for $e^x \sin x$ about $x = 0$ in two ways:

    a. By multiplying the Taylor series for e^x and that for $\sin x$.

    b. By using the `taylor` function directly on $e^x \sin x$.

27. Integrals that cannot be evaluated in closed form sometimes can be evaluated approximately by using a series representation for the integrand. For example, the following integral is used for some probability calculations (see Chapter 6, Section 6.2):

$$I = \int_0^1 e^{-x^2} dx$$

  a. Obtain the Taylor series for e^{-x^2} about $x = 0$ and integrate the first six nonzero terms in the series to find I. Use the seventh term to estimate the error.

  b. Compare your answer with that obtained with the MATLAB `erf(t)` function, defined as

$$\text{erf}(t) = \frac{2}{\sqrt{\pi}} \int_0^t e^{-t^2} dt$$

28.* Use MATLAB to compute the following limits:

  a. $\lim\limits_{x \to 1} \dfrac{x^2 - 1}{x^2 - x}$

  b. $\lim\limits_{x \to -2} \dfrac{x^2 - 4}{x^2 + 4}$

  c. $\lim\limits_{x \to 0} \dfrac{x^4 + 2x^2}{x^3 + x}$

29. Use MATLAB to compute the following limits:

  a. $\lim\limits_{x \to 0+} x^x$

  b. $\lim\limits_{x \to 0+} (\cos x)^{1/\tan x}$

  c. $\lim\limits_{x \to 0+} \left(\dfrac{1}{1 - x} \right)^{-1/x^2}$

  d. $\lim\limits_{x \to 0-} \dfrac{\sin x^2}{x^3}$

  e. $\lim\limits_{x \to 5-} \dfrac{x^2 - 25}{x^2 - 10x + 25}$

  f. $\lim\limits_{x \to 1+} \dfrac{x^2 - 1}{\sin(x - 1)^2}$

30. Use MATLAB to compute the following limits:

  a. $\lim\limits_{x \to \infty} \dfrac{x + 1}{x}$

  b. $\lim\limits_{x \to -\infty} \dfrac{3x^3 - 2x}{2x^3 + 3}$

31. Find the expression for the sum of the geometric series

$$\sum_{k=0}^{n-1} r^k$$

for $r \neq 1$.

32. A particular rubber ball rebounds to one-half its original height when dropped on a floor.

  a. If the ball is initially dropped from a height h and is allowed to continue to bounce, find the expression for the total distance traveled by the ball after the ball hits the floor for the nth time.

  b. If it is initially dropped from a height of 10 ft, how far will the ball have traveled after it hits the floor for the eighth time?

Section 8.4

33. The equation for the voltage y across the capacitor of an RC circuit is

$$RC \frac{dy}{dt} + y = v(t)$$

where $v(t)$ is the applied voltage. Suppose that $RC = 0.2$ s and that the capacitor voltage is initially 2 V. If the applied voltage goes from 0 to 10 V at $t = 0$, use MATLAB to determine and plot the voltage $y(t)$ for $0 \leq t \leq 1$ s.

34. The following equation describes the temperature $T(t)$ of a certain object immersed in a liquid bath of temperature $T_b(t)$:

$$10 \frac{dT}{dt} + T = T_b$$

Suppose the object's temperature is initially $T(0) = 70°F$ and the bath temperature is 170°F. Use MATLAB to answer the following questions:

  a. Determine $T(t)$.

  b. How long will it take for the object's temperature T to reach 168°F?

  c. Plot the object's temperature $T(t)$ as a function of time.

35.* This equation describes the motion of a mass connected to a spring with viscous friction on the surface

$$m\ddot{y} + c\dot{y} + ky = f(t)$$

where $f(t)$ is an applied force. The position and velocity of the mass at $t = 0$ are denoted by x_0 and v_0. Use MATLAB to answer the following questions:

  a. What is the free response in terms of x_0 and v_0 if $m = 3$, $c = 18$, and $k = 102$?

  b. What is the free response in terms of x_0 and v_0 if $m = 3$, $c = 39$, and $k = 120$?

36. The equation for the voltage y across the capacitor of an RC circuit is

$$RC \frac{dy}{dt} + y = v(t)$$

where $v(t)$ is the applied voltage. Suppose that $RC = 0.2$ s and that the capacitor voltage is initially 2 V. If the applied voltage is $v(t) = 10[2 - e^{-t} \sin(5\pi t)]$, use MATLAB to compute and plot the voltage $y(t)$ for $0 \leq t \leq 5$ s.

37. The following equation describes a certain dilution process, where $y(t)$ is the concentration of salt in a tank of fresh water to which salt brine is being added:

$$\frac{dy}{dt} + \frac{2}{10 + 2t}y = 4$$

Suppose that $y(0) = 0$. Use MATLAB to compute and plot $y(t)$ for $0 \leq t \leq 10$.

38. This equation describes the motion of a certain mass connected to a spring with viscous friction on the surface

$$3\ddot{y} + 18\dot{y} + 102y = f(t)$$

where $f(t)$ is an applied force. Suppose that $f(t) = 0$ for $t < 0$ and $f(t) = 10$ for $t \geq 0$.

  a. Use MATLAB to compute and plot $y(t)$ when $y(0) = \dot{y}(0) = 0$.

  b. Use MATLAB to compute and plot $y(t)$ when $y(0) = 0$ and $\dot{y}(0) = 10$.

39. This equation describes the motion of a certain mass connected to a spring with viscous friction on the surface

$$3\ddot{y} + 39\dot{y} + 120y = f(t)$$

where $f(t)$ is an applied force. Suppose that $f(t) = 0$ for $t < 0$ and $f(t) = 10$ for $t \geq 0$.

  a. Use MATLAB to compute and plot $y(t)$ when $y(0) = \dot{y}(0) = 0$.

  b. Use MATLAB to compute and plot $y(t)$ when $y(0) = 0$ and $\dot{y}(0) = 10$.

40. The equations for an armature-controlled dc motor follow. The motor's current is i and its rotational velocity is ω.

$$L \frac{di}{dt} = -Ri - K_e\omega + v(t)$$

$$I \frac{d\omega}{dt} = K_T i - c\omega$$

where L, R, and I are the motor's inductance, resistance, and inertia; K_T and K_e are the torque constant and back emf constant; c is a viscous damping constant; and $v(t)$ is the applied voltage.

Use the values $R = 0.8\ \Omega$, $L = 0.003$ H, $K_T = 0.05$ N·m/A, $K_e = 0.05$ V·s/rad, $c = 0$, and $I = 8 \times 10^{-5}$ kg · m^2.

Suppose the applied voltage is 20 V. Use MATLAB to compute and plot the motor's speed and current versus time for zero initial conditions. Choose a final time large enough to show the motor's speed becoming constant.

Section 8.5

41. The *RLC* circuit described in Problem 22 and shown in Figure P22 has the following differential equation model:

$$LC\ddot{v}_o + RC\dot{v}_o + v_o = RC\dot{v}_i(t)$$

Use the Laplace transform method to solve for the unit-step response of $v_o(t)$ for zero initial conditions, where $C = 10^{-5}$ F and $L = 5 \times 10^{-3}$ H. For the first case (a broadband filter), $R = 1000\ \Omega$. For the second case (a narrowband filter), $R = 10\ \Omega$. Compare the step responses of the two cases.

42. The differential equation model for a certain speed control system for a vehicle is

$$\ddot{v} + (1 + K_p)\dot{v} + K_I v = K_p \dot{v}_d + K_I v_d$$

where the actual speed is v, the desired speed is $v_d(t)$, and K_p and K_I are constants called the "control gains." Use the Laplace transform method to find the unit-step response (that is, $v_d(t)$ is a unit-step function). Use zero initial conditions. Compare the response for three cases:

a. $K_p = 9$, $K_I = 50$
b. $K_p = 9$, $K_I = 25$
c. $K_p = 54$, $K_I = 250$

43. The differential equation model for a certain position control system for a metal cutting tool is

$$\frac{d^3x}{dt^3} + (6 + K_D)\frac{d^2x}{dt^2} + (11 + K_p)\frac{dx}{dt} + (6 + K_I)x$$

$$= K_D\frac{d^2x_d}{dt^2} + K_p\frac{dx_d}{dt} + K_I x_d$$

where the actual tool position is x; the desired position is $x_d(t)$; and K_p, K_I, and K_D are constants called the control gains. Use the Laplace transform method to find the unit-step response (that is, $x_d(t)$ is a unit-step function). Use zero initial conditions. Compare the response for three cases:

a. $K_p = 30$, $K_I = K_D = 0$
b. $K_p = 27$, $K_I = 17.18$, $K_D = 0$
c. $K_p = 36$, $K_I = 38.1$, $K_D = 8.52$

44.* The differential equation model for the motor torque $m(t)$ required for a certain speed control system is

$$4\ddot{m} + 4K\dot{m} + K^2m = K^2v_d$$

where the desired speed is $v_d(t)$, and K is a constant called the control gain.
a. Use the Laplace transform method to find the unit-step response (that is, $v_d(t)$ is a unit-step function). Use zero initial conditions.
b. Use symbolic manipulation in MATLAB to find the value of the peak torque in terms of the gain K.

Section 8.6

45. Show that $\mathbf{R}^{-1}(a)\mathbf{R}(a) = \mathbf{I}$, where $\mathbf{I}$ is the identity matrix and $\mathbf{R}(a)$ is the rotation matrix given by (8.6–1). This equation shows that the inverse co-ordinate transformation returns you to the original coordinate system.

46. Show that $\mathbf{R}^{-1}(a) = \mathbf{R}(-a)$. This equation shows that a rotation through a negative angle is equivalent to an inverse transformation.

47.* Find the characteristic polynomial and roots of the following matrix:

$$\mathbf{A} = \begin{bmatrix} -6 & 2 \\ 3k & -7 \end{bmatrix}$$

48.* Use the matrix inverse and the left-division method to solve the following set for x and y in terms of c:

$$4cx + 5y = 43$$
$$3x - 4y = -22$$

49. The currents i_1, i_2, and i_3 in the circuit shown in Figure P49 are described by the following equation set if all the resistances are equal to R.

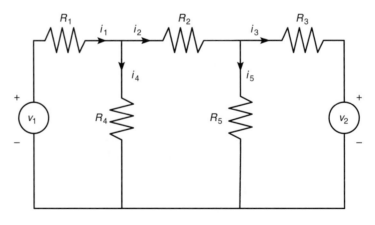

Figure P49

$$\begin{bmatrix} 2R & -R & 0 \\ -R & 3R & -R \\ 0 & R & -2R \end{bmatrix} \begin{bmatrix} i_1 \\ i_2 \\ i_3 \end{bmatrix} = \begin{bmatrix} v_1 \\ 0 \\ v_2 \end{bmatrix}$$

v_1 and v_2 are applied voltages; the other two currents can be found from $i_4 = i_1 - i_2$ and $i_5 = i_2 - i_3$.

a. Use both the matrix inverse method and the left-division method to solve for the currents in terms of the resistance R and the voltages v_1 and v_2.

b. Find the numerical values for the currents if $R = 1000\ \Omega$, $v_1 = 100$ V, and $v_2 = 25$ V.

50. The equations for the armature-controlled dc motor shown in Figure P50 follow. The motor's current is i, and its rotational velocity is ω.

$$L\frac{di}{dt} = -Ri - K_e\omega + v(t)$$

$$I\frac{d\omega}{dt} = K_Ti - c\omega$$

where L, R, and I are the motor's inductance, resistance, and inertia; K_T and K_e are the torque constant and back emf constant; c is a viscous damping constant; and $v(t)$ is the applied voltage.

a. Find the characteristic polynomial and the characteristic roots.

b. Use the values $R = 0.8\ \Omega$, $L = 0.003$ H, $K_T = 0.05$ N · m/A, $K_e = 0.05$ V · s/rad, and $I = 8 \times 10^{-5}$ kg · m². The damping constant c is often difficult to determine with accuracy. For these values find the expressions for the two characteristic roots in terms of c.

c. Using the parameter values in part b, determine the roots for the following values of c (in N · m · s): $c = 0$, $c = 0.01$, $c = 0.1$, and $c = 0.2$. For each case, use the roots to estimate how long the motor's speed will take to become constant; also discuss whether the speed will oscillate before it becomes constant.

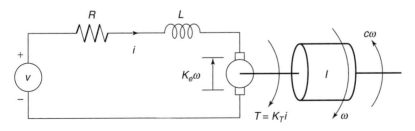

Figure P50

APPENDIX

Guide to Commands and Functions in This Text

Operators and special characters

Item	Description	Pages
+	Plus; addition operator.	6, 51
−	Minus; subtraction operator.	6, 51
*	Scalar and matrix multiplication operator.	6
.*	Array multiplication operator.	61
^	Scalar and matrix exponentiation operator.	6
.^	Array exponentiation operator.	61
\	Left-division operator.	6
/	Right-division operator.	6
.\	Array left-division operator.	61
./	Array right-division operator.	61
:	Colon; generates regularly spaced elements and represents an entire row or column.	9, 16, 40, 42
( )	Parentheses; encloses function arguments and array indices; overrides precedence.	6, 124
[ ]	Brackets; encloses array elements.	16, 126
{ }	Braces; encloses cell elements.	89
. . .	Ellipsis; line-continuation operator.	9
,	Comma; separates statements, and elements in a row of an array.	9, 10
;	Semicolon; separates columns in an array, and suppresses display.	9, 40, 41
%	Percent sign; designates a comment, and specifies formatting.	23
'	Quote sign and transpose operator.	27, 40, 42
.'	Nonconjugated transpose operator.	42
=	Assignment (replacement) operator.	8
@	Creates a function handle.	131

Logical and relational operators

Item	Description	Pages
==	Relational operator: equal to.	154
~=	Relational operator: not equal to.	154
<	Relational operator: less than.	154
<=	Relational operator: less than or equal to.	154
>	Relational operator: greater than.	154
>=	Relational operator: greater than or equal to.	154
&	Logical operator: AND.	157
&&	Short-circuit AND.	159
|	Logical operator: OR.	157
||	Short-circuit OR.	157
~	Logical operator: NOT.	157

Special variables and constants

Item	Description	Pages
ans	Most recent answer.	12
eps	Accuracy of floating-point precision.	12
i,j	The imaginary unit $\sqrt{-1}$.	12
Inf	Infinity.	12
NaN	Undefined numerical result (not a number).	12
pi	The number π.	12

Commands for managing a session

Item	Description	Pages
clc	Clears Command window.	9
clear	Removes variables from memory.	9
doc	Displays documentation.	32
exist	Checks for existence of file or variable.	9
global	Declares variables to be global.	131
help	Displays help text in the Command window.	32
helpwin	Displays help text in the Help Browser.	32
lookfor	Searches help entries for a keyword.	32
quit	Stops MATLAB.	9
who	Lists current variables.	9
whos	Lists current variables (long display).	9

System and file commands

Item	Description	Pages
cd	Changes current directory.	20
date	Displays current date.	140
dir	Lists all files in current directory.	20
load	Loads workspace variables from a file.	18, 145
path	Displays search path.	20
pwd	Displays current directory.	20
save	Saves workspace variables in a file.	18
type	Displays contents of a file.	32
uiimport	Activates the Import Wizard.	146
what	Lists all MATLAB files.	20
which	Displays pathname.	20
wk1read	Reads .wk1 spreadsheet file.	145
xlsread	Reads .xls spreadsheet file.	145

Input/output commands

Item	Description	Pages
disp	Displays contents of an array or string.	27
format	Controls screen display format.	13
input	Displays prompts and waits for input.	27
menu	Displays a menu of choices.	27
;	Suppresses screen printing.	9

Numeric display formats

Item	Description	Pages
format bank	Two decimal digits.	13
format compact	Suppresses some line feeds.	13
format long	16 decimal digits.	13
format long e	16 digits plus exponent.	13
format loose	Resets to less compact display mode.	13
format +	Positive, negative, or zero.	13
format rat	Rational approximation.	13
format short	Four decimal digits (default).	13
format short e	Five digits plus exponent.	13

Array functions

Item	Description	Pages
cat	Concatenates arrays.	49
find	Finds indices of nonzero elements.	160
length	Computes number of elements.	45
linspace	Creates regularly spaced vector.	45
logspace	Creates logarithmically spaced vector.	45
max	Returns largest element.	45
min	Returns smallest element.	45
norm	Computes vector geometric length.	45
size	Computes array size.	45
sort	Sorts each column.	45
sum	Sums each column.	45

Special matrices

Item	Description	Pages
eye	Creates an identity matrix.	67
ones	Creates an array of ones.	67
zeros	Creates an array of zeros.	67

Matrix functions for solving linear equations

Item	Description	Pages
det	Computes determinant of an array.	85, 383
inv	Computes inverse of a matrix.	85, 383
pinv	Computes pseudoinverse of a matrix.	85
rank	Computes rank of a matrix.	85
rref	Computes reduced row echelon form.	85

Exponential and logarithmic functions

Item	Description	Pages
exp(x)	Exponential; e^x.	121
log(x)	Natural logarithm; $\ln x$.	121
log10(x)	Common (base 10) logarithm; $\log x = \log_{10} x$.	121
sqrt(x)	Square root; $\sqrt{x}$.	121

Complex functions

Item	Description	Pages
abs(x)	Absolute value; $|x|$.	121
angle(x)	Angle of a complex number x.	121
conj(x)	Complex conjugate of x.	121
imag(x)	Imaginary part of a complex number x.	121
real(x)	Real part of a complex number x.	121

Numeric functions

Item	Description	Pages
ceil	Rounds to the nearest integer toward ∞.	121
fix	Rounds to the nearest integer toward zero.	121
floor	Rounds to the nearest integer toward $-\infty$.	121
round	Rounds toward the nearest integer.	121
sign	Signum function.	121

Trigonometric functions

Item	Description	Pages
Radian measure		
acos(x)	Inverse cosine; arccos $x = \cos^{-1}x$.	124
acot(x)	Inverse cotangent; arccot $x = \cot^{-1}x$.	124
acsc(x)	Inverse cosecant; arccsc $x = \csc^{-1}x$.	124
asec(x)	Inverse secant; arcsec $x = \sec^{-1}x$.	124
asin(x)	Inverse sine; arcsin $x = \sin^{-1}x$.	124
atan(x)	Inverse tangent; arctan $x = \tan^{-1}x$.	124
atan2(y,x)	Four-quadrant inverse tangent.	124
cos(x)	Cosine; cos x.	124
cot(x)	Cotangent; cot x.	124
csc(x)	Cosecant; csc x.	124
sec(x)	Secant; sec x.	124
sin(x)	Sine; sin x.	124
tan(x)	Tangent; tan x.	124
Degree measure		
asind(x)	Inverse sine; arcsin $x = \sin^{-1}x$.	18
sind(x)	Sine; sin x.	18

Hyperbolic functions

Item	Description	Pages
acosh(x)	Inverse hyperbolic cosine; $\cosh^{-1}x$.	126
acoth(x)	Inverse hyperbolic cotangent; $\coth^{-1}x$.	126
acsch(x)	Inverse hyperbolic cosecant; $\operatorname{csch}^{-1}x$.	126
asech(x)	Inverse hyperbolic secant; $\operatorname{sech}^{-1}x$.	126
asinh(x)	Inverse hyperbolic sine; $\sinh^{-1}x$.	126
atanh(x)	Inverse hyperbolic tangent; $\tanh^{-1}x$.	126
cosh(x)	Hyperbolic cosine; cosh x.	126
coth(x)	Hyperbolic cotangent; cosh x/sinh x.	126
csch(x)	Hyperbolic cosecant; 1/sinh x.	126
sech(x)	Hyperbolic secant; 1/cosh x.	126
sinh(x)	Hyperbolic sine; sinh x.	126
tanh(x)	Hyperbolic tangent; sinh x/cosh x.	126

Polynomial functions

Item	Description	Pages
conv	Computes product of two polynomials.	87
deconv	Computes ratio of polynomials.	87
eig	Computes the eigenvalues of a matrix.	333, 383
poly	Computes polynomial from roots.	87, 383
polyder	Differentiation of a polynomial.	318
polyfit	Fits a polynomial to data.	232, 239
polyint	Integration of a polynomial.	307
polyval	Evaluates polynomial.	87
roots	Computes polynomial roots.	87

String functions

Item	Description	Pages
lower	Converts string to all lowercase.	182

Logical functions

Item	Description	Pages
all	True if all elements are nonzero.	160
any	True if any elements are nonzero.	160
find	Finds indices of nonzero elements.	160
finite	True if elements are finite.	160
ischar	True if elements are a character array.	160
isempty	True if matrix is empty.	160
isinf	True if elements are infinite.	160
isnan	True if elements are undefined.	160
isnumeric	True if elements have numeric values.	160
isreal	True if all elements are real.	160
logical	Converts a numeric array to a logical array.	160
xor	Exclusive OR.	160

Miscellaneous mathematical functions

Item	Description	Pages
cross	Computes cross products.	69
dot	Computes dot products.	69
function	Creates a user-defined function.	129
nargin	Number of function input arguments.	167
nargout	Number of function output arguments.	168

Cell and structure functions

Item	Description	Pages
cell	Creates cell array.	90
fieldnames	Returns field names in a structure array.	93
isfield	Identifies a structure array field.	93
isstruct	Identifies a structure array.	93
rmfield	Removes a field from a structure array.	93
struct	Creates structure array.	93

Basic *xy* plotting commands

Item	Description	Pages
axis	Sets axis limits and other axis properties.	209
fplot	Intelligent plotting of functions.	209
ginput	Reads coordinates of the cursor position.	22
grid	Displays gridlines.	209
plot	Generates *xy* plot.	209, 216
print	Prints plot or saves plot to a file.	209
title	Puts text at top of plot.	209
xlabel	Adds text label to *x*-axis.	209
ylabel	Adds text label to *y*-axis.	209

Plot-enhancement commands

Item	Description	Pages
gtext	Enables label placement by mouse.	216
hold	Freezes current plot.	216
legend	Legend placement by mouse.	216
subplot	Creates plots in subwindows.	216
text	Places string in figure.	216

Specialized plot functions

Item	Description	Pages
bar	Creates bar chart.	219
loglog	Creates log-log plot.	219
open	Displays a report created by publish.	222
plotyy	Enables plotting on left and right axes.	219
polar	Creates polar plot.	219
publish	Creates reports with embedded graphics.	221
semilogx	Creates semilog plot (logarithmic abscissa).	219
semilogy	Creates semilog plot (logarithmic ordinate).	219
stairs	Creates stairs plot.	219
stem	Creates stem plot.	219

Three-dimensional plotting functions

Item	Description	Pages
contour	Creates contour plot.	254
mesh	Creates three-dimensional mesh surface plot.	254
meshc	Same as mesh with contour plot underneath.	254
meshgrid	Creates rectangular grid.	254
meshz	Same as mesh with vertical lines underneath.	254
plot3	Creates three-dimensional plots from lines and points.	250
surf	Creates shaded three-dimensional mesh surface plot.	254
surfc	Same as surf with contour plot underneath.	254
waterfall	Same as mesh with mesh lines in one direction.	254
zlabel	Adds text label to z-axis.	251

Program flow control

Item	Description	Pages
break	Terminates execution of a loop.	174
case	Provides alternate execution paths within switch structure.	181
continue	Passes control to the next iteration of a for or while loop.	174
else	Delineates alternate block of statements.	180
elseif	Conditionally executes statements.	180
end	Terminates for, while, and if statements.	180
for	Repeats statements a specific number of times.	180
if	Executes statements conditionally.	180
otherwise	Provides optional control within a switch structure.	182
switch	Directs program execution by comparing input with case expressions.	181
while	Repeats statements an indefinite number of times.	180

Optimization and root-finding functions

Item	Description	Pages
fminbnd	Finds the minimum of a function of one variable.	135
fminsearch	Finds the minimum of a multivariable function.	135
fzero	Finds the zero of a function.	135

Histogram functions

Item	Description	Pages
bar	Creates a bar chart.	219, 276
hist	Aggregates the data into bins.	276

Statistical functions

Item	Description	Pages
cumsum	Computes the cumulative sum across a row.	279
erf(x)	Computes the error function $\mathrm{erf}(x)$.	281
mean	Calculates the mean.	272
median	Calculates the median.	272
mode	Calculates the mode.	272
std	Calculates the standard deviation.	280
var	Calculates the variance.	280

Random number functions

Item	Description	Pages
rand	Generates uniformly distributed random numbers between 0 and 1; sets and retrieves the state.	285
randn	Generates normally distributed random numbers; sets and retrieves the state.	285
randperm	Generates random permutation of integers.	285

Interpolation functions

Item	Description	Pages
interp1	Linear and cubic-spline interpolation of a function of one variable.	293, 296
interp2	Linear interpolation of a function of two variables.	293
pchip	Piecewise cubic Hermite polynomial interpolation.	296
spline	Cubic-spline interpolation.	296
unmkpp	Computes the coefficients of cubic-spline polynomials.	296

Numerical integration functions

Item	Description	Pages
dblquad	Numerical integration of a double integral.	307
polyint	Integration of a polynomial.	307
quad	Numerical integration with adaptive Simpson's rule.	307
quadl	Numerical integration with Lobatto quadrature.	307
trapz	Numerical integration with the trapezoidal rule.	307
triplequad	Numerical integration of a triple integral.	307

Numerical differentiation functions

Item	Description	Pages
del 2	Computes the Laplacian from data.	318
diff(x)	Computes the differences between adjacent elements in the vector x.	318
gradient	Computes the gradient from data.	318
polyder	Differentiates a polynomial, a polynomial product, or a polynomial quotient.	318

ODE solvers

Item	Description	Pages
ode15s	Stiff, variable-order solver.	331
ode45	Nonstiff, medium-order solver.	331
odeset	Creates integrator options structure for ODE solvers.	331

LTI object functions

Item	Description	Pages
ss	Creates an LTI object in state-space form.	336
ssdata	Extracts state-space matrices from an LTI object.	336
tf	Creates an LTI object in transfer-function form.	336
tfdata	Extracts equation coefficients from an LTI object.	336

LTI ODE solvers

Item	Description	Pages
impulse	Computes and plots the impulse response of an LTI object.	337
initial	Computes and plots the free response of an LTI object.	337
lsim	Computes and plots the response of an LTI object to a general input.	337
step	Computes and plots the step response of an LTI object.	337

Predefined input functions

Item	Description	Pages
gensig	Generates a periodic sine, square, or pulse input.	343

Functions for creating and evaluating symbolic expressions

Item	Description	Pages
class	Returns the class of an expression.	361
digits	Sets the number of decimal digits used to do variable precision arithmetic.	361
double	Converts an expression to numeric form.	361
ezplot	Generates a plot of a symbolic expression.	361
findsym	Finds the symbolic variables in a symbolic expression.	361
latex	Converts a symbolic expression into a LaTeX typeset expression.	361
numden	Returns the numerator and denominator of an expression.	361
sym	Creates a symbolic variable.	361
syms	Creates one or more symbolic variables.	361
vpa	Sets the number of digits used to evaluate expressions.	361

Functions for manipulating symbolic expressions

Item	Description	Pages
collect	Collects coefficients of like powers in an expression.	362
expand	Expands an expression by carrying out powers.	362
factor	Factors an expression.	362
poly2sym	Converts a polynomial coefficient vector to a symbolic polynomial.	362
pretty	Displays an expression in a form that resembles typeset mathematics.	362
simple	Searches for the shortest form of an expression.	362
simplify	Simplifies an expression using Maple's simplification rules.	362
subs	Substitutes variables or expressions.	362
sym2poly	Converts an expression to a polynomial coefficient vector.	362

Symbolic solution of algebraic and transcendental equations

Item	Description	Pages
solve	Solves symbolic equations.	365

Symbolic calculus functions

Item	Description	Pages
diff	Returns the derivative of an expression.	366
dirac	Dirac delta function (unit impulse).	377
heaviside	Heaviside function (unit step).	377
int	Returns the integral of an expression.	366
limit	Returns the limit of an expression.	366
symsum	Returns the symbolic summation of an expression.	366
taylor	Returns the Taylor series of a function.	366
taylortool	Opens a graphical interface for analyzing Taylor series.	369

Symbolic solution of differential equations

Item	Description	Pages
dsolve	Returns a symbolic solution of a differential equation or set of equations.	371

Laplace transform functions

Item	Description	Pages
ilaplace	Returns the inverse Laplace transform.	380
laplace	Returns the Laplace transform.	380

Symbolic linear algebra functions

Item	Description	Pages
det	Returns the determinant of a matrix.	383
eig	Returns the eigenvalues (characteristic roots) of a matrix.	383
inv	Returns the inverse of a matrix.	383
poly	Returns the characteristic polynomial of a matrix.	383

B

APPENDIX

References

[Brown, 1994] Brown, T. L.; H. E. LeMay, Jr.; and B. E. Bursten. *Chemistry: The Central Science*, 6th ed. Upper Saddle River, NJ: Prentice-Hall, 1994.

[Felder, 1986] Felder, R. M. and R. W. Rousseau. *Elementary Principles of Chemical Processes*. New York: John Wiley & Sons, 1986.

[Kutz, 1999] Kutz, M., editor. *Mechanical Engineers' Handbook*. 2d ed. New York: John Wiley & Sons, 1999.

[Palm, 2005a] Palm, W. *Introduction to MATLAB 7 for Engineers*. New York: McGraw-Hill, 2005.

[Palm, 2005b] Palm, W. *System Dynamics*. New York: McGraw-Hill, 2005.

Answers to Selected Problems

Chapter 1

2. (a) -13.3333; (b) 0.6; (c) 15; (d) 1.0323

8. (a) $x + y = -3 - 2i$; (b) $xy = -13 - 41i$;
(c) $x/y = -1.72 + 0.04i$

18. $-15.685, 0.8425 \pm 3.4008i$

25. $L = 12.58$ m, perimeter $= 39.65$ m

Chapter 2

3.
$$\mathbf{A} = \begin{bmatrix} 0 & 6 & 12 & 18 & 24 & 30 \\ -20 & -10 & 0 & 10 & 20 & 30 \end{bmatrix}$$

7. (a) Length $= 3$, absolute value $= [2, 4, 7]$;
(b) Same as (a); (c) Length $= 3$, absolute
value $= [5.831, 5, 7.2801]$

12. (a)
$$\mathbf{A} + \mathbf{B} + \mathbf{C} = \begin{bmatrix} -4 & 2 \\ 22 & 15 \end{bmatrix}$$

(b) $\mathbf{A} - \mathbf{B} + \mathbf{C} = \begin{bmatrix} -16 & 12 \\ -2 & 19 \end{bmatrix}$

13. (a) $[1024, -128; 144, 32]$;
(b) $[4, -8; 4, 8]$;
(c) $[4096, -64; 216, -8]$

14. (a) Work done on each segment, in joules
(1 J $= 1$ N $\cdot$ m) is 800, 275, 525, 750, 1800;
(b) Total work done $= 4150$ J.

25.
$$\mathbf{AB} = \begin{bmatrix} -47 & -78 \\ 39 & 64 \end{bmatrix}$$

$$\mathbf{BA} = \begin{bmatrix} -5 & -3 \\ 48 & 22 \end{bmatrix}$$

28. 60 tons of copper, 67 tons of magnesium, 6 tons
of manganese, 76 tons of silicon, and 101 tons
of zinc

31. $M = 869$ N $\cdot$ m if $\mathbf{F}$ is in newtons and r is in
meters.

37. (a) $\mathbf{C} = \mathbf{B}^{-1}(\mathbf{A}^{-1}\mathbf{B} - \mathbf{A})$

(b)
$$\mathbf{C} = \begin{bmatrix} -0.6212 & -2.3636 \\ 1.197 & 2.1576 \end{bmatrix}$$

40. $x = 3c, y = -2c, z = c$

43. $T_1 = 19.8°C, T_2 = -7.0°C, T_3 = -9.7°C$. Heat
loss rate is 66.8 W.

46. The nonunique solution is $x = 1.38z + 4.92$,
$y = -0.077z - 1.38$, where z can have any value.

49. The exact and unique solution is $x = 8, y = 2$.

50. There is no exact solution. The least squares
solution is $x = 6.09, y = 2.26$.

54. $2.8x - 5.12$ with a remainder of $50.04x - 11.48$

55. 0.5676

Chapter 3

1. (*a*) 3, 3.1623, 3.6056;
   (*b*) 1.7321*i*, 0.2848 + 1.7553*i*, 0.5503 + 1.8174*i*;
   (*c*) 15 + 21*i*, 22 + 16*i*, 29 + 11*i*;
   (*d*) −0.4 − 0.2*i*, −0.4667 − 0.0667*i*,
   −0.5333 + 0.0667*i*

2. (*a*) $|xy| = 105$, $\angle xy = -2.6$ rad;
   (*b*) $|x/y| = 0.84$, $\angle x/y = -1.67$ rad

3. (*a*) 1.01 rad (58°); (*b*) 2.13 rad (122°);
   (*c*) −1.01 rad (−58°); (*d*) −2.13 rad (−122°)

7. $F_1 = 198$ N if $\mu = 0.3$, $F_2 = 100$ N, and $\beta = 130°$.

10. For the test values, $t = 7.46$ and 2.73 sec.

Chapter 4

1. (*a*) $z = 1$; (*b*) $z = 0$; (*c*) $z = 1$; (*d*) $z = 1$
2. (*a*) $z = 0$; (*b*) $z = 1$; (*c*) $z = 0$; (*d*) $z = 4$;
   (*e*) $z = 1$; (*f*) $z = 5$; (*g*) $z = 1$; (*h*) $z = 0$
3. (*a*) $z = [0, 1, 0, 1, 1]$;
   (*b*) $z = [0, 0, 0, 1, 1]$;
   (*c*) $z = [0, 0, 0, 1, 0]$;
   (*d*) $z = [1, 1, 1, 0, 1]$
8. (*a*) $z = [1, 1, 1, 0, 0, 0]$;
   (*b*) $z = [1, 0, 0, 1, 1, 1]$;
   (*c*) $z = [1, 1, 0, 1, 1, 1]$;
   (*d*) $z = [0, 1, 0, 0, 0, 0]$
10. (*a*) $7300; (*b*) $5600; (*c*) 1200 shares;
    (*d*) $15,800
26. (*a*) $x = 9$, $y = 16$ miles
31. 33 years
33. $W = 300$ N. If $W = 300$, the wire tensions are
    $T_i = 429, 471, 267, 233, 200$, and 100 N,
    respectively.
44. Weekly inventory for cases (*a*) and (*b*):

Week	1	2	3	4	5
Inventory (*a*)	50	50	45	40	30
Inventory (*b*)	30	25	20	20	10

Week	6	7	8	9	10
Inventory (*a*)	30	30	25	20	10
Inventory (*b*)	10	5	0	0	(<0)

Chapter 5

1. Production is profitable for $Q \geq 10^8$ gallons per
   year. The profit increases linearly with Q, so
   there is no upper limit on the profit.
3. To three significant digits, the roots are −0.480,
   1.13, and 3.83.

5. The left end is 47 m above the reference line.
   The right end is 110 m above the reference line.
10. 0.54 rad (31°).
14. The steady-state value of y is $y = 1$. $y = 0.98$
    at $t = 4/b$.
17. (*a*) The ball will rise 1.68 m and will travel
    9.58 m horizontally before striking the ground
    after 1.17 s.
26. (*a*) $y = 53.5x - 1354.5$;
    (*b*) $y = 3.58 \times 10^3 x^{-0.976}$;
    (*c*) $y = 2.06 \times 10^5 (10)^{-0.0067x}$
28. (*a*) $b = 1.2603 \times 10^{-4}$; (*b*) 836 years;
    (*c*) between 760 and 928 years ago
32. If unconstrained to pass through the origin,
    $f = 0.1999x - 0.0147$. If constrained to pass
    through the origin, $f = 0.1977x$.
34. $d = 0.0509v^2 + 1.1054v + 2.3571$, $J = 10.1786$,
    $S = 57,550$, $r^2 = 0.9998$
35. $y = 40 + 9.6x_1 - 6.75x_2$. Maximum percent
    error is 7.125 percent.

Chapter 6

7. (*a*) 99%; (*b*) 68%
11. (*a*) Mean pallet weight is 3000 lb, standard
    deviation is 10.95 lb; (*b*) 9 percent
18. Mean yearly profit = $64,609. Minimum
    expected profit = $51,340. Maximum expected
    profit = $79,440. Standard deviation of yearly
    profit = $5967.
22. The value at 5 P.M. is 22.5, the value at 9 P.M.
    is 16.5.

Chapter 7

1. 2360
7. 13.65 ft
10. 1363.4 m/s
25. (*a*) $v(t) = (f/500)(1 - e^{-t/2})$; (*b*) Steady-state
    speed is $f/500$ m/s. The speed is within 2 percent
    of this value after $t = 8$ s.
26. (*a*) $y(t) = C_1 e^{-3t} \sin 5t + C_2 e^{-3t} \cos 5t$;
    (*b*) $y(t) = C_1 e^{-8t} + C_2 e^{-5t}$
39. $\ddot{x}_1 + 7\dot{x}_1 + 14x_1 = 2u$

Chapter 8

3. (*a*) $60x^5 - 10x^4 + 108x^3 - 49x^2 + 71x - 24$;
   (*b*) 2546
4. $A = 1$, $B = -2a$, $C = 0$, $D = -2b$, $E = 1$,
   $F = r^2 - a^2 - b^2$

6. (a) $b = c \cos A \pm \sqrt{a^2 - c^2 \sin^2 A}$;

    (b) $b = 5.69$

8. (a) $x = \pm 10\sqrt{(4b^2 - 1)/(400b^2 - 1)}$, $y = \pm b\sqrt{99/(400b^2 - 1)}$;

    (b) $x = 0.9685$, $y = 0.4976$

11. Critical points: $x = 0$ and $x = 2$. Local minimum at $x = 0$. Inflection points at $x = 2$ and $x = 2/3$

17. $h = 15\sqrt{2}$

18. 49.68 m/s

28. (a) 2; (b) 0; (c) 0

35. (a) $y(t) = [0.6y(0) + 0.2v(0)]e^{-3t} \sin 5t + y(0)e^{-3t} \cos 5t$; (b) $y(t) = (1/3)[v(0) + 8y(0)]e^{-5t} - (1/3)[v(0) + 5y(0)]e^{-8t}$

44. (a) $m(t) = (K^2/4)te^{-Kt/2}$; (b) $m_{peak} = K/5.4366$

47. $s^2 + 13s + 42 - 6k$, $s = (-13 \pm \sqrt{1 + 24k})/2$

48. $x = 62/(16c + 15)$, $y = (129 + 88c)/(16c + 15)$

INDEX

Symbols

+ addition, 6, 51
− subtraction, 6, 51
* multiplication, 6
. * array multiplication, 61
^ exponentiation, 6
. ^ array exponentiation, 61
\ left division, 6
/ right division, 6
. \ array left division, 61
. / array right division, 61
: colon
array addressing, 42
array generation, 9, 16, 40

( ) parentheses
function arguments, 124
modifying precedence, 6
{ } braces; encloses cell elements, 89
[ ] brackets, 16
. . . ellipsis, 9
, comma
column separation, 9
statement separation, 10
; semicolon
display suppression, 9
row separation, 40, 41
% percent sign

comment designation, 23
' apostrophe
transpose, 40, 42
string designation, 27
. ' nonconjugated transpose, 42
= assignment or replacement operator, 8
= = equal to, 154
~= not equal to, 154
< less than, 154
<= less than or equal to, 154
> greater than, 154

>= greater than or equal to, 154
& AND, 157
&& short-circuit AND, 157
| OR, 197
|| short-circuit OR, 157
~ NOT, 157
>> MATLAB prompt, 3
@ creates a function handle, 131

MATLAB Commands

A

abs, 121
acos, 124
acosh, 126
acot, 124
acoth, 126
acsc, 124
acsch, 126
addpath, 20
all, 160
angle, 121
ans, 12
any, 160
asec, 124
asech, 126
asin, 124

asind, 18
asinh, 126
atan, 124
atan2, 124
atanh, 126
axis, 209

B

bar, 219
break, 174

C

case, 181
cat, 49
cd, 20

ceil, 121
cell, 90
cell disp, 90
cell plot, 90
class, 361
clc, 9
clear, 9
collect, 362
conj, 121
continue, 174
contour, 254
conv, 87
cos, 124
cosd, 18
cosh, 126
cot, 124

coth, 126
cross, 69
csc, 124
csch, 126
cumsum, 279

D

date, 140
dblquad, 307
deconv, 87
delz, 318
det, 85, 383
diff, 318, 366
digits, 361
dir, 20
dirac, 377

413

Topics